Antimicrobial Use in Companion Animals

Antimicrobial Use in Companion Animals

Guest Editor

George Valiakos

MDPI

Basel • Beijing • Wuhan • Barcelona • Belgrade • Novi Sad • Cluj • Manchester

Guest Editor
George Valiakos
Faculty of Veterinary Medicine
University of Thessaly
Karditsa
Greece

Editorial Office
MDPI AG
Grosspeteranlage 5
4052 Basel, Switzerland

This is a reprint of the Special Issue, published open access by the journal *Veterinary Sciences* (ISSN 2306-7381), freely accessible at: www.mdpi.com/journal/vetsci/special_issues/8785H55I8M.

For citation purposes, cite each article independently as indicated on the article page online and using the guide below:

Lastname, A.A.; Lastname, B.B. Article Title. *Journal Name* **Year**, *Volume Number*, Page Range.

ISBN 978-3-7258-2760-2 (Hbk)
ISBN 978-3-7258-2759-6 (PDF)
https://doi.org/10.3390/books978-3-7258-2759-6

Contents

About the Editor

George Valiakos

George Valiakos has been an Associate Professor at the Laboratory of Microbiology and Parasitology, Faculty of Veterinary Medicine, University of Thessaly since December 2017. He is a graduate of the Veterinary School of the Aristotle University of Thessaloniki, has a master's degree from the Department of Medicine of the University of Thessaly related to public health, and a doctorate from the Faculty of Veterinary Science of the same university. He is the author and co-author of more than 40 scientific publications, with over 900 references and a total h-index of 16 (Google Scholar). He has participated in more than 75 oral and poster announcements at national and international scientific conferences. His research interests include use of antimicrobials in veterinary medicine, antibiotic resistance, one health, and risk factor recognition for infectious diseases of animals.

Preface

Antimicrobial resistance (AMR) presents a serious challenge in today's world. The use of antimicrobials (AMU) significantly contributes to the emergence and spread of resistant bacteria. Historically, discussions surrounding AMU have primarily focused on quantitative aspects, with particular emphasis on food-producing animals, which account for the highest volume of antimicrobial consumption.

However, companion animals gain recognition as potential reservoirs and vectors for transmitting resistant microorganisms to both humans and other animals. The full extent of this transmission remains unclear, which is particularly concerning given the substantial and growing number of households with companion animals—approaching 200 million in Europe alone.

This situation highlights critical knowledge gaps in our understanding of risk factors and transmission pathways for AMR transfer between companion animals and humans. Moreover, there's a significant lack of information regarding AMU in everyday veterinary practices for companion animals. The exploration and development of alternative therapeutic approaches to antimicrobial treatments of companion animals represents also a research priority.

To address these pressing issues, this Reprint aims to compile and disseminate crucial additional knowledge. It serves as a platform for relevant research studies and reviews, shedding light on the complex interplay between AMU, AMR, and the role of companion animals in this global health challenge. This Reprint is especially addressed to companion animal veterinary practitioners as well as all researchers working on the field of AMR in both animals and humans, from a One Health perspective.

George Valiakos
Guest Editor

veterinary sciences

Article

Antibiotic Recommendations for Treatment of Canine Stromal Corneal Ulcers

Milan Joksimovic [1], Bradley A. Ford [2], Tatjana Lazic [1,3], Ivan Soldatovic [4], Sergey Luzetsky [3] and Sinisa Grozdanic [1,3,*]

1 Animal Eye Consultants of Iowa, 698 Boyson Road, Hiawatha, IA 52233, USA
2 Department of Pathology, Roy J. and Lucille A. Carver College of Medicine, University of Iowa, Iowa City, IA 52242, USA
3 Oculus–Ophthalmology Specialty Hospital, 11000 Belgrade, Serbia
4 Department of Biostatistics, College of Medicine, University of Belgrade, 11000 Belgrade, Serbia
* Correspondence: sgrozdan@gmail.com

Simple Summary: Infection of the cornea is among the most frequent causes for the loss of vision in dogs. The purpose of this study was to determine which particular antibiotics can be used immediately at the time of infection to eliminate bacteria from the infected region and prevent the loss of the eye. This study showed that combinations of antibiotics (amikacin and neopolybac or ofloxacin and neopolybac) are potentially the best first choice of treatment to eliminate the majority of commonly isolated bacteria from corneal infections in dogs.

Abstract: The aim of the study was to identify the aerobic bacterial isolates and determine corresponding antibiotic susceptibility profiles in vitro in canine clinical specimens with stromal corneal ulcers, with the goal of providing recommendations for first-line treatment with antibiotics. A total of 198 canine corneal stromal ulcer samples were studied between 2018 and 2021. A corneal swab was collected and cultured under aerobic conditions. Bacterial organisms were identified at the species level by MALDI-TOF mass spectrometry. Antibiotic susceptibility testing for commonly used topical and systemic antibiotics was performed by disk diffusion. Bacterial growth was obtained from 80% of samples. A variety of bacterial species were identified wherein the most common specimens were represented by *Staphylococcus pseudintermedius* (22%), *Staphylococcus epidermidis* (12%), *Staphylococcus capitis* (11%), and *Pseudomonas aeruginosa* (10%). Based on the overall antibiotic susceptibility data, neopolybac alone (96%) or a combination of neopolybac with either ofloxacin or amikacin (each 99%) showed the best coverage for commonly isolated bacterial organisms from canine corneal stromal ulcers. Results of this study support the use of the combined antibiotics as the first-line response for the treatment of canine corneal stromal ulcers. A statically significant increase in acquired bacterial resistance was detected during the longitudinal data observation.

Keywords: corneal; stromal; ulcer; bacterial; canines; antibiotic

Citation: Joksimovic, M.; Ford, B.A.; Lazic, T.; Soldatovic, I.; Luzetsky, S.; Grozdanic, S. Antibiotic Recommendations for Treatment of Canine Stromal Corneal Ulcers. *Vet. Sci.* **2023**, *10*, 66. https://doi.org/10.3390/vetsci10020066

Academic Editor: George Valiakos

Received: 28 November 2022
Revised: 10 January 2023
Accepted: 12 January 2023
Published: 17 January 2023

1. Introduction

The ocular surface is constantly exposed to a variety of environmental stimuli and contains different mechanisms which function as a first level of eye defense against possible pathogens. Bacteria often invade the damaged corneal surface, which in turn may lead to the acceleration of corneal tissue loss, resulting in structural integrity defects and potential loss of the eye [1,2]. The consequences could be vision-threatening and devastating for eye globe integrity if the corneal infection process is not immediately and aggressively treated or the causative bacterial organism is resistant to empiric antibiotic treatment [3]. The first step in treating corneal bacterial infections is empiric therapy based on epidemiological data and use of suggested antimicrobials [4]. While large epidemiological and corneal pathogen surveillance studies have been reported in humans, similar datasets are relatively sparsely reported in veterinary medicine, so the initial selection of the antibiotic treatment is frequently chosen based on personal preference and in-hospital ophthalmic drug availability [4–10].

A number of studies have evaluated a microbial community in canine corneal ulcers [5–15]. The most frequent bacterial groups identified are Gram-positive staphylococci and streptococci in addition to Gram-negative *Pseudomonas aeruginosa* [3,5,7–9,11–14,16,17].

The primary purpose of this manuscript was to perform in depth analysis of antimicrobial activity for commonly identified bacteria from canine corneal stromal ulcers and provide general guidelines for the immediate initiation of empiric antibiotic therapy, which may have the highest chance of being effective while waiting for the results of laboratory microbial identification and antibiotic susceptibility. Furthermore, we intended to evaluate trends in the antibiotic resistance development over a four-year period with a goal of providing predictive data for future topical antibiotic use for canine corneal stromal ulcers.

2. Materials and Methods

Canine corneal ulcer samples were harvested using a flocked swab kit and placed in the provided transport media (BD ESwab™ Collection Kit, COPAN ITALIA SpA, Brescia, Italy). All samples were collected 30 s after applying topical anesthetic on the ocular surface by gently rolling over the corneal surface for 10 s (Propracaine 0.05%, Akorn Pharmaceuticals, Lake Forest, IL, USA). The samples were then kept on ice packs until submitted to the laboratory or refrigerated at 4 °C and subsequently cultured after 1 to 5 days of collection.

Bacterial swabs were collected in the period from December 2018 to April 2021 from canine patients with corneal stromal ulcers presented to Animal Eye Consultants of Iowa in the state of Iowa, USA. All patients had a complete eye examination. The inclusion criteria for corneal stromal ulcers were presence of the corneal defect affecting at least 10% of the corneal stromal thickness with the clinical signs of cellular neutrophilic infiltrates with or without evidence of corneal melting. Half of each ESwab tube solution with a collected sample (approximately 0.5 mL) was cultured onto MacConkey agar (Hardy Diagnostics, Santa Maria, CA, USA), while the other half of the sample was cultured onto Chocolate agar (Hardy Diagnostics Hardy Diagnostics, Santa Maria, CA, USA). The plates were incubated at 37 °C in 5% CO_2 and examined at 24, 48, and 96 h after plating for bacterial colonies.

Antibiotic susceptibility testing was performed by Kirby–Bauer disc diffusion method in all isolates following Clinical and Laboratory Standards Institute (CLSI) guidelines (https://clsi.org/media/3481/m100ed30_sample.pdf, accessed on 10 January 2023 and https://clsi.org/media/2321/vet08ed4_sample.pdf, accessed on 10 January 2023). An ophthalmology antibiotic panel was developed based on the most frequently used and commercially available topical ophthalmic antibiotics in the midwestern US. Additionally, amoxicillin/clavulanic acid was added to the panel as this antimicrobial is frequently used as a systemic antibiotic after different ophthalmic surgeries. The topical antibiotic set included amikacin (30 μg), bacitracin (10 U), cefazolin (30 μg), cefoxitin (30 μg), gentamicin (10 μg), neomycin (30 μg), ofloxacin (5 μg), oxacillin (1 μg; used instead of cefoxitin in the case of *Staphylococcus pseudintermedius* and *Staphylococcus schleiferi* per CLSI guidelines; https://clsi.org/media/2321/vet08ed4_sample.pdf, accessed on 10 January 2023), polymyxin B (300 U), tetracycline (30 μg), and tobramycin (10 μg). The systemic antibiotic set included amoxicillin/clavulanic acid (20/10 μg), cephalexin (30 μg), ciprofloxacin (5 μg), clindamycin (2 μg), doxycycline (30 μg), enrofloxacin (5 μg), marbofloxacin (5 μg), penicillin G (10 U), and sulfamethoxazole/trimethoprim (1.25/23.75 μg). *Escherichia coli* ATCC 25922 and *Staphylococcus aureus* ATCC 25923 were the quality control organisms. All antibiotic discs and control cultures were provided by Hardy Diagnostics (Santa Maria, CA, USA) and Microbiologics INC (St Cloud, MN, USA), respectively.

Bacteria were identified to the species level by matrix-assisted laser desorption/ionization-time of flight (MALDI-TOF) analysis following manufacturer's instructions (Bruker, Madison, WI, USA). Briefly, a single colony no older than 5 days was taken from the culture plate with a toothpick. On the target plate, a thin bacterial layer was smeared onto a single spot and then the same specimen was placed onto the next spot to achieve a thinner bacterial layer. Each sample was covered with 1 μL of 100% formic acid and air dried and then 1 μL of HCCA matrix was added to each spot as instructed by manufacturer. Bacterial Test Standard, *Escherichia coli* ATCC 25922, *Enterococcus faecalis* ATCC 29212, and *Candida albicans* ATCC 10231 were used as quality controls (Microbiologics, INC; St Cloud, MN, USA). MALDI-TOF analysis was performed in a CLIA-certified diagnostic laboratory (Clinical Microbiology Laboratory, University of Iowa Hospitals & Clinics, Iowa City, IA, USA) using the Bruker BioTyper RUO Database which included continuously updated versions of the Compass reference library (in which veterinary isolates are well represented) as well as the optional mycobacterial and fungal libraries.

WHONET database software (World Health Organization) was used with 2022 CLSI breakpoints for dogs. If breakpoints were not available for dogs, other CLSI animal breakpoints were used followed by human CLSI breakpoints if no other animal breakpoints were available. These breakpoints were based on CLSI M100 Performance Standards for Antimicrobial Susceptibility Testing (https://clsi.org/media/3481/m100ed30_sample.pdf, accessed on 10 January 2023), CLSI VET08 Performance Standards for Antimicrobial Disk and Dilution Susceptibility Tests for Bacteria Isolated From Animals (https://clsi.org/media/2321/vet08ed4_sample.pdf, accessed on 10 January 2023), and Hardy Diagnostics Disk Diffusion Zone Diameter Chart (https://www.keyscientific.com/files/Other%20Manufacturers/Hardy%20Diagnostics/AST%20Discs/Hardy%20AST%20Disc%20Insert.pdf, accessed on 10 January 2023). WHONET database software was also used to manage and analyze microbiology laboratory data and antibiotic susceptibility test results. Hardy Diagnostics Disk Diffusion Zone Diameter Chart https://www.keyscientific.com/files/Other%20Manufacturers/Hardy%20Diagnostics/AST%20Discs/Hardy%20AST%20Disc%20Insert.pdf, accessed on 10 January 2023) was also used for polymyxin B breakpoints in *Pseudomonas aeruginosa* isolates (Resistant $\leq$ 11; Susceptible $\leq$ 12).

The Alere^{TM} PBP2A SA Culture Colony Test was performed to detect penicillin-binding protein 2A (PBP2A) in staphylococcal isolates according to manufacturer's instructions (Alere Scarborough, Inc., 10 Southgate Road, Scarborough, ME 04074, USA). In the case of *Staphylococcus pseudintermedius* and *Staphylococcus schleiferi* the methicillin resistance was confirmed with oxacillin disks per CLSI VET08 guidelines (https://clsi.org/media/2321/vet08ed4_sample.pdf, accessed on 10 January 2023).

All bacterial isolates were separated into susceptible and resistant categories according to the interpretive criteria above. Susceptible and intermediate levels of response were assigned to the susceptible class for the purposes of antibiogram creation [5,6,10].

Each antibiotic was classified into the antibiotic categories of aminoglycosides (amikacin, gentamicin, neomycin, and tobramycin), polypeptides/polymyxins (bacitracin, polymyxin B), anti-staphylococcal β-lactams (cephamycins, oxacillin, cefoxitin), tetracyclines (tetracycline), non-extended spectrum cephalosporins (1st and 2nd generation cephalosporins, cefazolin), penicillin and β-lactamase inhibitors (amoxicillin/clavulanic acid), and fluoroquinolones (ofloxacin). Each clinical isolate was classified by group based upon its susceptibility data according to resistance pattern as not multidrug-resistant (Not MDR), multidrug-resistant (MDR), or possible extensively multidrug-resistant (possible XDR). The MDR group was defined as resistance to at least one antibiotic in three or more antibiotic categories. XDR was defined as resistant to at least one antibiotic in all but two or fewer antibiotic categories (i.e., bacterial isolates remain susceptible to only one or two categories) as previously proposed [18,19]. Additionally, intrinsic resistance of an isolate to a particular antibiotic was excluded from this analysis as previously suggested [20].

Statistical analyses were performed using a paired *t*-test and contingency table analyses (chi-square and Fisher's exact tests) for the indicated observed parameters with commercial software as described in the manuscript (Prism, version 5.0; GraphPad, San Diego, CA, USA).

3. Results

3.1. Bacterial Growth from Patient Samples

A total of 187 dogs (198 eyes) with corneal stromal ulcers were subjected to sample collection and a total of 198 samples were plated; 159/198 (80.3%) of plated samples demonstrated bacterial growth, while 39/198 (19.7%) yielded no growth. A total of 167 isolates were collected. Regarding prior antibiotic exposure, 134/198 (67.7%) samples were collected from patients having previous antibiotic treatment, while 64/198 (32.3%) samples were collected from patients with no previous antibiotic treatment; 101 of 134 (75.4%) plated samples resulted in the growth of isolates, while 33/134 (24.6%) yielded no growth in the group of patients with previous antibiotic treatment. Regarding patients with no previous antibiotic treatment, 54/64 (84.4%) plated samples returned growth of isolates while 10/64 (15.6%) resulted in no isolate growth.

3.2. Distribution of Bacterial Species in Patients Diagnosed with Corneal Stromal Ulcers

The most common bacterial species identified in the corneal stromal samples was *Staphylococcus pseudintermedius* present in 22% of samples. *Staphylococcus epidermidis, Staphylococcus capitis,* and *Pseudomonas aeruginosa* were present in 12%, 11%, and 10% of samples, respectively. *Enteric Gram-negative rods, coagulase negative staphylococci,* and *Streptococcus canis* were present in 7%, 5%, and 5% of samples, respectively (Table 1).

Table 1. Bacterial species distribution from corneal stromal ulcer samples. Distribution of incidence of the specific microorganism detection in canine patients. Data are presented as a percentage of total isolates (n = 167). [a] *Citrobacter freundii, Enterobacter cloacae, Escherichia coli,* and *Serratia marcescens*; [b] *Staphylococcus* spp. (*S. auricularis, S. hominis, S. saprophyticus, S. simulans, S. warneri, and S. xylosus*) *excluding S. pseudintermedius, S. capitis, S. epidermidis, Staphylococcus intermedius and S.* aureus. [c] *Gram-positive* and *Gram-negative bacteria* (this group was identified by Gram labeling since MALDI-TOF analysis yielded no identification) and *unknown* (this group was also not identified by MALDI-TOF analysis). These two groups likely represent multiple bacterial species on the target plate spot or no hit in the MALDI-TOF data base. [d] Presence of each isolate of 1% or less represented by *Acinetobacter johnsonii, Actinomyces* sp., *Bacillus pumilus, Bacillus* sp., *Exiguobacterium* sp., *Klebsiella oxytoca, Kocuria* sp., *Microbacterium* sp. *Micrococcus luteus, Moraxella canis, Pasteurella canis, Pseudomonas* sp., *Psychrobacter* sp., *Staphylococcus aureus, Streptococcus lutetiensis,* and *Streptococcus salivarius.*

Organism	**(n = 167)**
Staphylococcus pseudintermedius	22%
Staphylococcus epidermidis	12%
Staphylococcus capitis	11%
Pseudomonas aeruginosa	10%
enteric Gram-negative rods [a]	7%
coagulase negative staphylococci [b]	5%
Streptococcus canis	5%
Corynebacterium sp.	2%
Enterococcus faecalis	2%
Streptococcus sp.	2%
Rothia sp.	2%
Enterococcus faecium	1%
Staphylococcus aureus	1%
Staphylococcus intermedius	1%
Other [c]	5%
Other [d]	10%

3.3. Antibiotic Susceptibility Pattern of Bacteria Isolated from Corneal Stromal Ulcers

To gain insight into which antibiotics to use in treating current canine corneal stromal ulcers, we analyzed resistance profiles of isolates relative to a single or combination antibiotic between two time points, 2018–2019 and 2020–2021. Based on overall antibiotic susceptibility data, neopolybac alone (96%) or a combination of neopolybac with either ofloxacin or amikacin (each 99%) showed the best antibiotic coverage for commonly isolated bacterial organisms from canine corneal stromal ulcers (Figure 1). No statistically significant difference ($p = 0.1637$, paired t-test) was observed in bacterial resistance to a single or combination antibiotic between these two time points (Figure 1). For all samples tested with topical antibiotics, bacterial species were most frequently resistant to polymyxin B, oxacillin, cefoxitin, and cefazolin. The least resistance was detected against amikacin, gentamicin, and ofloxacin (Table 2). When acquired resistance was analyzed, a similar trend of resistance was observed (Table 2).

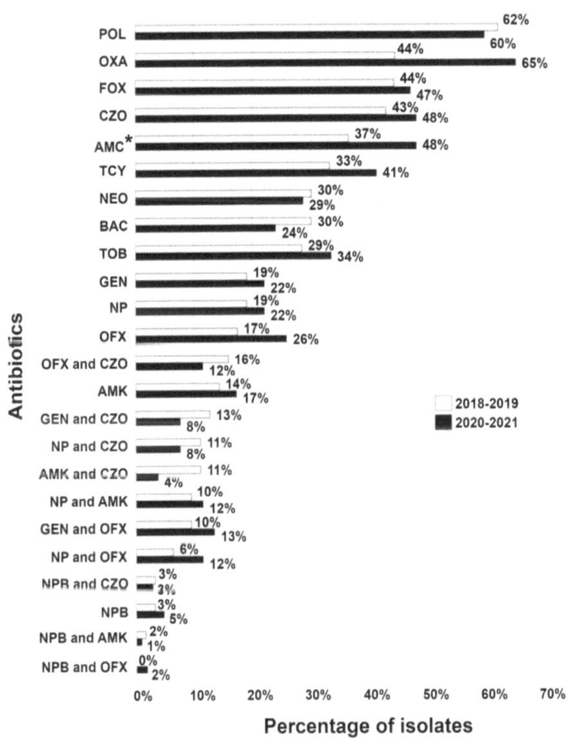

Figure 1. Percentage of total resistant isolates to a single or combination antibiotics from 2018–2019 and 2020–2021. Percentage (%) of overall resistant bacterial species isolated from patients with corneal stromal ulcers (combined intrinsic and acquired resistance). For the period 2018–2019, the number of isolates is 63 for each antibiotic/antibiotic combination except for oxacillin, which numbered 9. For the period 2020–2021, the number of isolates is 104 for each antibiotic/antibiotic combination except for oxacillin, which numbered 20. * = although amoxicillin/clavulanic acid is not used topically, data for this antibiotic are presented here since it is often used postoperatively as a systemic antibiotic. Abbreviations: AMC, amoxicillin/clavulanic acid; AMK, amikacin; BAC, bacitracin; CZO, cefazolin; FOX, cefoxitin; GEN, gentamicin; NEO, neomycin; NP, neomycin and polymyxin B; NPB, neomycin, polymyxin B, and bacitracin; OFX, ofloxacin; OXA, oxacillin; POL, polymyxin B; TCY, tetracycline; TOB, tobramycin.

Table 2. Resistance profile of isolates from corneal stromal ulcer samples from 2018–2021. Percentage (%) of resistant bacterial species from patient samples presented as combined resistance (TOTAL; intrinsic and acquired resistance together), intrinsic resistance only (INTRINSIC), or acquired resistance only (ACQUIRED). The number of isolates is 167 for each antibiotic in the TOTAL column exceptfor oxacillin, which numbered 29. * = although amoxicillin/clavulanic acid is not used topically, data for this antibiotic are presented here since it is often used postoperatively as a systemic antibiotic.

Antibiotic	Total	Intrinsic	Acquired
polymyxin B	60%	1%	59%
oxacillin	59%	0%	59%
cefoxitin	46%	7%	39%
cefazolin	46%	11%	35%
amoxicillin/clavulanic acid *	44%	10%	34%
tetracycline	38%	5%	33%
tobramycin	32%	5%	27%
neomycin	29%	10%	19%
bacitracin	26%	0%	26%
ofloxacin	23%	0%	23%
gentamicin	21%	4%	17%
amikacin	16%	6%	10%

For all samples tested with systemic antibiotics, bacterial species were predominantly resistant to penicillin G (74%), cephalexin (65%), clindamycin (59%), amoxicillin/clavulanic acid (44%), ciprofloxacin (41%), sulfamethoxazole/trimethoprim (41%), and doxycycline (34%), while the least resistance was seen against enrofloxacin (21%), and marbofloxacin (15%).

3.4. *Increase in Acquired Resistance in Isolates from Corneal Stromal Ulcers*

To gain insight into the temporal dynamics of acquired resistance, we analyzed isolates between two time points, 2018–2019 and 2020–2021. In the period 2018–2019, the highest percentage of isolates was resistant to polymyxin B, oxacillin, and cefoxitin with a similar trend in the period 2020–2021 (Figure 2). The lowest percentage of isolates was resistant to amikacin, gentamicin, ofloxacin, and neomycin in the period 2018–2019 with a comparable tendency in the period 2020–2021. Strikingly, a temporal increase in acquired resistance was statistically significant ($p = 0.0025$, paired t-test) from 2018–2019 to 2020–2021 (Figure 2).

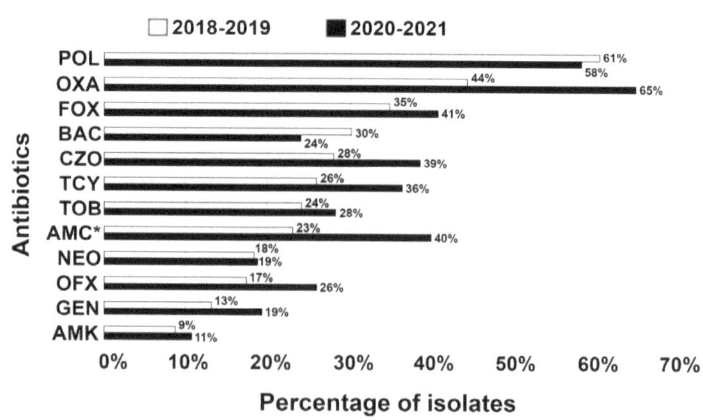

Figure 2. Increase in acquired resistance in isolates from corneal stromal ulcers between 2018–2019 and 2020–2021. Percentage (%) of resistant bacterial species from patient samples excluding intrinsic

resistance of isolates. The number of isolates ranges from 9 to 63 for the period 2018–2019 and from 20 to 104 for the period 2020–2021. * = although amoxicillin/clavulanic acid is not used topically, data for this antibiotic are presented here since it is often used postoperatively as a systemic antibiotic. Abbreviations: AMC, amoxicillin/clavulanic acid; AMK, amikacin; BAC, bacitracin; CZO, cefazolin; FOX, cefoxitin; GEN, gentamicin; NEO, neomycin; OFX, ofloxacin; OXA, oxacillin; POL, polymyxin B; TCY, tetracycline; TOB, tobramycin.

3.5. Distribution of Bacterial Species in Patients Diagnosed with Corneal Stromal Ulcers Relative to Previous Patient's Antibiotic Treatment

The most common bacterial species identified in the corneal stromal samples relative to the patient's previous antibiotic treatments were Staphylococcus pseudintermedius, Staphylococcus epidermidis, Staphylococcus capitis, and Pseudomonas aeruginosa (Table 3). When compared to the overall data in Table 1, most of isolates were present in comparable percentages. Accordingly, no statistically significant difference ($p = 0.3367$, paired t-test) was observed in the distribution of bacterial species relative to the patient's previous antibiotic treatments.

Table 3. Bacterial species distribution from corneal stromal ulcer samples relative to the patient's previous antibiotic treatments. Distribution of isolates in canine patients. Data are presented as a percentage of total isolates ($n = 61$ and $n = 106$). [a] *Citrobacter freundii, Enterobacter cloacae, Escherichia coli,* and *Serratia marcescens;* [b] *Staphylococcus* spp. (*S. auricularis, S. hominis, S. saprophyticus, S. simulans, S. warneri,* and *S. xylosus*) excluding *S. pseudintermedius, S. capitis, S. epidermidis, Staphylococcus intermedius* and *S. aureus.* [c] Gram-positive and Gram-negative bacteria (this group was identified by Gram labeling since MALDI-TOF analysis yielded no identification) and unknown (this group was also not identified by MALDI-TOF analysis). These two groups likely represent multiple bacterial species on the target plate spot or no hit in the MALDI-TOF data base. [d] Presence of each isolate of 1% or less represented by *Acinetobacter johnsonii, Actinomyces* sp., *Bacillus pumilus, Bacillus* sp., *Exiguobacterium* sp., *Klebsiella oxytoca, Kocuria* sp., *Microbacterium* sp. *Micrococcus luteus, Moraxella canis, Pasteurella canis, Pseudomonas* sp., *Psychrobacter* sp., *Streptococcus lutetiensis,* and *Streptococcus salivarius.*

Organism	No Previous Antibiotic Treatment ($n = 61$)	Previous Antibiotic Treatment ($n - 106$)
Staphylococcus pseudintermedius	34%	15%
Staphylococcus epidermidis	2%	18%
Staphylococcus capitis	8%	13%
Pseudomonas aeruginosa	15%	8%
enteric Gram-negative rods [a]	7%	7%
coagulase negative staphylococci [b]	7%	5%
Streptococcus canis	5%	6%
Corynebacterium sp.	0%	3%
Enterococcus faecalis	2%	2%
Streptococcus sp.	3%	1%
Rothia sp.	2%	3%
Enterococcus faecium	0%	2%
Staphylococcus aureus	0%	2%
Staphylococcus intermedius	3%	0%
Other [c]	3%	7%
Other [d]	10%	10%

3.6. Susceptibility Profile of Isolates from Corneal Stromal Ulcers Relative to Patient's Previous Antibiotic Treatments

To examine a trend of acquired resistance relative to previous the patient's antibiotic treatments, we analyzed isolates within an approximately three-year period from 2018 to 2021. In the group with no previous exposure to antibiotics, the highest percentage of isolates was resistant to oxacillin, polymyxin B, cefoxitin, and tetracycline with a similar trend in the group previously treated with antibiotics (Figure 3). The lowest percentage of isolates was resistant to amikacin and ofloxacin in both groups. Furthermore, no statistically significant changes ($p = 0.0977$, paired t-test) in acquired resistance were observed relative to the patient's previous antibiotic treatments within the examined three-year period (Figure 3). Furthermore, we analyzed total isolate resistance against the most commonly prescribed antibiotics by local non-specialty veterinary practices, ofloxacin and tobramycin, relative to the patient's previous antibiotic treatments. In either case, no statistically significant changes in resistance were observed relative to previous patient's antibiotic treatments (ofloxacin; odds ratio = 1.744; CI= 0.3717–8.647, $p = 0.7163$; tobramycin (odds ratio = 0.2857; CI = 0.05942–1.240, $p = 0.1919$; contingency table analyses (chi-square and Fisher's exact tests)).

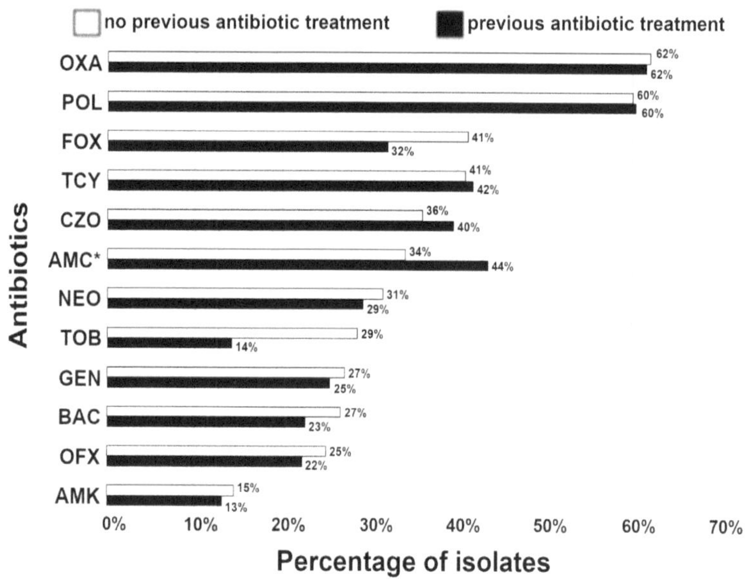

Figure 3. Acquired resistance in isolates from corneal stromal ulcers regarding the patient's previous antibiotic treatments. Percentage (%) of resistant bacterial species against antibiotics excluding intrinsic resistance of isolates. The number of isolates range from 16 to 60 for the patients without previous antibiotic treatments and from 13 to 106 for the patients with previous antibiotic treatments. * = although amoxicillin/clavulanic acid is not used topically, data for this antibiotic are presented here since it is often used postoperatively as a systemic antibiotic. Abbreviations: AMC, amoxicillin/clavulanic acid; AMK, amikacin; BAC, bacitracin; CZO, cefazolin; FOX, cefoxitin; GEN, gentamicin; NEO, neomycin; OFX, ofloxacin; OXA, oxacillin; POL, polymyxin B; TCY, tetracycline; TOB, tobramycin.

3.7. Resistance Profile Based on the Percentage of Isolates Resistant to Multiple Antibiotics in Canine Corneal Stromal Ulcers

The highest percentage of bacteria was not resistant to any tested antibiotic (24/167; 14%) followed by bacteria resistant to one antibiotic (22/167; 13%), while the vast majority of isolates (73%) were resistant to two or more tested antibiotics (Figure 4). Furthermore, no statistically significant difference (p = 0.1855, paired t-test) was observed in the percentage of isolates resistant to multiple antibiotics relative to the patient's previous antibiotic treatments. Some highly aggressive isolates showed antibiotic resistance to nine or more tested antibiotics (Figure 4), such as a particular case of *Staphylococcus pseudintermedius* (Figure 5).

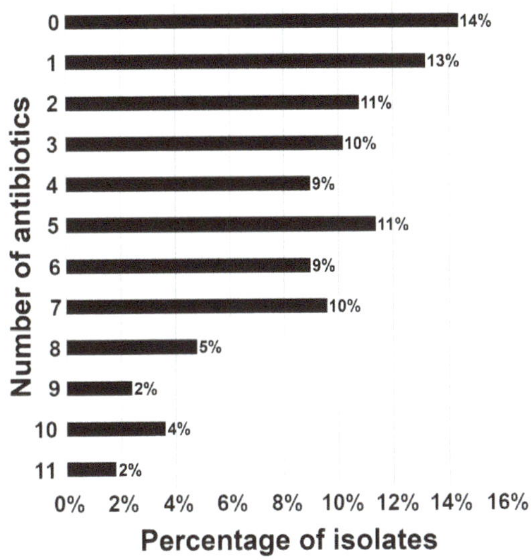

Figure 4. Distribution of resistant isolates against multiple antibiotics. Percentage of the isolates that are resistant to the corresponding number of antibiotics. Bacterial samples were isolated from canine corneal stromal ulcers (number of isolates = 167).

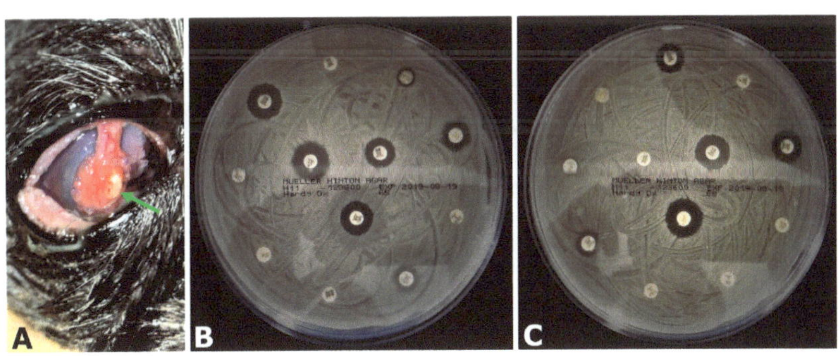

Figure 5. Aggressive corneal infection by *Staphylococcus pseudintermedius*. (**A**) Patient cornea was severely affected with a focal zone of conjunctival graft destruction at 4 o'clock position in the paracentral graft region (green arrow). Patient was treated with topical ofloxacin and cefazolin and systemic amoxicillin/clavulanic acid immediately after surgery. (**B**,**C**) Antibiotic susceptibility testing performed for topical (**A**) and systemic antibiotic sets (**B**). The isolate was resistant to the multiple topical (amikacin, cefazolin, cefoxitin, gentamicin, neomycin, ofloxacin, polymyxin B, tetracycline, and tobramycin) and systemic antibiotics (amoxicillin/clavulanic acid, cephalexin, ciprofloxacin, clindamycin, doxycycline, enrofloxacin, marbofloxacin, penicillin G, and sulfamethoxazole/trimethoprim). The isolate was also penicillin-binding protein 2A–positive and susceptible only to topical bacitracin.

3.8. Distribution of Multidrug-Resistant (MDR) Bacteria in Clinical Corneal Ulcers

Overall, in the corneal stromal ulcer samples, 62% (103/167) were not MDR isolates, 20% (33/167) were MDR isolates, while 18% (31/167) were possible XDR isolates from 2018–2021. Regarding the patient's previous antibiotic treatments, no statistically significant difference (p = 0.4325, paired t-test) was observed in the distribution of multidrug-resistant bacteria. In the period 2018–2019, 67% (42/63) were not MDR isolates and 19% (12/63) were MDR isolates, while 14% (9/63) were possible XDR isolates. In the period 2020–2021, 59% (61/104) were not MDR isolates and 20% (21/104) were MDR isolates, while 21% (22/104) were possible XDR isolates. Moreover, no statistically significant difference (p = 0.2777, paired t-test) was observed in multidrug resistance between these two time points.

3.9. Methicillin-Resistant Staphylococcus spp.

To assess methicillin resistance in *Staphylococcus* species, the isolates were tested with cefoxitin disks [21]. A few species were also tested with the penicillin-binding protein 2A (PBP2A) antibody test. Overall, in the 2018–2021 period, 49% (40/81) of isolates were methicillin-resistant. Relative to the patient's previous antibiotic treatments, similar percentages of resistance were detected. In the group with no previous antibiotic treatment, 54% (15/28) of isolates were methicillin resistant. In the group with previous antibiotic treatment 47% (25/53) of isolates were methicillin resistant. Over time, a substantial increase in the number of methicillin-resistant *Staphylococcus* species were detected from 2018–2019 (39%; 11/28) to 2020–2021 (55%; 29/53).

4. Discussion

In this study, the overall bacterial growth from patients with corneal stromal ulcers (80.3%) was generally higher than previously reported rates (57–71%) [5,7,8,10,13,22,23]. The higher positive culture rate in our study may reflect the use of improved elution swabs (BD ESwab™ Collection Kit) and possibly be due to increased yield of plating half of the volume of each collection swab onto a culture plate rather than a subset of the total volume adsorbed onto the swab.

The most common bacterial isolates from the stromal corneal ulcers in our study were *Staphylococcus* spp. accounting for 50% of all bacterial species, consistent with or slightly above levels reported in previous studies that were performed in various geographical locations (Table 4) [5,6,8,9,11,13,15,22,24–26]. In contrast, studies from Australia and UK reported *Pseudomonas aeruginosa* and *Streptococcus* spp. as the most commonly isolated bacteria from canine ulcers (Table 4) [7,10]. This discrepancy could be potentially explained by regional differences of bacterial species in various geographical locations due to local climate factors [4,9], wherein weather conditions in the midwestern US are not as warm and humid as in the southeastern US, Brazil, or Australia (Table 4).

Based on the overall antibiotic susceptibility data, neopolybac alone (96%) or a combination of neopolybac with either ofloxacin or amikacin showed the best coverage for commonly isolated bacterial organisms from canine corneal stromal ulcers in line with a previous report [15]. Considering that canine stromal corneal ulcers may be extremely aggressive, an immediate and aggressive initiation of antibiotic therapy with commercially available ophthalmic antibiotics (neopolybac and ofloxacin) may be the prudent strategy while waiting for the results of the microbial identification and susceptibility from the affected patient.

Data from this study described a trend of increased resistance to polymyxin B and ofloxacin when compared to previous studies [5,10,12]. However, a comparison of data from this study to the recent report [15], both performed at the same general geographical location (midwestern US), revealed substantial differences in topical susceptibility profiles for polymyxin B, bacitracin, and cefazolin (Table 5). This discrepancy can be partly explained by different methods used in these two studies (Kirby–Bauer disc diffusion method vs. minimum inhibitory concentration (MIC) susceptibility testing). Since the reliability of bacterial resistance to polymyxin B assessed by the Kirby–Bauer disc diffusion method is still questionable [27,28], reported polymyxin B data should be carefully scrutinized when making the clinical judgement on the choice of antibiotic. The same logic can be applied for bacitracin as this antibiotic, together with polymyxin B, belongs to the same antibiotic group of polypeptides. The trend of increased resistance to ofloxacin may reflect the acquisition of mutations through mobile genetic elements as reported in the case of *Pseudomonas aeruginosa* [29].

Table 4. Distribution of the most common bacterial species from canine corneal ulcers across various geographical locations.

Location	Author, Year	*Staphylococcus* spp.	*Pseudomonas aeruginosa*	*Streptococcus* spp.
Australia	Hindley et al., 2015 [7]	18%	31%	31%
Taiwan	Lin et al., 2007 [25]	49%	8%	7%
Thailand	Ekapopphan et al., 2018 [11]	46%	21%	8%
UK	Tsvetanova et al., 2020 [10]	14%	40%	28%
Switzerland	Suter et al., 2018 [8]	41%	11%	26%
Brazil	Prado et al., 2005 [24]	57%	5%	11%
Brazil	Varges et al., 2009 [26]	59%	-	-
Southeast US	Tolar et al., 2006 [6]	33%	21%	17%
Southeast US	McKeever, 2021 [5]	34%	18%	28%
Midwest US	Jinks et al., 2020 [12]	36%	10%	34%
Midwest US	Hewitt et al., 2020 [15]	32%	12%	19%
Midwest US	this study, 2022	50%	10%	7%

Table 5. Comparison of susceptibility profiles of isolates in the midwestern US between Hewitt et al. and this study.

	Topical Antibiotic Susceptibility	
Antibiotic	**Hewitt et al.**	**This Study**
amikacin	77%	84%
amikacin and cefazolin	79%	93%
bacitracin	7%	74%
cefazolin	8%	54%
gentamicin	74%	79%
gentamicin and cefazolin	76%	90%
gentamicin and ofloxacin	87%	88%
neomycin	76%	71%
neopoly	76%	79%
neopolybac	77%	96%
ofloxacin	53%	77%
ofloxacin and cefazolin	55%	87%
polymyxin B	0%	40%
tobramycin	57%	68%
	Systemic Antibiotic Susceptibility	
amoxicillin/clavulanic acid	78%	56%
cephalexin	23%	35%
clindamycin	61%	41%
doxycycline	56%	66%
enrofloxacin	64%	79%
marbofloxacin	75%	85%
penicillin	35%	26%
trimethoprim/sulfamethoxazole	53%	59%

In our study, a patient's previous antibiotic treatments do not affect isolate resistance to tobramycin in general. However, resistance to tobramycin appears to be increased between two time points, in line with a previous report performed at the same general geographical location [15] (Table 5; midwestern US). In contrast, the other study performed in the southeastern US demonstrated a substantially higher increase in and the percentage of resistant isolates to tobramycin [6]. As previously discussed, this inconsistency could be due to local climate factors in conjunction with the regional variation of bacterial species [4,9].

In the cases where severe corneal neovascularization is present or conjunctival pedicle graft surgery was performed so iatrogenic blood supply can be provided to the corneal ulcer region, treatments with topical medications can be complemented by systemic antibiotics. Based on data reported in this study, a systemic fluoroquinolone antibiotic (marbofloxacin, enrofloxacin) should be the first choice for treatment of corneal stromal ulcers, which is in line with a recently published report from the US Midwest on corneal ulcers [15]. In comparison to the earlier report from the southeastern US [6], in this study there was a tendency of increased resistance to enrofloxacin and ciprofloxacin, although this difference may be caused by regional differences in a distribution of bacterial species at different geographical locations and local climate factors [4,9].

In this study, we report a statistically significant increase in acquired resistance in isolates from 2018–2019 to 2020–2021. Our data not only point to this alarming trend but also indicate a presence of detectable deteriorating changes in antimicrobial susceptibility within a relatively short three-year period. However, the presence or absence of previous antibiotic treatments does not appear to influence an overall status of acquired bacterial resistance.

In our study, we detected 8% of isolates which showed antibiotic resistance to nine or more tested antibiotics. This pattern is of particular concern and qualifies corresponding isolates as potentially very aggressive pathogens causing corneal pathology poorly responsive to medical and surgical treatments.

The surge in antibiotic resistance is an alarming concern not only in global health care but also in animal ophthalmology [30]. In this study, over a third of isolates from clinical corneal stromal ulcers belong to the MDR group. However, none of *Pseudomonas aeruginosa* specimens belong to the MDR class, consistent with a previous report [15]. This study did not detect a statistically significant MDR increase between two time points 2018–2019 and 2020–2021 in contrast to the previous report, suggesting an MDR increase over time period of 2016–2020 [15].

Methicillin resistance of *Staphylococcus* spp. isolates is a serious concern in human and veterinary medicine due to the cross-species infectious behavior of these bacteria [31,32]. In this study, a half of *Staphylococcus* spp. isolates were methicillin resistant, which is in line with a previous report [33] with some isolates showing extremely aggressive clinical behavior (Figure 5).

In conclusion, the current study reports four important findings directly relevant to antibiotic treatments of canine corneal stromal ulcers: (1) clinical corneal stromal isolates showed increased acquired resistance within a three-year period; (2) many isolates were resistant to a large number of antibiotics; (3) over a third of analyzed specimens belong to the multidrug resistance group; and (4) some clinical isolates showed resistance to a combination of up to four antibiotics. Similar data have been recently reported in the ARMOR study from human corneal isolates; however, analysis of the resistance trend did not show gradual progression over a period of 10 years [4].

Analogous to earlier studies of antibiotic susceptibility in animals, key limitations of this study are the fact that CLSI interpretive criteria and breakpoints for particular bacterial species and antibiotic combinations are based on systemic minimum inhibitory concentration of antibiotics, since specific standards for corneal infections were never developed in human or veterinary medicine. Consequently, complete reliability of the Kirby–Bauer disc diffusion system as a method to assess corneal infections can be fully evaluated when these standards are developed. Until specific ophthalmology antibiotic standards become available, this study may provide a general guideline when initially choosing empirical therapies for treating canine corneal stromal ulcers while waiting for the patient-specific antibiotic susceptibility profile.

5. Conclusions

The results of this study support the use of the combined antibiotics as the first-line response for the treatment of canine corneal stromal ulcers. Neopolybac alone or a combination of neopolybac with either ofloxacin or amikacin is recommended as the initial antibiotic treatment while waiting for the patient-specific antibiotic susceptibility profile.

Author Contributions: Conceptualization: M.J. and S.G. Methodology: M.J., B.A.F., I.S., S.L. and S.G. Validation: M.J., B.A.F., T.L., I.S., S.L. and S.G. Formal analysis: M.J., I.S. and S.G. Investigation: M.J., T.L. and S.G. Resources: B.A.F., T.L., I.S. and S.G. Writing—original draft preparation: M.J. and S.G. Writing—review and editing: M.J., B.A.F., T.L., I.S., S.L. and S.G. Visualization: M.J., I.S. and S.G. Supervision: S.G. Project administration: M.J. and S.G. Funding acquisition: T.L. and S.G. All authors have read and agreed to the published version of the manuscript.

Funding: This research received no external funding.

Institutional Review Board Statement: The study was conducted in accordance with the Declaration of Helsinki and approved by the Animal Eye Consultants of Iowa Hospital Board. The study is performed as a retrospective study with analysis of database data without client animal identifiers, so IACUC protocol was not required.

Informed Consent Statement: An informed consent was provided to the owners prior to collecting samples.

Data Availability Statement: The data presented in this study are available on request from the corresponding author. The data are not publicly available due to privacy concerns (client identification information).

Conflicts of Interest: The authors declare no conflict of interest.

References

1. Rogers, C.M.; Scott, E.M.; Sarawichitr, B.; Arnold, C.; Suchodolski, J.S. Evaluation of the bacterial ocular surface microbiome in ophthalmologically normal dogs prior to and following treatment with topical neomycin-polymyxin-bacitracin. *PLoS ONE* **2020**, *15*, e0234313. [CrossRef]
2. Leis, M.L.; Costa, M.O. Initial description of the core ocular surface microbiome in dogs: Bacterial community diversity and composition in a defined canine population. *Vet. Ophthalmol.* **2019**, *22*, 337–344. [CrossRef]
3. Ledbetter, E.C.; Franklin-Guild, R.J.; Edelmann, M.L. Capnocytophaga keratitis in dogs: Clinical, histopathologic, and microbiologic features of seven cases. *Vet. Ophthalmol.* **2018**, *21*, 638–645. [CrossRef]
4. Asbell, P.A.; Sanfilippo, C.M.; Sahm, D.F.; DeCory, H.H. Trends In Antibiotic Resistance Among Ocular Microorganisms in the United States From 2009 to 2018. *JAMA Ophthalmol.* **2020**, *138*, 439–450. [CrossRef]
5. McKeever, J.M.; Ward, D.A.; Hendrix, D.V.H. Comparison of antimicrobial resistance patterns in dogs with bacterial keratitis presented to a veterinary teaching hospital over two multi-year time periods (1993–2003 and 2013–2019) in the Southeastern United States. *Vet. Ophthalmol.* **2021**, *24*, 653–658. [CrossRef]
6. Tolar, E.L.; Hendrix, D.V.; Rohrbach, B.W.; Plummer, C.E.; Brooks, D.E.; Gelatt, K.N. Evaluation of clinical characteristics and bacterial isolates in dogs with bacterial keratitis: 97 cases (1993–2003). *J. Am. Vet. Med. Assoc.* **2006**, *228*, 80–85. [CrossRef]
7. Hindley, K.E.; Groth, A.D.; King, M.; Graham, K.; Billson, F.M. Bacterial isolates, antimicrobial susceptibility, and clinical characteristics of bacterial keratitis in dogs presenting to referral practice in Australia. *Vet. Ophthalmol.* **2016**, *19*, 418–426. [CrossRef]
8. Suter, A.; Voelter, K.; Hartnack, S.; Spiess, B.M.; Pot, S.A. Septic keratitis in dogs, cats, and horses in Switzerland: Associated bacteria and antibiotic susceptibility. *Vet. Ophthalmol.* **2018**, *21*, 66–75. [CrossRef]
9. Wang, L.; Pan, Q.; Zhang, L.; Xue, Q.; Cui, J.; Qi, C. Investigation of bacterial microorganisms in the conjunctival sac of clinically normal dogs and dogs with ulcerative keratitis in Beijing, China. *Vet. Ophthalmol.* **2008**, *11*, 145–149. [CrossRef]
10. Tsvetanova, A.; Powell, R.M.; Tsvetanov, K.A.; Smith, K.M.; Gould, D.J. Melting corneal ulcers (keratomalacia) in dogs: A 5-year clinical and microbiological study (2014–2018). *Vet. Ophthalmol.* **2021**, *24*, 265–278. [CrossRef]
11. Ekapopphan, D.; Srisutthakarn, A.; Moonarmart, W.; Buddhirongawatr, R.; Bangphoomi, N. Identification and antimicrobial susceptibility of microorganisms isolated from severe corneal ulcers of dogs in Thailand. *J. Vet. Med. Sci.* **2018**, *80*, 1259–1265. [CrossRef] [PubMed]

12. Jinks, M.R.; Miller, E.J.; Diaz-Campos, D.; Mollenkopf, D.F.; Newbold, G.; Gemensky-Metzler, A.; Chandler, H.L. Using minimum inhibitory concentration values of common topical antibiotics to investigate emerging antibiotic resistance: A retrospective study of 134 dogs and 20 horses with ulcerative keratitis. *Vet. Ophthalmol.* **2020**, *23*, 806–813. [CrossRef] [PubMed]
13. Ledbetter, E.C.; Hendricks, L.M.; Riis, R.C.; Scarlett, J.M. In vitro fluoroquinolone susceptibility of *Pseudomonas aeruginosa* isolates from dogs with ulcerative keratitis. *Am. J. Vet. Res.* **2007**, *68*, 638–642. [CrossRef] [PubMed]
14. LoPinto, A.J.; Mohammed, H.O.; Ledbetter, E.C. Prevalence and risk factors for isolation of methicillin-resistant Staphylococcus in dogs with keratitis. *Vet. Ophthalmol.* **2015**, *18*, 297–303. [CrossRef] [PubMed]
15. Hewitt, J.S.; Allbaugh, R.A.; Kenne, D.E.; Sebbag, L. Prevalence and Antibiotic Susceptibility of Bacterial Isolates From Dogs With Ulcerative Keratitis in Midwestern United States. *Front. Vet. Sci.* **2020**, *7*, 583965. [CrossRef]
16. Ledbetter, E.C.; Mun, J.J.; Kowbel, D.; Fleiszig, S.M. Pathogenic phenotype and genotype of *Pseudomonas aeruginosa* isolates from spontaneous canine ocular infections. *Investig. Ophthalmol. Vis. Sci.* **2009**, *50*, 729–736. [CrossRef] [PubMed]
17. Leigue, L.; Montiani-Ferreira, F.; Moore, B.A. Antimicrobial susceptibility and minimal inhibitory concentration of *Pseudomonas aeruginosa* isolated from septic ocular surface disease in different animal species. *Open Vet. J.* **2016**, *6*, 215–222. [CrossRef]
18. Magiorakos, A.P.; Srinivasan, A.; Carey, R.B.; Carmeli, Y.; Falagas, M.E.; Giske, C.G.; Harbarth, S.; Hindler, J.F.; Kahlmeter, G.; Olsson-Liljequist, B.; et al. Multidrug-resistant, extensively drug-resistant and pandrug-resistant bacteria: An international expert proposal for interim standard definitions for acquired resistance. *Clin. Microbiol. Infect. Off. Publ. Eur. Soc. Clin. Microbiol. Infect. Dis.* **2012**, *18*, 268–281. [CrossRef]
19. Schwarz, S.; Silley, P.; Simjee, S.; Woodford, N.; van Duijkeren, E.; Johnson, A.P.; Gaastra, W. Editorial: Assessing the antimicrobial susceptibility of bacteria obtained from animals. *J. Antimicrob. Chemother.* **2010**, *65*, 601–604. [CrossRef]
20. Sweeney, M.T.; Lubbers, B.V.; Schwarz, S.; Watts, J.L. Applying definitions for multidrug resistance, extensive drug resistance and pandrug resistance to clinically significant livestock and companion animal bacterial pathogens. *J. Antimicrob. Chemother.* **2018**, *73*, 1460–1463. [CrossRef]
21. Bonjean, M.; Hodille, E.; Dumitrescu, O.; Dupieux, C.; Nkoud Mongo, C.; Allam, C.; Beghin, M.; Paris, M.; Borrel, O.; Chardon, H.; et al. Disk Diffusion Testing for Detection of Methicillin-Resistant Staphylococci: Does Moxalactam Improve upon Cefoxitin? *J. Clin. Microbiol.* **2016**, *54*, 2905–2909. [CrossRef]
22. Auten, C.R.; Urbanz, J.L.; Dees, D.D. Comparison of bacterial culture results collected via direct corneal ulcer vs conjunctival fornix sampling in canine eyes with presumed bacterial ulcerative keratitis. *Vet. Ophthalmol.* **2020**, *23*, 135–140. [CrossRef]
23. Gerding, P.A., Jr.; McLaughlin, S.A.; Troop, M.W. Pathogenic bacteria and fungi associated with external ocular diseases in dogs: 131 cases (1981–1986). *J. Am. Vet. Med. Assoc.* **1988**, *193*, 242–244.
24. Prado, M.R.; Rocha, M.F.; Brito, E.H.; Girão, M.D.; Monteiro, A.J.; Teixeira, M.F.; Sidrim, J.J. Survey of bacterial microorganisms in the conjunctival sac of clinically normal dogs and dogs with ulcerative keratitis in Fortaleza, Ceará, Brazil. *Vet. Ophthalmol.* **2005**, *8*, 33–37. [CrossRef]
25. Lin, C.T.; Petersen-Jones, S.M. Antibiotic susceptibility of bacterial isolates from corneal ulcers of dogs in Taiwan. *J. Small Anim. Pract.* **2007**, *48*, 271–274. [CrossRef]
26. Varges, R.; Penna, B.; Martins, G.; Martins, R.; Lilenbaum, W. Antimicrobial susceptibility of Staphylococci isolated from naturally occurring canine external ocular diseases. *Vet. Ophthalmol.* **2009**, *12*, 216–220. [CrossRef]
27. Behera, B.; Mathur, P.; Das, A.; Kapil, A.; Gupta, B.; Bhoi, S.; Farooque, K.; Sharma, V.; Misra, M.C. Evaluation of susceptibility testing methods for polymyxin. *Int. J. Infect. Dis.* **2010**, *14*, e596–e601. [CrossRef]
28. Jerke, K.H.; Lee, M.J.; Humphries, R.M. Polymyxin Susceptibility Testing: A Cold Case Reopened. *Clin. Microbiol. Newsl.* **2016**, *38*, 69–77. [CrossRef]
29. Khan, M.; Summers, S.; Rice, S.A.; Stapleton, F.; Willcox, M.D.P.; Subedi, D. Acquired fluoroquinolone resistance genes in corneal isolates of *Pseudomonas aeruginosa*. *Infect. Genet. Evol. J. Mol. Epidemiol. Evol. Genet. Infect. Dis.* **2020**, *85*, 104574. [CrossRef]
30. Soimala, T.; Lubke-Becker, A.; Hanke, D.; Eichhorn, I.; Fessler, A.T.; Schwarz, S.; Eule, J.C. Molecular and phenotypic characterization of methicillin-resistant *Staphylococcus pseudintermedius* from ocular surfaces of dogs and cats suffering from ophthalmological diseases. *Vet. Microbiol.* **2020**, *244*, 108687. [CrossRef]
31. Rich, M. Staphylococci in animals: Prevalence, identification and antimicrobial susceptibility, with an emphasis on methicillin-resistant Staphylococcus aureus. *Br. J. Biomed. Sci.* **2005**, *62*, 98–105. [CrossRef] [PubMed]
32. Malik, S.; Peng, H.; Barton, M.D. Antibiotic resistance in staphylococci associated with cats and dogs. *J. Appl. Microbiol.* **2005**, *99*, 1283–1293. [CrossRef] [PubMed]
33. Jones, R.D.; Kania, S.A.; Rohrbach, B.W.; Frank, L.A.; Bemis, D.A. Prevalence of oxacillin- and multidrug-resistant staphylococci in clinical samples from dogs: 1,772 samples (2001–2005). *J. Am. Vet. Med. Assoc.* **2007**, *230*, 221–227. [CrossRef] [PubMed]

veterinary sciences

MDPI

Systematic Review

Fosfomycin Resistance in Bacteria Isolated from Companion Animals (Dogs and Cats)

Marios Lysitsas [1], Irene Chatzipanagiotidou [2], Charalambos Billinis [1] and George Valiakos [1,*]

[1] Faculty of Veterinary Science, University of Thessaly, 43100 Karditsa, Greece; lysitsas91@gmail.com (M.L.); billinis@uth.gr (C.B.)
[2] Department of Biochemistry and Biotechnology, University of Thessaly, 41500 Larissa, Greece
* Correspondence: georgevaliakos@uth.gr

Simple Summary: Fosfomycin is an antibiotic with renewed interest over the last years, especially in human medicine, as it possesses some advantageous properties and a broad spectrum of bactericidal activity. Moreover, the emerging issue of multi-resistance against traditional antibiotics obligates specialists to turn to new alternative agents. However, cases of Fosfomycin-resistant strains are being detected at a rising rate worldwide, among humans and animals. The objective of this review is to collect, present and analyze studies related to Fosfomycin resistance in isolates from companion animals and specifically dogs and cats. Variable articles were collected and data for the relevant strains were scanned and evaluated. Since Fosfomycin is an agent not routinely used by veterinarians, the detection of Fosfomycin-resistant strains in canine and feline samples indicates possible dissemination of these strains among humans, pets, and the environment, reinforced by other factors. Concerning the origin, the species, and the resistance patterns of the related bacteria, useful conclusions were drawn, about the presence, the spreading, and possibly the causes of Fosfomycin resistance among companion animals and between them and their environment.

Abstract: Fosfomycin is an old antibacterial agent, which is currently used mainly in human medicine, in uncomplicated Urinary Tract Infections (UTIs). The purpose of this review is to investigate the presence and the characteristics of Fosfomycin resistance in bacteria isolated from canine or feline samples, estimate the possible causes of the dissemination of associated strains in pets, and underline the requirements of prospective relevant studies. Preferred Reporting Items for Systematic Reviews (PRISMA) guidelines were used for the search of current literature in two databases. A total of 33 articles were finally included in the review. Relevant data were tracked down, assembled, and compared. Referring to the geographical distribution, Northeast Asia was the main area of origin of the studies. *E. coli* was the predominant species detected, followed by other Enterobacteriaceae, Staphylococci, and *Pseudomonas* spp. *FosA* and *fosA3* were the more frequently encountered Antimicrobial Resistance Genes (ARGs) in the related Gram-negative isolates, while *fosB* was regularly encountered in Gram-positive ones. The majority of the strains were multidrug-resistant (MDR) and co-carried resistance genes against several classes of antibiotics and especially β-Lactams, such as $bla_{CTX\text{-}M}$ and *mecA*. These results demonstrate the fact that the cause of the spreading of Fosfomycin-resistant bacteria among pets could be the extended use of other antibacterial agents, that promote the prevalence of MDR, epidemic strains among an animal population. Through the circulation of these strains into a community, a public health issue could arise. Further research is essential though, for the comprehensive consideration of the issue, as the current data are limited.

Keywords: Fosfomycin resistance; dog; cat; companion animals; pets; PRISMA guidelines; ARGs

check for updates

Citation: Lysitsas, M.; Chatzipanagiotidou, I.; Billinis, C.; Valiakos, G. Fosfomycin Resistance in Bacteria Isolated from Companion Animals (Dogs and Cats). *Vet. Sci.* **2023**, *10*, 337. https://doi.org/10.3390/vetsci10050337

Academic Editor: Jianhua Liu

Received: 31 March 2023
Revised: 25 April 2023
Accepted: 6 May 2023
Published: 9 May 2023

1. Introduction

1.1. Properties and Mode of Action of Fosfomycin

Fosfomycin is a hydrophilic, low molecular mass (138 Da) derivative of a phosphoric acid, which was first discovered in Spain in 1969, under the name phosphonomycin [1]. It has been isolated from cultures of *Streptomyces* spp. [2,3]. It remains the only antibiotic in its class, and thus its role as an alternative therapeutic option is unquestionably important [4].

Fosfomycin's bactericidal effect is achieved by blocking the first step of peptidoglycan synthesis, which is the structural unit of the bacterial cell wall, resulting in the inhibition of its biogenesis and consequently lysis of the bacterial cell [5]. The transport of Fosfomycin to the interior of the bacteria is performed through specific permeases (membrane transport proteins that subserve the diffusion of specific molecules intracellularly), which are the glycerol-3-phosphate transporter (GlpT) and the glucose-6-phosphate transporter (UhpT) [4]. Once diffused intracellularly, it inhibits the UDP-N-acetylglucosamine enolpyruvyl transferase (MurA) enzyme, which is responsible for catalyzing the formation of N-acetylmuramic acid, peptidoglycan's precursor. Therefore, it inhibits cell wall formation by blocking the first step of its biosynthesis [5].

Fosfomycin is an antibacterial agent of broad spectrum and very low toxicity. Additionally, it has a very low protein binding (0,5%) and is highly distributed throughout the body, including inflamed tissues and pus [6]. Several studies confirm that Fosfomycin has the ability to penetrate into tissues where antibiotics frequently demonstrate low penetration ability; therefore, it has a significant potential for usage against many difficult-to-treat infections [3,7]. Another major advantage is its diminished cross-resistance property, as its mechanism of action is unique, resulting in an absence of crossed resistances with other classes of antibiotics [4].

1.2. Usage in Human and Veterinary Medicine

Oral Fosfomycin is primarily used in human medicine, in cases of uncomplicated urinary tract infections (UTIs) and prostatitis caused by multidrug-resistant (MDR) Gram-negative bacteria. Intravenous Fosfomycin has been mainly used in combination with other agents against various types of complicated, severe MDR infections [8].

In recent years, however, the use of Fosfomycin has increased spectacularly due to the considerable incidence of MDR microorganisms, against which it constitutes (alone or in combination) a treatment alternative [3,9]. It has a broad spectrum of in vitro activity against a variety of pathogens, such as methicillin-resistant *Staphylococcus aureus* (MRSA), drug-resistant Enterobacteriaceae, and *Pseudomonas aeruginosa* including extended-spectrum-β-lactamase (ESBL)-producing and carbapenem-resistant (CR) organisms [10–12]. Different studies also confirm the in vivo and in vitro synergistic action of Fosfomycin in combination with other antibiotics, against MDR *Acinetobacter baumannii* [13–15] and Vancomycin-resistant Enterococci (VRE) strains [16,17].

In veterinary medicine, Fosfomycin is used mostly against infections caused by a number of Gram-positive and Gram-negative pathogens, including *E. coli*, in intensively bred piglets and broiler chickens [3,18]. Especially in piglets, Fosfomycin is effective in the treatment of several stress-associated or secondary (due to the dissemination of viruses) bacterial infections [18]. It is more widely used in farms in South and Central America. Even though it is occasionally used by veterinarians for decades and in several countries, it Is still considered a second-line antibiotic. It is not approved for veterinary use in many countries, and therefore knowledge of its pharmacokinetic and pharmacodynamic properties among professionals is rather insufficient [19]. Moreover, it is currently categorized as a critically important antimicrobial for human medicine and thus its use is rather disapproved, especially in European Union [20,21]. Nevertheless, a small number of protocols have been suggested, with promiscuous results, in several livestock and domestic species such as broiler chickens [22,23], dogs [19,24], pigs [25,26], cattle [27], and horses [28]. Contrarily, acute renal insufficiency and failure have been referred to after administration in cats [29,30].

1.3. Fosfomycin Resistance

Even though Fosfomycin seems to be still effective against a great variety of bacteria, several resistance mechanisms have been described. The main mechanisms are described in Table 1.

Table 1. Main mechanisms of bacterial resistance against Fosfomycin.

Mechanism of Resistance	Description	Related Genes	References
Modification or overexpression of *MurA* gene	A modification of MurA gene, could alter amino acid sequence in Fosfomycin's binding site, conferring resistance [28,29]. Furthermore, a resistant phenotype can also be achieved by increased synthesis of the enzyme through overexpression of the MurA gene.	*MurA*	[31,32]
Reduced permeability	• Mutations affecting metabolic pathways of the membrane transporters responsible for the uptake of Fosfomycin (GlpT and UhpT) or their substrates, glycerol-3-P and glucose-6-P. • Mutations affecting intracellular levels of cAMP-receptor protein (CRP), which affects the action of GlpT and UhpT.	Variable related genes	[31,33,34]
Fosfomycin modifying enzymes	*FosA* enzyme: a glutathione S-transferase that inactivates Fosfomycin by the addition of glutathione. Mn^{+2} and K^{+} are used as cofactors. Mostly found in Gram-negative bacteria such as Enterobacteriaceae and *Pseudomonas*. Several subtypes of *fosA* have been identified: *fosA3, fosA4, fosA5, fosA6, fosA8, fosA9, fosA10*, $fosA^{EC}$ and $fosA^{SH}$ [1].	*fosA, fosA3, fosA4, fosA5, fosA6, fosA8, fosA9, fosA10*, $fosA^{EC}$, $fosA^{SH}$	[35–39]
	FosB enzyme: differs from *fosA* in being a Mg^{+2} dependent enzyme and using l-cysteine or possibly bacillithiol, as the physiologic thiol donor. Additionally, an extracytoplasmic sigma factor *SigW*, seems to be essential for its expression. It is routinely detected in Gram-positive bacteria, (*Staphylococcus* spp., *Enterococcus* spp., *Bacillus subtilis*).	*fosB*	[36,40,41]
	FosX enzyme: a Mn^{2+}-dependent epoxide hydrolase, which catalyzes the addition of a water molecule to C1 position of Fosfomycin's oxirane ring and as a result breaks it and inactivates its antibacterial properties. It can be regularly found in specific species, such as *Listeria monocytogenes, Clostridium botulinum*, and *Brucella melitensis*.	*fosX*	[31,42]
	FomA and *fomB* enzymes: kinases that modify Fosfomycin by phosphorylation and thus detoxify it intracellularly. They are encountered in Fosfomycin-producing bacteria, such as *Streptomyces* spp.	*fomA, fomB*	[43]
	FosC enzyme: a kinase that converts Fosfomycin to Fosfomycin monophosphate, conferring intrinsic resistance in *Pseudomonas syringae*, through the expression of a chromosomally encoded gene.	*fosC*	[44]
Efflux pumps	*Tet38* (when overexpressed) and *AbaF* pumps in *Staphylococcus aureus and Acinetobacter baumannii*, respectively, contribute to Fosfomycin resistance, possibly by acting as efflux transporters of the agent.	*Tet38, abaF*	[45,46]

[1] $FosA^{EC}$ and $fosA^{SH}$ were initially reported as *fosA2* and *fosA7* genes, chromosomal variants of *fosA* of *E. cloace* and *S. enterica* serovar Heidelberg, respectively. They were later named $fosA^{EC}$ and $fosA^{SH}$ in the newly proposed nomenclature [35].

1.4. Objectives of the Review

The objective of this review is to collect, present and analyze the results of studies related to Fosfomycin resistance in bacteria isolated from dogs and cats, on an international level. Moreover, the objective is to evaluate the available data for the associated bacterial species, their phenotypic and molecular resistance profiles, and their dissemination in animal populations, in order to determine possible causes for the prevalence of Fosfomycin resistance in pets.

The final objective is to underline possible emerging concerns regarding public health and, considering these concerns, recommend potential requirements for future research in this area of interest, as the current data are rather insufficient for a circumstantial comprehension of the issue.

2. Materials and Methods

The Preferred Reporting Items for Systematic Reviews (PRISMA) guidelines were evaluated for this study [47]. The individual steps of the process are presented in Figure 1.

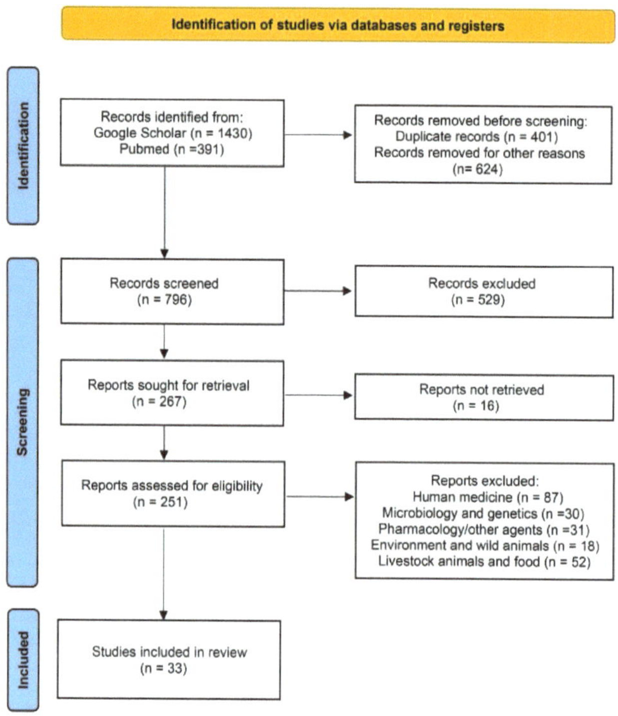

Figure 1. Identification of studies regarding Fosfomycin resistance in companion animals via databases using PRISMA guidelines [47].

Initially, we searched for reviews on the subject of Fosfomycin resistance in companion animals at an international level in the following databases: Google Scholar and PubMed. In these databases, 1821 studies were found using the keywords Fosfomycin and resistance and extra keywords, such as companion animals, dog, cat, canine, feline, and pet, alone or in variable combinations.

The selected studies were published in peer-reviewed journals, websites of organizations, books, and dissertations mostly in the English language. The first step was a screening based on the titles of the articles. We excluded all those not related such as duplicates and studies referring to human medicine, pharmacodynamics, pharmacokinetics and other antibacterial agents, microbiology and genetics, wild animals, water, and environment, and studies referring to food, livestock, and other domestic animals except dogs and cats. During the second selection phase, the abstracts of the reviewed studies were examined fully and independently, in order to identify their relevance according to the information that was searched. Generic information was collected from each article, such as the author, year of publication, the country where the study was conducted, its design and unit of interest, and number of subjects.

Specifically, a total of 1821 manuscripts were finally found: 391 in PubMed and 1430 in Google Scholar. A total of 1035 publications were first excluded as their title was fully irrelevant or they were duplicates or they did not provide an abstract or they were written exclusively in a different language. Subsequently, of the remaining 796 articles whose abstracts were examined, 529 were rejected because their abstracts were irrelevant to the scope of this review, according to the criteria referred to previously. For these reasons, 267 manuscripts remained and 16 could not be retrieved. Therefore, 251 studies were left to be examined and, among them, we rejected 87 as they only concerned human medicine, 30 as they only concerned microbiology and genetics; 31 as they only concerned pharmacodynamics, pharmacokinetics, or different antibacterial agents; 18 as they only concerned wild animals, water or the environment; and 52 as they only concerned food, livestock, or other domestic animals. Finally, 33 manuscripts were used in this review.

The public information extracted from each of the selected articles is presented in Table 2. The country/area of isolation, the bacterial species, the ARGs (if referred), the origin of the sample (dogs, cats, both), the number of fosfomycin-resistant isolates reported (with associated genes recognized in parenthesis), and the date/time period of isolation are listed. Statistical analysis included descriptive analysis of the prevalence of fosfomycin-resistant bacterial strains reported in the various manuscripts significant, comparing the findings among bacterial species, genes detected, and countries reporting results.

Table 2. Public information about the studies included in this review.

Country/Area	Bacterial Species	ARGs	Sample Origin	Fosfomycin Resistant/Total	Date/Period	Reference
USA	*E. coli*	NS [1]	Dogs, cats	3/275	2008–2010	[48]
France	MRSA	*fosB*	Dogs, cats	19/23 (19 *fosB*)	2006–2010	[49]
China	*E. coli*	*fosA3*	Dogs, cats	33/323 (29 *fosA3*)	2006–2010	[50]
China	*E. coli*	*fosA3*	Dog	1/1 (1 *fosA3*)	2008	[51]
China	*E. coli*	*fosA3*	Dogs, cats	12/766 (11 *fosA3*)	2008–2010	[52]
USA, Canada	MRSP [2]	*fosB*	Dogs	7/31 (?7 *fosB*)	NR [3]	[53]
France	*P. aeruginosa*	NS	Dogs	22/46	2008–2011	[54]
Germany	*Ac. baumannii*	NR	Dogs, cats	25/25	2000–2013	[55]
China	Enterobacteriaceae (*E. coli*, *Pr. mirabilis*, *E. fergusonii*, *C. freundii* *E. aerogenes*, *Kl. oxytoca*, *Kl. pneumoniae*	*fosA3*, *fosA*	Dogs, cats, pet owners	19/171 (16 *fosA3*: 8 *E. coli*, 4 *Pr. mirabilis*, 3 *E. fergusonii*, 1 *C. freundii* and 3 *fosA*: 1 *E. aerogenes*, 1 *Kl. oxytoca*, 1 *Kl. pneumoniae*)	2013	[56]
Japan	*P. aeruginosa*	NR	Dogs, cats	71/200	NR	[57]
Japan	Enterobacteriaceae (381 *E. coli*, 50 *Kl. pneumoniae*, 56 *P. mirabilis*)	*fosA3*	Dog, cat	3/487 (3 *fosA3*: 2 *E. coli*, 1 *Kl. pneumoniae*)	2016	[58]
Canada	Enterobacteriaceae (*Kl. pneumoniae*, *E. cloacae*)	*fosA*	Dogs	3/47 (3 *fosA*: 2 *Kl. pneumoniae*, 1 *E. cloacae*)	2015–2016	[59]
Germany	*Staphylococcus cohnii* subsp. *Cohnii*	NR	Dog	1/1	2015–2016	[60]
India	XDR [4] *E. coli*	NR	Dog	1/1	NR	[61]
Brazil	*P. aeruginosa*	*fosA*	Dog	1/1 (1 *fosA*)	2016	[62]
China	*E. coli*	NR	Dog	1/1	2013	[63]
Ecuador	*E. coli*	NR	Dogs	10/23	2017	[64]
Canada	*E. coli*, *St. pseudintermedius*	NF [5] (Absence of *fosA*, *fosA3*, *fosB*, *fosC2*)	Dogs	7/274, 4/113	2013–2016	[65]

Table 2. *Cont.*

Country/Area	Bacterial Species	ARGs	Sample Origin	Fosfomycin Resistant/Total	Date/ Period	Reference
Australia	Enterobacteriaceae, *P. aeruginosa*	*fosA*	Small Animal Hospital environment	65/656, 23/59	2016–2017	[66]
Brazil	*Kl. pneumoniae* (susceptible in the AST)	*fosA*	Dog	1/1 (1 *fosA*)	2018	[67]
Brazil	*Kl. pneumoniae* (susceptible in the AST)	*fosA*	Dog	1/1 (1 *fosA*)	2019	[68]
Taiwan	*Enterobacter cloacae*	*fosA3*, *fosA*	Dogs, cats	8/19 (8 *fosA*, 3 co-carried *fosA3*)	2010–2013	[69]
Ecuador	*E. coli*	*fosA3*	Dog	1/1 (1 *fosA3*)	2016	[70]
USA	*Salmonella* spp.	*fosA7* (*fosA^SH^*) [6]	Dogs	2/27 (2 *fosA7*)	2013–2014	[71]
China	*Kl. pneumoniae*	*fosA*	Dogs, cats	105/105 (105 *fosA*)	2017–2019	[72]
Germany (China) [7]	*Salmonella enterica* serovar Telelkebir	*fosA7* (*fosA^SH^*) [6]	Dog	1/1 (1 *fosA7*)	2007	[73]
Hungary	*Staphylococcus aureus*	*fosB*, *murA* and *glpT* modification	Dog and owner	4/27 (2 *fosB*, 2 *murA* and *glpT*)	NR	[74]
USA (China) [7]	*Micrococcus luteus*	*murA*	Dog	1/1 (1 *murA*)	2019	[75]
China	*E. coli (mcr-1)*	*fosA3*	Dogs, cats	7/7 (7 *fosA3*)	2021	[76]
Caribbean	*Klebsiella pneumoniae*	*fosA*	Dogs, cats	2/2 (2 *fosA*)	2011–2018	[77]
South Africa	*E.coli* [MIC:(R), DD:(S)] [8]	NR	Dog	1/1	NR	[78]
USA	*Staphylococcus aureus*	*fosB*	Dogs, cats	42/53 (42 *fosB*)	2017–2020	[79]
China	*Salmonella enterica* serovar Dublin	*fosX*	Dogs	NR	2018	[80]

[1] NS: Not searched. [2] MRSP: Methicillin-Resistant *Staphylococcus pseudintermedius*. [3] NR: Not referred. [4] XDR: Extensively Drug-Resistant. [5] NF: Not found. [6] *FosA7* was later named $FosA^{SH}$ in a newly proposed nomenclature [39]. [7] Germany and the USA were the countries where the bacteria were isolated, while China was the country of origin of these studies [73,75]. [8] Resistance when examined with the Minimum Inhibitory Concentration (MIC) method and susceptibility when examined with the Disc Diffusion (DD) method.

3. Results

3.1. Geographical Distribution

This distribution of the isolates included in this review is visualized in Figure 2. Regarding the continental distribution, Asia, North, South America, and Europe hold the grand majority, whereas Africa and Oceania are hardly represented.

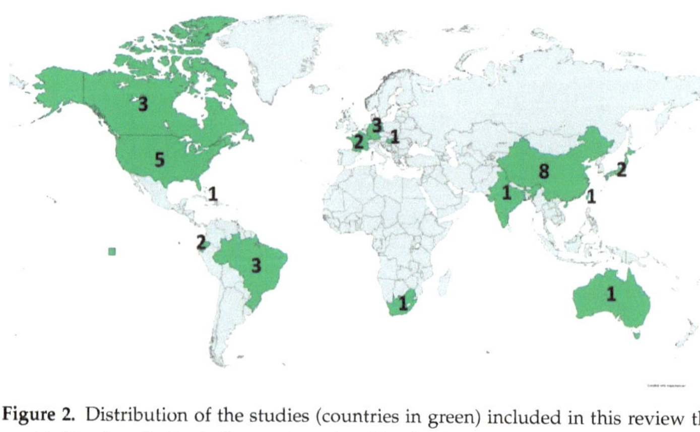

Figure 2. Distribution of the studies (countries in green) included in this review throughout the world. Numbers indicate the number of studies that investigated isolates from the corresponding country. (https://www.mapchart.net/world.html, accessed on 9 February 2023).

3.2. Relevant Findings of the Studies Per Country of Isolates Origin

3.2.1. China

In a study implemented in Guangdong province [50], 323 *E. coli* strains were tested for Fosfomycin resistance. They were isolated from living and diseased pets (dogs and cats), from 2006 to 2010. A total of 33 isolates (10.2%) were resistant, while 29 of them carried the ARG *fosA3*. The presence of resistant bacteria while the antibacterial agent was not administered to any of these pets, could be a result of the co-selection of the *fosA3* gene with other ARGs, after the extended use of aminoglycosides and/or cephalosporines and dissemination through the horizontal transfer of plasmids [50].

One of the strains of the previous study, recovered from a dog in Guangzhou, in 2008, was further investigated. Plasmid DNA was fully sequenced and it was proved that the *fosA3* gene was located in a Multiresistance Region (MRR) of a plasmid, together with aminoglycoside-resistance gene *rmtB* and ESBL gene $bla_{CTX\text{-}M\text{-}65}$ [51]. This enhances the hypothesis that the dissemination of resistance to Fosfomycin could be achieved through epidemic plasmids, under the pressure of other antibacterial agents, which are routinely used in companion animals.

During the period of September 2008–December 2010, a total of 2106 fecal animal samples were collected from the area of Hong Kong and, among them, 368 were from stray dogs and 398 from stray cats. Nine Fosfomycin-resistant strains were found in dogs and three in cats and the *fosA3* gene was detected in all but one canine isolate. Additionally, an interesting fact was that Fosfomycin-resistant isolates were more likely to be MDR than susceptible ones. Finally, the results, in concordance with the previous studies, demonstrated the dissemination of *fosA3*-mediated resistance among MDR bacteria, in several domestic and wild animals [52].

In another study, 171 samples of pets and their owners were collected in a veterinary Teaching Hospital in Beijing, China, from March to June 2013. All of the samples were inoculated in media supplemented with Fosfomycin in order to isolate resistant strains. Nineteen resistant Enterobacteriaceae isolates were detected. Among them, 16 strains were *fosA3* positive (8 *E. coli*, 4 *P. mirabilis*, 3 *E. fergusonii*, and 1 *C. freundii)* and three were *fosA* positive (*Enterobacter aerogenes*, *Klebsiella oxytoca*, *Klebsiella pneumoniae*). All of them were MDR. Two different genetic environments of the *fosA3* gene were detected, both related to plasmids previously studied, from bacteria of human and swine origin. Therefore, the danger of transmission of *fosA3*-carrying plasmids or clones between humans and animals is indicated [56].

An *E. coli* strain isolated from a dog, in an Animal Teaching Hospital (Beijing), in 2013, during a surveillance study with samples originating from companion animals, was found positive for the $bla_{NDM\text{-}1}$ gene, that confers resistance to carbapenems. This isolate was also highly resistant to Fosfomycin among other antibacterial agents [63].

From July 2017 to October 2019, 5359 samples of companion animals were collected in Beijing and Tianjin, from which 105 *Klebsiella pneumoniae* strains were isolated. Even though Fosfomycin was not among the antibiotics tested in the AST, the *fosA* gene was detected in all the isolates [72].

In Shanghai, 79 samples were collected from 31 hospitalized pets in a veterinary hospital, from May to July 2021. Seven *E. coli* isolates positive for the colistin-resistance gene *mcr-1* were detected. All of them carried the *fosA3* gene and also exhibited an MDR phenotype [76]. The spread of the *mcr-1* gene in companion animals was associated with plasmid transmission, horizontal and vertical. The *fosA3* gene was also related, as it was co-carried in the same plasmid with the *mcr-1* in all isolates, and both Colistin and Fosfomycin are agents of limited use in dogs and cats.

Finally, in a study conducted in Zhejiang, canine and feline fecal samples were collected and *Salmonella enterica* isolates were obtained. A high percentage of these bacteria was MDR and several ARGs were identified. Among them, Fosfomycin-resistance gene *fosX* was detected in a number of *Salmonella* Dublin strains [80].

3.2.2. USA

In a study, 200 clinical and 75 experimental strains of *E. coli* were isolated from samples of canine and feline origin, from March 2008 to December 2010. The minimum inhibitory concentration of Fosfomycin and six other classes of antibacterial agents was determined. Only 3/275 isolates were non-susceptible to Fosfomycin, while according to the antimicrobial susceptibility testing (AST) phenotype, they were categorized as: resistant to no drugs (NDR): 47 strains, a single drug class resistance (SDR): 88 strains, 3–4 drug classes resistance (MDR): 112 strains and 5–6 drug classes resistant (Extensively Drug Resistant, XDR): 18 strains. These data suggested that, even though there was a very limited rate of non-susceptibility, Fosfomycin was undoubtedly very effective against *E. coli* related to disease in dogs and cats. Moreover, a large number of MDR and XDR strains were identified. These in vitro results though, require further research for dosing, clinical safety, and efficacy to be determined [48].

Fecal samples of 554 dogs from seven animal shelters across Texas were collected from May 2013 to December 2014 and tested for the detection of *Salmonella* spp. The 27 *Salmonella enterica* isolates were sequenced and screened for ARGs. Two of them (serotypes Heidelberg and Derby) carried the Fosfomycin-resistance gene *fosA7* [71].

In a study from China, a comprehensive comparative analysis of 66 genomes of *Micrococcus luteus*, downloaded from the National Center for Biotechnology Information (NCBI) GenBank database was performed. Among them, a strain isolated from a sample of canine ocular discharge, originating from the USA (New Hampshire) carried the *murA* gene, associated with Fosfomycin resistance. This gene was identified in all 66 genomes [75].

Finally, *Staphylococcus aureus* isolates from clinical animal specimens were collected in New England, from September 2017 through March 2020. Using whole genome sequencing, the distribution of acquired genes related to antibiotic resistance was searched in 53 genomes of canine or feline origin, among other tasks. The Fosfomycin-resistance gene *fosB* was identified in 42 genomes of canine or feline origin [79].

3.2.3. Canada

A total of 542 Enterobacteriaceae strains were isolated from canine clinical samples from two diagnostic laboratories in Ontario, from November 2015 to October 2016. They were subjected to AST and PCR screening for ARGs. The genome of all *bla* $_{CTX\text{-}M}$ positive bacteria was sequenced (a total of 47 isolates). Fosfomycin-resistance gene *fosA* was detected in two *Klebsiella pneumoniae* and one *Enterobacter cloacae* isolate [59].

R. M. Courtice [65] searched the prevalence of Fosfomycin resistance in 387 canine urine isolates, including 274 *E. coli* and 113 *S. pseudintermedius*. Of these strains, 11 were resistant (seven *E. coli* and four *S. pseudintermedius* strains). Among other results, a statistically significant relationship between methicillin and Fosfomycin resistance in *Staphylococcus pseudintermedius* was observed, as three of four Methicillin-Resistant *Staphylococcus pseudintermedius* (MRSP) detected among the 113 isolates were also Fosfomycin-resistant. It is noted that this should be estimated with caution, due to the small number of MRSP isolates. Additionally, as ARGs *fosA*, *fosA3*, *fosB*, and *fosC* were not identified in any of the resistant strains, it was suggested that an alternative mechanism such as mutations in *murA*, *glpT*, *uhpA*, or *uhpT* genes could be the cause of the resistance.

3.2.4. USA and Canada

Isolates of MRSP, from Canada (21 strains) and the USA (10 strains), were screened for Fosfomycin resistance and for the presence of the ARG *fosB*. Only seven strains were resistant while 27 of them (all of the resistant and 20 of the susceptible) carried the gene. It was estimated that this could be a result of repression or a non-functional *fosB* and the need for further study of Fosfomycin resistance in MRSPs was noted [53].

3.2.5. Germany

In a study in Germany, 223 clinical isolates of *Acinetobacter baumannii* were obtained from veterinary clinics between 2000 and 2013. They were screened for carbapenem non-susceptibility and AST for several antibiotics was also performed. A subgroup of 25 strains was afterward tested against some extra antibacterial agents including Fosfomycin. All of them (25/25) were Fosfomycin-resistant [55].

Between May 2015 and March 2016, bacteria were obtained from samples of companion animals. Antimicrobial susceptibility testing was performed. Staphylococci were screened for Methicillin Resistance and Enterobacteriaceae for ESBL production, by phenotypic methods and PCR. Four Staphylococci isolates were resistant to Fosfomycin; three were derived from rabbits and a *Staphylococcus cohnii* subsp. *cohnii* isolate originated from a canine buccal sample [60].

In another study that took place in China, a *Salmonella enterica* serovar Telelkebir clinical isolate and 120 available genomes obtained from databases, were investigated for variable factors, such as their relatedness, ARGs, and virulence factors. Among them, there was an isolate from a companion animal in Germany (2007), which carried Fosfomycin-resistance gene *fosA7* [73].

3.2.6. Brazil

In December 2016, a *Pseudomonas aeruginosa* isolate was recovered from the ear canal of a dog in a veterinary clinic in Brazil. This strain harbored many ARGs for several classes of antibacterial agents including β-Lactams and Carbapenems, Aminoglycosides, Tetracyclines, Folate Pathway Inhibitors, Phenicols, and Fosfomycin. More specifically, concerning Fosfomycin resistance, the *fosA* gene was detected. Furthermore, as the same strain was isolated from the pet owner, who has a recent event of hospitalization and the house environment, it was proved that household dissemination occurred. This fact indicated the danger of transmission of hospital-acquired MDR pathogens from humans to companion animals and the circulation of them in a household environment [62].

Another case of an MDR isolate, carrying a Fosfomycin-resistance gene (*fosA*), occurred in 2018. In a veterinary hospital, a *Klebsiella pneumoniae* strain was obtained from a urinary tract infection (UTI) of a 6-year-old dog. Antimicrobial susceptibility testing and genomic DNA sequencing were performed. Antimicrobial resistance genes were identified against several classes of antibiotics and *fosA* was one of them, although by disc diffusion method, a susceptible phenotype for Fosfomycin had been demonstrated [67].

In March 2019, in a veterinary teaching hospital in southeast Brazil, a *Klebsiella pneumoniae* strain was isolated from a dog with otitis. Antimicrobial susceptibility testing and whole genome sequencing were performed. *FosA* was detected among variable ARGs, after resistome analysis, even though the strain was phenotypically susceptible to the agent [68].

3.2.7. France

In a study conducted in France, coagulase-positive Staphylococci isolated from canine and feline samples, from 2006 to 2010, were tested for Methicillin resistance. A group of 23 *MRSA* was obtained. Subsequently, susceptibility testing and molecular typing were carried out. In 19 of 23 strains, Fosfomycin-resistance gene *fosB* was identified. Most of these strains were MDR and related to human MRSA clones [49].

From 2008 to 2011, 68 *Pseudomonas aeruginosa* clinical isolates of variable animals (dogs, dairy cows, horses) were collected. Genotyping and AST were performed. A rate of 47.8% (22/46) of Fosfomycin resistance was observed in canine isolates. Additionally, among these isolates, a common combination of resistances noticed was the β-Lactams-Aminoglycosides-Fosfomycin [54].

3.2.8. Japan

From September 2014 to February 2015, 200 *Pseudomonas aeruginosa* strains were obtained from samples of dogs and cats, in veterinary hospitals located in several areas of Japan. Antimicrobial susceptibility testing and screening for aminoglycoside-resistance enzymes and metallo-β-lactamases were performed. Resistance against Fosfomycin was observed in 35.5% of the isolates, the highest among the tested agents, as rates of resistance were generally low [57].

In another study, isolates from specimens of dogs and cats were collected between May and September 2016, in clinical settings and hospitals throughout Japan. A total of 381 *E. coli*, 50 *Klebsiella pneumoniae*, and 56 *Proteus mirabilis* were obtained. AST and a multilevel molecular analysis were performed. Plasmid-mediated Fosfomycin-resistance gene *fosA3* was detected in two *E. coli* strains, co-harboring $bla_{CTX\text{-}M\text{-}55}$ plus $bla_{CMY\text{-}2}$ and $bla_{CTX\text{-}M\text{-}27}$ genes, respectively, and in a *Klebsiella pneumoniae* strain co-harboring $bla_{SHV\text{-}12}$, $bla_{DHA\text{-}1}$, and *armA* [58].

3.2.9. Ecuador

In Quito, 50 canine fecal samples were collected in August 2017 and inoculated in ceftriaxone-supplemented agar, to enhance the growth of strains resistant to the agent. A total of 23 *E. coli* isolates were obtained. AST and genotyping characterization of the bacteria were subsequently carried out. Resistance for Fosfomycin was observed at 43% of the strains (10/23). Additionally, all of them were MDR, and a very high resistance rate was detected for most of the antibiotics tested [64].

For another study, 42 samples were collected from backyard animals in June 2016. They were plated to selective media in order to isolate Colistin-resistant strains. Three isolates were obtained, two of canine origin. Among other tasks, the Fosfomycin-resistance gene *fosA3* was detected [70].

3.2.10. India

An XDR-resistant *E. coli* was isolated from the scrotal fluid of a 3-year-old Labrador, in a veterinary hospital. Antimicrobial susceptibility testing and phenotypic and molecular screening for ARGs were accomplished and the strain was resistant against almost every tested agent (except Tigecycline), including Fosfomycin. Variable ARGs for β-Lactams, including a bla_{NDM1} carbapenemase-producing gene, were detected [61].

3.2.11. Australia

A study was carried out in Melbourne, between November 2016 and May 2017. Samples of environmental sources of an animal Hospital were collected. The microbial population was evaluated and screened for ARGs, before further molecular investigation. Antimicrobial resistance genes against Fosfomycin were detected in Enterobacteriaceae species from all four sources of the samples (Intensive Care Unit (ICU) cages, Laundry Trolley, Mop Bucket, and Office Corridor) and in *Pseudomonas* spp. from ICU cages. *FosA* detection is specifically referred to in *Klebsiella* and *Enterobacter* species, located on chromosomal sequences [66].

3.2.12. Taiwan

A group of 19 *Enterobacter cloacae* isolates was obtained from UTIs of companion animals (10 dogs and 9 cats) in a Veterinary Teaching Hospital. Antimicrobial resistance genes for Fosfomycin and co-existing resistance genes were screened, and AST and conjugation experiments were also performed. The *fosA* gene was present in eight strains, three of which co-carried the *fosA3* gene. *FosA* was likely located on the chromosome, while *fosA3* was on mobile genetic elements [69].

3.2.13. Hungary

During a study, a group of 102 dogs and 84 owners was sampled in Budapest and 14 other towns. Subsequently, 27 *S. aureus* and 58 *S. pseudintermedius* isolates were obtained. AST and molecular investigation of the strains occurred. In a pair of *S. aureus* isolates, from a dog and its owner, the *fosB* gene was present. In another pair of *S. aureus* isolates, from a dog and another owner, amino acid changes in *murA* transferase and *glpT* transporter were detected. These changes can confer Fosfomycin resistance by different mechanisms [74].

3.2.14. Caribbean

For a study carried out on St. Kitts, a small Caribbean island, 82 *Klebsiella* spp. strains were collected from several sources (human, canine, feline, equine, vervet). A whole genome sequence analysis was accomplished. The *fosA* gene was detected in a *Klebsiella pneumoniae* of canine origin and one of feline origin. It is noted in the study that this gene belongs to the core genome of *Klebsiella* and therefore it should not be regarded as acquired [77].

3.2.15. South Africa

In Pretoria, an *E. coli* strain was isolated from a deceased dog. Antimicrobial susceptibility testing was performed by both disc diffusion and MIC. The isolate was MDR and possessed virulence factors for two pathotypes. It was proved to be Fosfomycin-resistant in the MIC test, even though in the disk diffusion test it demonstrated a susceptible phenotype [78].

3.3. *Microorganisms and ARGs of Interest*

The isolated bacteria related to Fosfomycin Resistance in the selected articles and the associated ARGs are presented in Table 3.

Table 3. Microorganisms referred to in the selected studies.

Microorganisms	Number of References	Fosfomycin-Resistant Isolates *	Related Fosfomycin ARGs **
E. coli	13	86	*fosA3* (58)
Kl. pneumoniae	7	113	*fosA* (112), *fosA3* (1)
P. aeruginosa	4	117	*fosA* (24)
St. aureus	3	65	*fosB* (63), *murA* (2), *glpT* (2)
Salmonella spp.	3	3	*fosA*^{SH} (3), *fosX*
St. pseudintermedius	2	11	*fosB* (7)
Enterobacter cloacae	2	9	*fosA* (9), *fosA3* (3)
E. aerogenes	1	1	*fosA* (1)
Kl. oxytoca	1	1	*fosA* (1)
E. fergusonii	1	3	*fosA3* (3)
Pr. mirabilis	1	4	*fosA3* (4)
C. freundii	1	1	*fosA3* (1)
Micrococcus luteus	1	1	*murA* (1)
Ac. baumannii	1	25	NR
Staphylococcus cohnii	1	1	NR

* study [66] was excluded from the analysis regarding Enterobacteriaceae, as they did not refer to specific numbers of isolates investigated per bacterial species. Study [80] does not refer to the specific number of *Salmonella* Dublin isolates carrying the *fosX* ARG. ** the number of isolates that were detected to carry each ARG is referred to in parentheses.

E. coli is more frequently referred to (13/33, 39% of studies), followed by *Kl. pneumoniae* (7/33, 21% of studies), while the Enterobacteriaceae family is strongly represented by many more of its members (8/33, 24% of studies). These bacteria are mostly associated (when searched) with the presence of *fosA* (including *fosA*^{SH}) and *fosA3* genes. Staphylococci and *Pseudomonas aeruginosa* also make their appearance more than once (6/33 18% and 4/33, 12% of studies, respectively), usually related to *fosB* and *fosA* genes, respectively. Finally, to our knowledge, there is only one relevant reference for *Micrococcus luteus* and for *Acinetobacter baumannii*.

In reference to the molecular basis of the resistance, the *fosA* gene is the most frequently detected ARG with nine references plus two references of *fosA*^{SH} (*fosA7*). *FosA3* is present in eight of the studies, almost all of them (7/8) located in southeast Asia and Japan. *FosB* is identified in four articles, from Staphylococci (*S. aureus* and *S. pseudintermedius*) in

Europe and North America (two articles, respectively). There are only two references to the detection of the *murA* gene and one of the *glpT* and *fosX*, in the selected literature.

3.4. Resistome and Phenotypic Resistance against Other Antibacterial Agents in Fosfomycin-Resistant Strains

The selected studies contain a large amount of information about susceptibility testing and different mechanisms of resistance of the associated bacteria, such as acquired, intrinsic, silent, and protoresistance (little/no activity until mutated). All these mechanisms are described by the term resistome, as it has been previously described [81]. In order to evaluate the resistome of the isolates related to Fosfomycin resistance, data of their ARGs and phenotypic susceptibility testing were collected and presented in Table 4. Some articles have been excluded due to difficulties in the classification of the available information.

Table 4. ARGs and resistance against other antibacterial agents, in the isolates included in this review.

Country /Area	Bacterial Species	Fosfomycin Related ARGs	Other ARGs with ≥50% Prevalence among Fos-Resistant Isolates [1]	Agents with ≥50% Resistance Rates among Fos-Resistant Isolates [1,2]	Reference
France	*S. aureus*	*fosB*	*mecA, blaZ, aadD*	ENR, ERY, FOX, KAN, LIN, PEN, SPI, TOB	[49]
China	*E. coli*	*fosA3*	*bla*$_{CTX-M}$, *rmtB*	AMK, CHL, CTX, GEN, TET	[50]
China	*E. coli*	*fosA3*	*bla*$_{CTX-M}$, *rmtB*	—	[51]
China	*E. coli*	*fosA3*	*bla*$_{CTX-M}$	CHL, CIP, COT, GEN, NAL, TET	[52]
USA, Canada	*S. pseudi-ntermedius*	*fosB*	*mecA*	b-lactams [3]	[53]
Germany	*A. baumannii*	NR	—	CXM, CFD, PIT, SAM	[55]
China	*E. coli, Pr. mirabilis, E. fergusonii, C. freundii*	*fosA3*	*bla*$_{CTX-M}$	AMK, AMP, FAZ, CHL, CIP, GEN, FFC, KAN	[56]
Japan	*E. coli, Kl. pneumoniae*	*fosA3*	ESBL, *pAmpC*	—	[58]
Canada	*Kl. pneumoniae, E. cloacae*	*fosA*	*bla*$_{CTX-M-15}$, *aac(3)-IIa, strA, strB, aac(6')Ib-cr, bla*$_{OXA-1}$, *bla*$_{SHV-83}$, *bla*$_{TEM-1-B}$, *qnrB1, sul2, drfA14, tetA*	—	[59]
Germany	*Staphylococcus cohnii*	NR	*mecA*	OXA	[60]
India	*E. coli*	NR	*bla*$_{CTX-M}$, *bla*$_{AmpC}$*bla*$_{TEM}$, *bla*$_{NDM-1}$, *sul1*	AMK, AMC, AZT, CAZ, CFD, CFM, CHL, CIP, COL, CRO, CTX, CTC, CTR, ERT, FEP, FOX, GAT, GEN, IMP, MER, MOX, NOR, NIT, OFL, PMB, SXT, TET, TOB, VAN	[61]
Brazil	*P. aeruginosa*	*fosA*	*bla*$_{VIM-2}$, *bla*$_{PAO}$, *bla*$_{OXA-4}$, *bla*$_{OXA-50}$, *aadA2, aac(3)-Id, aph(3)-IIb, catB7, cmlA1, sul1, dfrB5, tetG*	AMK, AMC, CAZ, CIP, CHL, CRO, CTX, FEP, FOX, GEN, IMP, MER, NAL, PIT, STX, TET, TIC	[62]
China	*E. coli*	NR	*bla*$_{NDM-1}$, *drfA17, sul1, aadA5*	CAZ, CTX, CIP, ERT, GEN, IMP, MER, PIP, TET	[63]
Ecuador	*E. coli*	NR	*bla*$_{CTX-M}$	AZT, CAZ, CHL, CIP, CTX, DOX, FEP, LEV, NAL, NOR, TET	[64]
Canada	*S. pseudinter-medius*	NF	*mecA*	PEN, OXA, AMP, CLI	[65]
Brazil	*Kl. pneumoniae*	*fosA*	*bla*$_{CTX-M-15}$, *bla*$_{SHV}$, *bla*$_{OXA-1}$, *aph(3'')-Ib, aph(6)-Id, aac (3)-IIa, tetA, aac(6')-Ib-cr, qnrB1, oqxA and oqxB, dfrA*	AMC, CAZ, CIP, CTX, ENR FEP, FUR, GEN, LEV, NOR, OFL, TET	[67]
Brazil	*Kl. pneumoniae*	*fosA*	*bla*$_{CTX-M-15}$, *bla*$_{SHV}$, *bla*$_{OXA-1}$, *1, aadA2, aph(3')-Ia], mphA, catB3, aac(6')Ib-cr, oqxA, oqxB], sul1, tetA, dfrA12, GyrA, ParC*	AMC, AZT, CIP, CRO, ENR, FEP, FUR, LEV, NAL, SXT, TET	[68]
Taiwan	*Enterobacter cloacae*	*fosA3, fosA*	*bla* $_{TEM}$	AMP, SUD	[69]
Ecuador	*E. coli*	*fosA3*	*mcr-1.1, bla*$_{CTX-M-3}$, *bla*$_{TEM-206}$, *bla*$_{TEM-1B}$, *tetA, GyrA, ParC*	CIP, COL, CRO, FEP, SAM	[70]
China	*Kl. pneumoniae*	*fosA*	*bla*$_{SHV}$, *oqxA, oqxB, sul*	AMC, DOX, FFC, SXT	[72]
Germany (China)	*Salmonella enterica*	*fosA7* (*fosA*^{SH})	*aac(6')-Iaa_1*	—	[73]
Hungary	*Staphylococcus aureus*	*fosB, murA, glpT*	*blaZ*	PEN	[74]

Table 4. *Cont.*

Country /Area	Bacterial Species	Fosfomycin Related ARGs	Other ARGs with ≥50% Prevalence among Fos-Resistant Isolates [1]	Agents with ≥50% Resistance Rates among Fos-Resistant Isolates [1,2]	Reference
China	*E. coli*	*fosA3*	*aac(3)-IVa, aph(3′)-IIa, aph(3′)-Ia, and aph(4)-Ia,* $bla_{CTX-M-65}$, bla_{TEM-1B}, *floR, drfA14, mcr-1, sul2, qnrS1, mdfA*	COL, CTX, FFC	[76]
South Africa	*E.coli*	NR	—	AMC, AMP, CEP, ENR, FUR, NEO, PEN	[78]

[1] In case of one relevant strain in an article, ARGs and phenotypic resistances are referred to this isolate. [2] Antibacterial agents: AMC: Amoxicillin-clavulanate, AMK: Amikacin, AMP: Ampicillin, AZT: Aztreonam, CAZ: Ceftazidime, CEP: Cephalothin, CFM: Cefixime, CHL: Chloramphenicol, CFD: Cefpodoxime, CIP: Ciprofloxacin, COL: Colistin, CPR: Cefoperazone, CRO: Ceftriaxone, CTC: Cefotaxime-clavulanate, CTX: Cefotaxime, CXM: Cefuroxime, DOX: Doxycycline, ENR, Enrofloxacin, ERT: Ertapenem, ERY: Erythromycin, FEP: Cefepime, FFC: Florfenicol, FOX: Cefoxitin, FUR: Ceftiofur, GAT: Gatifloxacin, GEN: Gentamicin, IMP: Imipenem, KAN: Kanamycin, LEV: Levofloxacin, LIN: Lincomycin, MOX: Moxalactam, NAL: Nalidixic acid, NEO: Neomycin, NOR: Norfloxacin, NIT: Nitrofurantoin, OFL: Ofloxacin, OXA: Oxacillin, PEN: Penicillin PMB: Polymyxin B, PIP: Piperacillin, PIT: Piperacillin-tazobactam, SAM: Ampicillin-sulbactam, SPI: Spiramycin, SXT: Sulfamethoxazole-Trimethoprim, SUD: Sulfadiazine, TET: Tetracycline, TIC: Ticarcillin, TOB: Tobramycin, VAN: Vancomycin. [3] Except Ceftaroline.

The majority of the correlated strains are MDR bacteria, as indicated by the number of antibacterial agents with >50% resistance rate among them, presented in Table 4. Furthermore, there is a notable presence of methicillin resistance in the related Staphylococci (4/5 articles) and a notable presence of ESBL genes in the related Enterobacteriaceae, with bla_{CTX-M} gene detected in at least 11 studies and more β-Lactamase encoding genes such as bla_{SHV}, bla_{OXA}, bla_{TEM}, even bla_{NDM-1} referred in variable occasions. Other classes of antibiotics are also included several times in the resistome of the isolates, such as aminoglycosides, quinolones, tetracyclines, folate pathway inhibitors, phenicols, polymyxins, etc.

4. Discussion

In this review, several cases of Fosfomycin resistance in bacteria from companion animals were presented, even though the available data are relatively limited and difficult to evaluate at present. However, the emergence of higher resistance rates would definitely constitute a challenge for veterinarians and a concern for public health, as this antibiotic demonstrates potential as an alternative agent in the upcoming era of bacterial multi-resistance [3,6,10].

Moreover, the dissemination of Fosfomycin resistance is already a matter of concern in human medicine. The *fosA3* gene is rather endemic in Southeast Asia. The detection of this resistance gene in plasmids of MDR clones, co-carrying other ARGs, indicates the danger of higher rates of resistance, under the pressure of antibacterial agents widely used, especially β-Lactams [5,82]. Importation of isolates carrying similar plasmids in Europe has also been reported [83], demonstrating the danger of universal spreading.

Regarding companion animals, even if the use of Fosfomycin is relatively limited in many countries, cases of phenotypic resistance and correlated ARGs are referred to in the studies included in this review. Pets, especially dogs and cats, could contribute to the prevalence of resistant isolates in a community, considering their household accommodation, their close contact with their owners, their large numbers in urban areas, and the possibility of hospital-acquired MDR infections in veterinary hospitals. The danger of a pet–owner circulation of resistant bacteria is a subject of research in some of the included articles [56,62,70,74] and a matter of concern in the grand majority of them. Additionally, hospital-acquired infections are present in small animal practices, among veterinary hospitals of variable sizes and it is estimated that their frequency is going to increase [84].

Moreover, the real occurrence of Fosfomycin resistance in pets is undoubtedly expected to be more significant, considering that several facts could influence its current sub-detection. Specifically, Fosfomycin is a rarely used agent in the AST of isolates from companion animals in many countries, especially in bacteria not originated from UTIs, and therefore a phenotypic resistance could be frequently missed. ARGs for Fosfomycin are also not usually screened in molecular assays and, consequently, they are not detected. A characteristic of the limited available data is that only one of the articles included in this review originated from Africa [78] and the references from Europe come only from three countries (France, Germany, Hungary) [49,54,55,60,73,74]. More relevant studies have been conducted in Southeast Asia (Figure 2). Furthermore, only a few of the 33 included studies were particularly aiming to search for phenotypic Fosfomycin resistance or correlated ARGs [48,50–53,56,65,69], whereas, in most cases, the detection was random during investigations.

Difficulties may occur in regard to the interpretation of Fosfomycin resistance. The MIC and the disk diffusion methods for Fosfomycin have their own specifications and should be evaluated with caution [85]. Furthermore, in the included studies, bacteria carrying ARGs for Fosfomycin demonstrated a susceptible AST in some cases [53,67,68], while there is an ambiguous AST, with an isolate observed to be susceptible by the disk diffusion method and proved to be resistant by the MIC method [78]. Additionally, the existing clinical breakpoints refer exclusively to human medicine with only a few exceptions [86], and the clinical trials in companion animals are limited [18]. Therefore, only an estimation can be formulated about the effectiveness of the drug in vivo, by the evaluation of the in vitro susceptibility testing.

As a result, the lack of sufficient relevant data and the demanding assessment of the existing information were limiting factors for this current review.

However, a shaky interest in Fosfomycin might emerge lately, even in veterinary medicine. This can be demonstrated by the fact that all of the relative articles included in this study have been completed over the last two decades and published after 2010. As a consequence, a more comprehensive evaluation of Fosfomycin resistance in companion animals is expected to be possible during the following years.

Nevertheless, some interesting facts can be estimated, even from the current data. Initially, Fosfomycin resistance in dogs and cats has mostly been searched in specific areas of the World, such as Southeast Asia, America, and part of Europe. The lack of relevant data from other areas is not necessarily a result of full susceptibility, as it is usually not included in the AST. Moreover, the interpretation of the resistance is a challenging task and in vitro AST tests or molecular assays might not always represent the clinical effectiveness of the drug.

Enterobacteriaceae are the isolates mainly associated with Fosfomycin resistance in dogs and cats (Table 3). This was anticipated, considering their wide dissemination, their significance as pathogens, and the increased interest in their resistance mechanisms in the current research fields. *FosA* and *fosA3* are the dominant relevant ARGs in Gram-negative bacteria, located in chromosomal DNA and plasmids, respectively. The location of *fosA3* gene in mobile genetic elements creates concerns about the wide dissemination of the resistance through the transmission of these elements among different bacterial strains [50–52,56,58,69,70,76]. Acquired *fosA3*-mediated resistance is the dominant mechanism detected in *E. coli* isolates, while it is only occasionally detected [56,58] in other bacterial species of the Enterobacteriaceae family. Furthermore, *fosA3* and $bla_{CTX\text{-}M}$ are regularly co-carried [50–52,56,70,76] and possibly co-transferred through common mobile genetic elements. *FosB* is the main ARG identified in Gram-positive cocci [49,53,74,79].

Concerning the prevalence of resistance among the total number of isolates included in the selected studies, *Acinetobacter baumannii* (25/25 isolates, 100%) and *Pseudomonas* spp. (117/306 isolates, 38%) are exhibiting the highest resistance rates, even though they are more infrequently detected than Enterobacteriaceae. This fact is unsurprising, as these species demonstrate reduced susceptibility through inherent mechanisms [3,10].

The connection between Fosfomycin resistance and MDR strains is beyond a doubt a fact, especially in Enterobacteriaceae and Staphylococci, as is clearly indicated by Table 4. Particularly, ARGs and phenotypic resistances for β-Lactams are detected in the grand majority of the studies. In combination with the infrequent use of Fosfomycin, this fact indicates the possible prevalence of Fosfomycin-resistant bacteria under the pressure of wide usage of antibacterial agents belonging to this class. Several other antibiotics make their appearance regularly in Table 4, suggesting that more routinely used agents could also have an impact, subserving the dominance of resistant strains. Aminoglycosides and tetracyclines are the most frequently observed classes. This co-selection issue is highlighted in a number of the included studies [50–52,76]. Moreover, recent studies have identified the colonization of pets by MDR bacteria after antibiotic treatment with commonly used drugs, such as β-Lactams and Fluoroquinolones, and thus the requirement for increased surveillance efforts [87,88].

Consideration of the aforementioned facts underlines the necessity of future research. The agenda of the prospective studies could include surveillance studies in countries/areas where there are no current data for Fosfomycin resistance in veterinary samples, in order to determine the presence of resistant bacteria and their phenotypic and genotypic characteristics. Additionally, a correlation of the presence of resistant strains, with previous antibiotic treatments of the animal, evaluating all the available data (such as the number of treatments during the preceding time period, classes of antibiotics received, duration of the therapy, etc.), would be of major significance in the assessment of antibiotic usage as a predisposing factor. Furthermore, molecular investigation of the related isolates for the verification of their resistome and the detection of mobile genetic elements, where the Fosfomycin ARGs may be located, is essential in order to identify and evaluate the mechanisms of their dissemination. Another important project is a circumstantial molecular investigation of the Fosfomycin ARGs, in order to verify the exact factors that could provoke the presence or absence of a resistant phenotype and determine the levels of resistance.

Data provided by research in these sectors could undoubtedly contribute to an in-depth comprehension of the phenomenon and indicate the requirements for possible surveillance and control measures.

5. Conclusions

There is a renewed interest in Fosfomycin in the last two decades, as high rates of non-susceptibility against the traditionally used factors appear. Its desirable properties and wide spectrum of bactericidal activity reinforce its potential as an alternative agent. However, Fosfomycin resistance emerges worldwide. The appearance of resistant isolates in companion animals, where the drug has not been widely used, is an even more disturbing fact, indicating the wide dissemination of MDR strains and the danger of circulation of these strains among pets, their owners, and the environment. The results of this review demonstrate the presence of relevant strains in dogs and cats and the fact that the cause of their spreading, could be the extended use of other, routinely used antibacterial agents, that promote the prevalence of MDR, epidemic strains among an animal population. As the relative data are yet limited, further and more extensive research is essential in order to identify the various aspects of the phenomenon and evaluate possible preventive or control measures.

Author Contributions: Conceptualization, M.L and I.C.; methodology and investigation, G.V., M.L. and I.C.; writing—original draft preparation, M.L. and I.C.; writing—review and editing, G.V. and C.B. All authors have read and agreed to the published version of the manuscript.

Funding: This research received no external funding.

Institutional Review Board Statement: Not applicable.

Informed Consent Statement: Not applicable.

Data Availability Statement: No new data were created or analyzed in this study. Data sharing is not applicable to this article.

Conflicts of Interest: The authors declare no conflict of interest.

References

1. Hendlin, D.; Stapley, E.O.; Jackson, M.; Wallick, H.; Miller, A.K.; Wolf, F.J.; Miller, T.W.; Chaiet, L.; Kahan, F.M.; Foltz, E.L.; et al. Phosphonomycin, a New Antibiotic Produced by Strains of *Streptomyces*. *Science* **1969**, *166*, 122–123. [CrossRef] [PubMed]
2. Zhanel, G.G.; Walkty, A.J.; Karlowsky, J.A. Fosfomycin: A First-Line Oral Therapy for Acute Uncomplicated Cystitis. *Can. J. Infect. Dis. Med. Microbiol.* **2016**, *2016*, 2082693. [CrossRef]
3. Falagas, M.E.; Vouloumanou, E.K.; Samonis, G.; Vardakas, K.Z. Fosfomycin. *Clin. Microbiol. Rev.* **2016**, *29*, 321–347. [CrossRef] [PubMed]
4. Díez-Aguilar, M.; Cantón, R. New microbiological aspects of fosfomycin. *Rev. Esp. Quimioter.* **2019**, *32* (Suppl. S1), 8–18. [PubMed]
5. Silver, L.L. Fosfomycin: Mechanism and Resistance. *Cold Spring Harb. Perspect. Med.* **2017**, *7*, a025262. [CrossRef]
6. Dijkmans, A.C.; Zacarías, N.V.O.; Burggraaf, J.; Mouton, J.W.; Wilms, E.; van Nieuwkoop, C.; Touw, D.J.; Stevens, J.; Kamerling, I.M.C. Fosfomycin: Pharmacological, Clinical and Future Perspectives. *Antibiotics* **2017**, *6*, 24. [CrossRef]
7. Parker, S.; Lipman, J.; Koulenti, D.; Dimopoulos, G.; Roberts, J.A. What Is the Relevance of Fosfomycin Pharmacokinetics in the Treatment of Serious Infections in Critically Ill Patients? A Systematic Review. *Int. J. Antimicrob. Agents* **2013**, *42*, 289–293. [CrossRef]
8. Bassetti, M.; Graziano, E.; Berruti, M.; Giacobbe, D.R. The Role of Fosfomycin for Multidrug-Resistant Gram-Negative Infections. *Curr. Opin. Infect. Dis.* **2019**, *32*, 617–625. [CrossRef]
9. Candel, F.J.; Cantón, R. Uso Actual de La Fosfomicina: Del Laboratorio a La Práctica Clínica. *Enfermedades Infecc. Microbiol. Clínica* **2019**, *37*, 1–3. [CrossRef]
10. Michalopoulos, A.S.; Livaditis, I.G.; Gougoutas, V. The Revival of Fosfomycin. *Int. J. Infect. Dis.* **2011**, *15*, e732–e739. [CrossRef]
11. Karaiskos, I.; Giamarellou, H. Multidrug-Resistant and Extensively Drug-Resistant Gram-Negative Pathogens: Current and Emerging Therapeutic Approaches. *Expert Opin. Pharmacother.* **2014**, *15*, 1351–1370. [CrossRef] [PubMed]
12. Zhao, M.; Bulman, Z.P.; Lenhard, J.R.; Satlin, M.J.; Kreiswirth, B.N.; Walsh, T.J.; Marrocco, A.; Bergen, P.J.; Nation, R.L.; Li, J.; et al. Pharmacodynamics of Colistin and Fosfomycin: A 'Treasure Trove' Combination Combats KPC-Producing Klebsiella Pneumoniae. *J. Antimicrob. Chemother.* **2017**, *72*, 1985–1990. [CrossRef] [PubMed]
13. Ku, N.S.; Lee, S.-H.; Lim, Y.-S.; Choi, H.; Ahn, J.Y.; Jeong, S.J.; Shin, S.J.; Choi, J.Y.; Choi, Y.H.; Yeom, J.-S.; et al. In Vivo Efficacy of Combination of Colistin with Fosfomycin or Minocycline in a Mouse Model of Multidrug-Resistant Acinetobacter Baumannii Pneumonia. *Sci. Rep.* **2019**, *9*, 17127. [CrossRef] [PubMed]
14. Saelim, W.; Changpradub, D.; Thunyaharn, S.; Juntanawiwat, P.; Nulsopapon, P.; Santimaleeworagun, W. Colistin plus Sulbactam or Fosfomycin against Carbapenem-Resistant *Acinetobacter Baumannii*: Improved Efficacy or Decreased Risk of Nephrotoxicity? *Infect. Chemother.* **2021**, *53*, 128. [CrossRef]
15. Singkham-in, U.; Chatsuwan, T. In Vitro Activities of Carbapenems in Combination with Amikacin, Colistin, or Fosfomycin against Carbapenem-Resistant Acinetobacter Baumannii Clinical Isolates. *Diagn. Microbiol. Infect. Dis.* **2018**, *91*, 169–174. [CrossRef]
16. Hall Snyder, A.D.; Werth, B.J.; Nonejuie, P.; McRoberts, J.P.; Pogliano, J.; Sakoulas, G.; Yim, J.; Singh, N.; Rybak, M.J. Fosfomycin Enhances the Activity of Daptomycin against Vancomycin-Resistant Enterococci in an In Vitro Pharmacokinetic-Pharmacodynamic Model. *Antimicrob. Agents Chemother.* **2016**, *60*, 5716–5723. [CrossRef]
17. Yan, Y.; Yang, G.; Li, Y.; Mao, J.; Wang, S.; Zhang, N.; Liu, H.; Huang, X. Factorial Design and Post-Antibiotic Sub-MIC Effects of Linezolid Combined with Fosfomycin against Vancomycin-Resistant Enterococci. *Ann. Transl. Med.* **2022**, *10*, 148. [CrossRef]
18. Pérez, D.; Tapia, M.; Soraci, A. Fosfomycin: Uses and Potentialities in Veterinary Medicine. *Open Vet. J.* **2014**, *5*, 26. [CrossRef]
19. Harada, K.; Shimizu, T.; Kawaguchi, K.; Furuhashi, T.; Ishihara, G. Urinary Pharmacokinetic and Pharmacodynamic Profiles of Fosfomycin against Extended-Spectrum β-Lactamase-Producing *Escherichia coli* with Canine Ex Vivo Modeling: A Pilot Study. *Antibiotics* **2020**, *9*, 230. [CrossRef]
20. European Medicines Committee. *Categorisation of Antibiotics Used in Animals Promotes Responsible Use to Protect Public and Animal Health*; European Medicines Agency: Amsterdam, The Netherlands, 2020.
21. World Health Organization. *Critically Important Antimicrobials for Human Medicine*, 6th ed.; World Health Organization: Geneva, Switzerland, 2019; ISBN 978-92-4-151552-8.
22. Aramayona, J.J.; Bregante, M.A.; Solans, C.; Rueda, S.; Fraile, L.J.; Garcia, M.A. Pharmacokinetics of fosfomycin in chickens after a single intravenous dose and tissue levels following chronic oral administration. *Vet Res.* **1997**, *28*, 581–588.
23. Soraci, A.L.; Pérez, D.S.; Tapia, M.O.; Martinez, G.; Dieguez, S.N.; Buronfosse-Roque, F.; Colusi, A.; Romano, O. Pharmacocinétique et biodisponibilité de fosfomycine chez le poulet de chair. *Ec. Natl. Vet. Toulouse Rev. Med. Vet.* **2011**, *162*, 358–363.
24. Gutierrez, O.L.; Ocampo, C.L.; Aguilera, J.R.; Luna, J.; Sumano, L.H. Pharmacokinetics of Disodium-Fosfomycin in Mongrel Dogs. *Res. Vet. Sci.* **2008**, *85*, 156–161. [CrossRef] [PubMed]
25. Soraci, A.L.; Pérez, D.S.; Martínez, G.; Amanto, F.; Tapia, M.O.; Dieguez, S.; Fernández Paggi, M.B. Fosfomycin Concentrations in Epithelial Lining Fluid in Weaning Piglets: Fosfomycin Concentrations in ELF. *J. Vet. Pharmacol. Ther.* **2012**, *35*, 406–409. [CrossRef] [PubMed]
26. Prez, D.; Soraci, A.; Tapia, M. Pharmacokinetics and Bioavailability of Calcium Fosfomycin in Post Weaning Piglets after Oral Administration. *Int. J. Agro Vet. Med. Sci.* **2012**, *6*, 424. [CrossRef]
27. Sumano, L.H.; Ocampo, C.L.; Gutierrez, O.L. Intravenous and Intramuscular Pharmacokinetics of a Single-Daily Dose of Disodium Fosfomycin in Cattle, Administered for 3 days. *J. Vet. Pharmacol. Ther.* **2007**, *30*, 49–54. [CrossRef] [PubMed]

28. Zozaya, D.H.; Gutirrez, O.L.; Ocampo, C.L.; Sumano, L.H. Pharmacokinetics of a Single Bolus Intravenous, Intramuscular and Subcutaneous Dose of Disodium Fosfomycin in Horses. *J. Vet. Pharmacol. Ther.* **2008**, *31*, 321–327. [CrossRef]
29. Fukata, T.; Imai, N.; Shibata, S. Acute Renal Insufficiency in Cats after Fosfomycin Administration. *Vet. Rec.* **2008**, *163*, 337–338. [CrossRef]
30. Sakai, Y.; Yoshida, T.; Shibutani, M.; Ohmori, K. Acute Renal Failure in an Adult Cat Following Oral Administration of Fosfomycin. *J. Feline Med. Surg. Open Rep.* **2018**, *4*, 205511691878660. [CrossRef]
31. Castañeda-García, A.; Blázquez, J.; Rodríguez-Rojas, A. Molecular Mechanisms and Clinical Impact of Acquired and Intrinsic Fosfomycin Resistance. *Antibiotics* **2013**, *2*, 217–236. [CrossRef]
32. Couce, A.; Briales, A.; Rodríguez-Rojas, A.; Costas, C.; Pascual, Á.; Blázquez, J. Genomewide Overexpression Screen for Fosfomycin Resistance in Escherichia Coli: MurA Confers Clinical Resistance at Low Fitness Cost. *Antimicrob. Agents Chemother.* **2012**, *56*, 2767–2769. [CrossRef]
33. Xu, S.; Fu, Z.; Zhou, Y.; Liu, Y.; Xu, X.; Wang, M. Mutations of the Transporter Proteins GlpT and UhpT Confer Fosfomycin Resistance in Staphylococcus Aureus. *Front. Microbiol.* **2017**, *8*, 914. [CrossRef] [PubMed]
34. Leite, G.C.; Perdigão-Neto, L.V.; Ruedas Martins, R.C.; Rizek, C.; Levin, A.S.; Costa, S.F. Genetic Factors Involved in Fosfomycin Resistance of Multidrug-Resistant Acinetobacter Baumannii. *Infect. Genet. Evol.* **2021**, *93*, 104943. [CrossRef] [PubMed]
35. Ito, R.; Mustapha, M.M.; Tomich, A.D.; Callaghan, J.D.; McElheny, C.L.; Mettus, R.T.; Shanks, R.M.Q.; Sluis-Cremer, N.; Doi, Y. Widespread Fosfomycin Resistance in Gram-Negative Bacteria Attributable to the Chromosomal *FosA* Gene. *mBio* **2017**, *8*, e00749-17. [CrossRef] [PubMed]
36. Falagas, M.E.; Athanasaki, F.; Voulgaris, G.L.; Triarides, N.A.; Vardakas, K.Z. Resistance to Fosfomycin: Mechanisms, Frequency and Clinical Consequences. *Int. J. Antimicrob. Agents* **2019**, *53*, 22–28. [CrossRef] [PubMed]
37. Poirel, L.; Vuillemin, X.; Kieffer, N.; Mueller, L.; Descombes, M.-C.; Nordmann, P. Identification of FosA8, a Plasmid-Encoded Fosfomycin Resistance Determinant from Escherichia Coli, and Its Origin in Leclercia Adecarboxylata. *Antimicrob. Agents Chemother.* **2019**, *63*, e01403-19. [CrossRef]
38. ten Doesschate, T.; Abbott, I.J.; Willems, R.J.L.; Top, J.; Rogers, M.R.C.; Bonten, M.M.; Paganelli, F.L. In Vivo Acquisition of Fosfomycin Resistance in Escherichia Coli by FosA Transmission from Commensal Flora. *J. Antimicrob. Chemother.* **2019**, *74*, 3630–3632. [CrossRef]
39. Huang, Y.; Lin, Q.; Zhou, Q.; Lv, L.; Wan, M.; Gao, X.; Wang, C.; Liu, J.-H. Identification of FosA10, a Novel Plasmid-Mediated Fosfomycin Resistance Gene of Klebsiella Pneumoniae Origin, in Escherichia Coli. *Infect. Drug Resist.* **2020**, *13*, 1273–1279. [CrossRef]
40. Gaballa, A.; Newton, G.L.; Antelmann, H.; Parsonage, D.; Upton, H.; Rawat, M.; Claiborne, A.; Fahey, R.C.; Helmann, J.D. Biosynthesis and Functions of Bacillithiol, a Major Low-Molecular-Weight Thiol in Bacilli. *Proc. Natl. Acad. Sci. USA* **2010**, *107*, 6482–6486. [CrossRef]
41. Cao, M.; Bernat, B.A.; Wang, Z.; Armstrong, R.N.; Helmann, J.D. FosB, a Cysteine-Dependent Fosfomycin Resistance Protein under the Control of ς ^{W}, an Extracytoplasmic-Function ς Factor in *Bacillus Subtilis*. *J. Bacteriol.* **2001**, *183*, 2380–2383. [CrossRef]
42. Fillgrove, K.L.; Pakhomova, S.; Schaab, M.R.; Newcomer, M.E.; Armstrong, R.N. Structure and Mechanism of the Genomically Encoded Fosfomycin Resistance Protein, FosX, from *Listeria Monocytogenes*. *Biochemistry* **2007**, *46*, 8110–8120. [CrossRef]
43. Kobayashi, S.; Kuzuyama, T.; Seto, H. Characterization of the *FomA* and *FomB* Gene Products from *Streptomyces Wedmorensis*, Which Confer Fosfomycin Resistance on *Escherichia Coli*. *Antimicrob. Agents Chemother.* **2000**, *44*, 647–650. [CrossRef] [PubMed]
44. García, P.; Arca, P.; Evaristo Suárez, J. Product of FosC, a Gene from Pseudomonas Syringae, Mediates Fosfomycin Resistance by Using ATP as Cosubstrate. *Antimicrob. Agents Chemother.* **1995**, *39*, 1569–1573. [CrossRef] [PubMed]
45. Truong-Bolduc, Q.C.; Wang, Y.; Hooper, D.C. Tet38 Efflux Pump Contributes to Fosfomycin Resistance in Staphylococcus Aureus. *Antimicrob. Agents Chemother.* **2018**, *62*, e00927-18. [CrossRef] [PubMed]
46. Sharma, A.; Sharma, R.; Bhattacharyya, T.; Bhando, T.; Pathania, R. Fosfomycin Resistance in *Acinetobacter Baumannii* Is Mediated by Efflux through a Major Facilitator Superfamily (MFS) Transporter—AbaF. *J. Antimicrob. Chemother.* **2017**, *72*, 68–74. [CrossRef]
47. Page, M.J.; McKenzie, J.E.; Bossuyt, P.M.; Boutron, I.; Hoffmann, T.C.; Mulrow, C.D.; Shamseer, L.; Tetzlaff, J.M.; Akl, E.A.; Brennan, S.E.; et al. The PRISMA 2020 Statement: An Updated Guideline for Reporting Systematic Reviews. *BMJ* **2021**, *372*, n71. [CrossRef]
48. Hubka, P.; Boothe, D.M. In Vitro Susceptibility of Canine and Feline Escherichia Coli to Fosfomycin. *Vet. Microbiol.* **2011**, *149*, 277–282. [CrossRef]
49. Haenni, M.; Saras, E. A USA300 Variant and Other Human-Related Methicillin-Resistant Staphylococcus Aureus Strains Infecting Cats and Dogs in France. *J. Antimicrob. Chemother.* **2012**, *67*, 326–329. [CrossRef]
50. Hou, J.; Huang, X.; Deng, Y.; He, L.; Yang, T.; Zeng, Z.; Chen, Z.; Liu, J.-H. Dissemination of the Fosfomycin Resistance Gene *FosA3* with CTX-M β-Lactamase Genes and *RmtB* Carried on IncFII Plasmids among Escherichia Coli Isolates from Pets in China. *Antimicrob. Agents Chemother.* **2012**, *56*, 2135–2138. [CrossRef]
51. He, L.; Partridge, S.R.; Yang, X.; Hou, J.; Deng, Y.; Yao, Q.; Zeng, Z.; Chen, Z.; Liu, J.-H. Complete Nucleotide Sequence of PHN7A8, an F33:A-:B-Type Epidemic Plasmid Carrying BlaCTX-M-65, FosA3 and RmtB from China. *J. Antimicrob. Chemother.* **2013**, *68*, 46–50. [CrossRef]

52. Ho, P.L.; Chan, J.; Lo, W.U.; Law, P.Y.; Li, Z.; Lai, E.L.; Chow, K.H. Dissemination of Plasmid-Mediated Fosfomycin Resistance *FosA3* among Multidrug-Resistant *Escherichia Coli* from Livestock and Other Animals. *J. Appl. Microbiol.* **2013**, *114*, 695–702. [CrossRef]
53. DiCicco, M.; Weese, S.; Neethirajan, S.; Rousseau, J.; Singh, A. Fosfomycin Susceptibility of Canine Methicillin-Resistant Staphylococcus Pseudintermedius Isolates. *Res. Vet. Sci.* **2014**, *96*, 251–253. [CrossRef] [PubMed]
54. Haenni, M.; Hocquet, D.; Ponsin, C.; Cholley, P.; Guyeux, C.; Madec, J.-Y.; Bertrand, X. Population Structure and Antimicrobial Susceptibility of Pseudomonas Aeruginosa from Animal Infections in France. *BMC Vet. Res.* **2015**, *11*, 9. [CrossRef] [PubMed]
55. Ewers, C.; Klotz, P.; Leidner, U.; Stamm, I.; Prenger-Berninghoff, E.; Göttig, S.; Semmler, T.; Scheufen, S. OXA-23 and IS Aba1 –OXA-66 Class D β-Lactamases in Acinetobacter Baumannii Isolates from Companion Animals. *Int. J. Antimicrob. Agents* **2017**, *49*, 37–44. [CrossRef] [PubMed]
56. Yao, H.; Wu, D.; Lei, L.; Shen, Z.; Wang, Y.; Liao, K. The Detection of Fosfomycin Resistance Genes in Enterobacteriaceae from Pets and Their Owners. *Vet. Microbiol.* **2016**, *193*, 67–71. [CrossRef]
57. Yukawa, S.; Tsuyuki, Y.; Sato, T.; Fukuda, A.; Usui, M.; Tamura, Y. Antimicrobial Resistance of *Pseudomonas Aeruginosa* Isolated from Dogs and Cats in Primary Veterinary Hospitals in Japan. *Jpn. J. Infect. Dis.* **2017**, *70*, 461–463. [CrossRef]
58. Maeyama, Y.; Taniguchi, Y.; Hayashi, W.; Ohsaki, Y.; Osaka, S.; Koide, S.; Tamai, K.; Nagano, Y.; Arakawa, Y.; Nagano, N. Prevalence of ESBL/AmpC Genes and Specific Clones among the Third-Generation Cephalosporin-Resistant Enterobacteriaceae from Canine and Feline Clinical Specimens in Japan. *Vet. Microbiol.* **2018**, *216*, 183–189. [CrossRef]
59. Zhang, P.L.C.; Shen, X.; Chalmers, G.; Reid-Smith, R.J.; Slavic, D.; Dick, H.; Boerlin, P. Prevalence and Mechanisms of Extended-Spectrum Cephalosporin Resistance in Clinical and Fecal Enterobacteriaceae Isolates from Dogs in Ontario, Canada. *Vet. Microbiol.* **2018**, *213*, 82–88. [CrossRef]
60. Kaspar, U.; von Lützau, A.; Schlattmann, A.; Roesler, U.; Köck, R.; Becker, K. Zoonotic Multidrug-Resistant Microorganisms among Small Companion Animals in Germany. *PLoS ONE* **2018**, *13*, e0208364. [CrossRef]
61. Pruthvishree, B.S.; Kumar, O.R.V.; Sivakumar, M.; Tamta, S.; Sunitha, R.; Sinha, D.K.; Singh, B.R. Molecular Characterization of Extensively Drug Resistant (XDR), Extended Spectrum Beta-Lactamases (ESBL) and New Delhi Metallo Beta-Lactamase-1 (BlaNDM1) Producing Escherichia Coli Isolated from a Male Dog—A Case Report. *Escherichia Coli. Vet. Arhiv.* **2018**, *88*, 139–148. [CrossRef]
62. Fernandes, M.R.; Sellera, F.P.; Moura, Q.; Carvalho, M.P.N.; Rosato, P.N.; Cerdeira, L.; Lincopan, N. Zooanthroponotic Transmission of Drug-Resistant *Pseudomonas Aeruginosa*, Brazil. *Emerg. Infect. Dis.* **2018**, *24*, 1160–1162. [CrossRef]
63. Cui, L.; Lei, L.; Lv, Y.; Zhang, R.; Liu, X.; Li, M.; Zhang, F.; Wang, Y. Bla NDM-1 -Producing Multidrug-Resistant Escherichia Coli Isolated from a Companion Dog in China. *J. Glob. Antimicrob. Resist.* **2018**, *13*, 24–27. [CrossRef] [PubMed]
64. Ortega-Paredes, D.; Haro, M.; Leoro-Garzón, P.; Barba, P.; Loaiza, K.; Mora, F.; Fors, M.; Vinueza-Burgos, C.; Fernández-Moreira, E. Multidrug-Resistant Escherichia Coli Isolated from Canine Faeces in a Public Park in Quito, Ecuador. *J. Glob. Antimicrob. Resist.* **2019**, *18*, 263–268. [CrossRef]
65. Courtise, R.M. Characterisation of Antimicrobial Resistance among Canine Urinary Isolates in Western Canada. Master's Thesis, Department of Veterinary Microbiology, University of Saskatchewan, Saskatoon, SK, Canada, 2018; pp. 110–136.
66. Kamathewatta, K.I.; Bushell, R.N.; Young, N.D.; Stevenson, M.A.; Billman-Jacobe, H.; Browning, G.F.; Marenda, M.S. Exploration of Antibiotic Resistance Risks in a Veterinary Teaching Hospital with Oxford Nanopore Long Read Sequencing. *PLoS ONE* **2019**, *14*, e0217600. [CrossRef] [PubMed]
67. Sartori, L.; Sellera, F.P.; Moura, Q.; Cardoso, B.; Cerdeira, L.; Lincopan, N. Multidrug-Resistant CTX-M-15-Positive Klebsiella Pneumoniae ST307 Causing Urinary Tract Infection in a Dog in Brazil. *J. Glob. Antimicrob. Resist.* **2019**, *19*, 96–97. [CrossRef] [PubMed]
68. Sellera, F.P.; Lopes, R.; Monte, D.F.M.; Cardoso, B.; Esposito, F.; dos Anjos, C.; da Silva, L.C.B.A.; Lincopan, N. Genomic Analysis of Multidrug-Resistant CTX-M-15-Positive Klebsiella Pneumoniae Belonging to the Highly Successful ST15 Clone Isolated from a Dog with Chronic Otitis. *J. Glob. Antimicrob. Resist.* **2020**, *22*, 659–661. [CrossRef]
69. Chen, L.; Ou, B.; Zhang, M.; Chou, C.-H.; Chang, S.-K.; Zhu, G. Coexistence of Fosfomycin Resistance Determinant *FosA* and *FosA* 3 in *Enterobacter Cloacae* Isolated from Pets with Urinary Tract Infection in Taiwan. *Microb. Drug Resist.* **2021**, *27*, 415–423. [CrossRef] [PubMed]
70. Loayza-Villa, F.; Salinas, L.; Tijet, N.; Villavicencio, F.; Tamayo, R.; Salas, S.; Rivera, R.; Villacis, J.; Satan, C.; Ushiña, L.; et al. Diverse Escherichia Coli Lineages from Domestic Animals Carrying Colistin Resistance Gene Mcr-1 in an Ecuadorian Household. *J. Glob. Antimicrob. Resist.* **2020**, *22*, 63–67. [CrossRef]
71. Cummings, K.J.; Mitchell, P.K.; Rodriguez-Rivera, L.D.; Goodman, L.B. Sequence Analysis of *Salmonella Enterica* Isolates Obtained from Shelter Dogs throughout Texas. *Vet. Med. Sci.* **2020**, *6*, 975–979. [CrossRef] [PubMed]
72. Zhang, Z.; Lei, L.; Zhang, H.; Dai, H.; Song, Y.; Li, L.; Wang, Y.; Xia, Z. Molecular Investigation of Klebsiella Pneumoniae from Clinical Companion Animals in Beijing, China, 2017–2019. *Pathogens* **2021**, *10*, 271. [CrossRef]
73. Qiu, Y.; Nambiar, R.B.; Xu, X.; Weng, S.; Pan, H.; Zheng, K.; Yue, M. Global Genomic Characterization of Salmonella Enterica Serovar Telelkebir. *Front. Microbiol.* **2021**, *12*, 704152. [CrossRef]
74. Sahin-Tóth, J.; Kovács, E.; Tóthpál, A.; Juhász, J.; Forró, B.; Bányai, K.; Havril, K.; Horváth, A.; Ghidán, Á.; Dobay, O. Whole Genome Sequencing of Coagulase Positive Staphylococci from a Dog-and-Owner Screening Survey. *PLoS ONE* **2021**, *16*, e0245351. [CrossRef] [PubMed]

75. Li, Y.; Sun, Z.-Z.; Rong, J.-C.; Xie, B.-B. Comparative Genomics Reveals Broad Genetic Diversity, Extensive Recombination and Nascent Ecological Adaptation in Micrococcus Luteus. *BMC Genom.* **2021**, *22*, 124. [CrossRef] [PubMed]
76. Lin, H.; Chen, W.; Zhou, R.; Yang, J.; Wu, Y.; Zheng, J.; Fei, S.; Wu, G.; Sun, Z.; Li, J.; et al. Characteristics of the Plasmid-Mediated Colistin-Resistance Gene Mcr-1 in Escherichia Coli Isolated from a Veterinary Hospital in Shanghai. *Front. Microbiol.* **2022**, *13*, 1002827. [CrossRef] [PubMed]
77. Butaye, P.; Stegger, M.; Moodley, A.; Damborg, P.; Williams, A.; Halliday-Simmonds, I.; Guardabassi, L. One Health Genomic Study of Human and Animal Klebsiella Pneumoniae Isolated at Diagnostic Laboratories on a Small Caribbean Island. *Antibiotics* **2021**, *11*, 42. [CrossRef]
78. Jonker, A.; Gouws, J.; Kapp, E.R.; Henning, A. Isolation of a Multidrug-Resistant *Escherichia Coli* Pathotype Stx2:Cnf1:Cnf2:Eae as a Potential Cause of Hemorrhagic Diarrhea and Secondary Septicemia in a Dog. *J. Vet. Diagn. Investig.* **2022**, *34*, 339–342. [CrossRef]
79. Bruce, S.A.; Smith, J.T.; Mydosh, J.L.; Ball, J.; Needle, D.B.; Gibson, R.; Andam, C.P. Shared Antibiotic Resistance and Virulence Genes in Staphylococcus Aureus from Diverse Animal Hosts. *Sci. Rep.* **2022**, *12*, 4413. [CrossRef]
80. Teng, L.; Liao, S.; Zhou, X.; Jia, C.; Feng, M.; Pan, H.; Ma, Z.; Yue, M. Prevalence and Genomic Investigation of Multidrug-Resistant Salmonella Isolates from Companion Animals in Hangzhou, China. *Antibiotics* **2022**, *11*, 625. [CrossRef]
81. Kim, D.-W.; Cha, C.-J. Antibiotic Resistome from the One-Health Perspective: Understanding and Controlling Antimicrobial Resistance Transmission. *Exp. Mol. Med.* **2021**, *53*, 301–309. [CrossRef]
82. Jiang, Y.; Shen, P.; Wei, Z.; Liu, L.; He, F.; Shi, K.; Wang, Y.; Wang, H.; Yu, Y. Dissemination of a Clone Carrying a FosA3-Harbouring Plasmid Mediates High Fosfomycin Resistance Rate of KPC-Producing Klebsiella Pneumoniae in China. *Int. J. Antimicrob. Agents* **2015**, *45*, 66–70. [CrossRef]
83. Mendes, A.C.; Rodrigues, C.; Pires, J.; Amorim, J.; Ramos, M.H.; Novais, Â.; Peixe, L. Importation of Fosfomycin Resistance *FosA3* Gene to Europe. *Emerg. Infect. Dis.* **2016**, *22*, 346–348. [CrossRef]
84. Stull, J.W.; Weese, J.S. Hospital-Associated Infections in Small Animal Practice. *Vet. Clin. N. Am. Small Anim. Pract.* **2015**, *45*, 217–233. [CrossRef] [PubMed]
85. CLSI. *Performance Standards for Antimicrobial Susceptibillity Testing. Supplement M100*, 32nd ed.; Clinical and Laboratory Standards Institute: Wayne, PA, USA, 2022; ISBN 978-1-68440-134-5.
86. CLSI. *Performance Standards for Antimicrobial Disk and Dilution Susceptibility Tests for Bacteria Isolated from Animals. Supplement VET01S*, 6th ed.; Clinical and Laboratory Standards Institute: Wayne, PA, USA, 2023; ISBN 978-1-68440-167-3.
87. Schmidt, V.M.; Pinchbeck, G.; McIntyre, K.M.; Nuttall, T.; McEwan, N.; Dawson, S.; Williams, N.J. Routine Antibiotic Therapy in Dogs Increases the Detection of Antimicrobial-Resistant Faecal Escherichia Coli. *J. Antimicrob. Chemother.* **2018**, *73*, 3305–3316. [CrossRef] [PubMed]
88. Menard, J.; Goggs, R.; Mitchell, P.; Yang, Y.; Robbins, S.; Franklin-Guild, R.J.; Thachil, A.J.; Altier, C.; Anderson, R.; Putzel, G.G.; et al. Effect of Antimicrobial Administration on Fecal Microbiota of Critically Ill Dogs: Dynamics of Antimicrobial Resistance over Time. *Anim. Microbiome* **2022**, *4*, 36. [CrossRef] [PubMed]

veterinary sciences

MDPI

Article

Antibiotic Resistance Genes Carried by Commensal *Escherichia coli* from Shelter Cats in Italy

Delia Gambino [1], Francesco Giuseppe Galluzzo [1,*], Luca Cicero [1,*], Roberta Cirincione [1], Erika Mannino [1], Veronica Fiore [1], Daniela Proverbio [2], Eva Spada [2], Giovanni Cassata [1] and Valeria Gargano [1]

[1] Istituto Zooprofilattico Sperimentale della Sicilia "A. Mirri", 90129 Palermo, Italy; deliagamb@gmail.com (D.G.); robertacirincione@gmail.com (R.C.); manninoerika@gmail.com (E.M.); veronicafiore@hotmail.it (V.F.); giovanni.cassata@izssicilia.it (G.C.); valeria.gargano@izssicilia.it (V.G.)

[2] Department of Veterinary Medicine and Animal Sciences (DIVAS), University of Milan, 26900 Lodi, Italy; daniela.proverbio@unimi.it (D.P.); eva.spada@unimi.it (E.S.)

* Correspondence: francesco.galluzzo@izssicilia.it (F.G.G.); luca.cicero@izssicilia.it (L.C.)

Simple Summary: The epidemic of antimicrobial resistance is a widespread health challenge that deserves a One Health approach. Bacteria resistant to antimicrobials and their resistance genes can be transferred from food-producing animals and pets to humans and vice versa. Many studies have shown that resistant bacteria are emerging in companion animals and that a number of resistance genes are being shared between pets and humans. Even stray cats, which have contact with humans and share the urban environment with them, can therefore act as reservoirs of antimicrobial resistance for humans and their pets. Therefore, to investigate the implication of these animals as disseminators of antibiotic resistance, we phenotypically and genotypically assessed the resistance of commensal *E. coli* isolated from stray cat feces. The *E. coli* analyzed were resistant to ampicillin, tetracyclines and sulfisoxazole and carried genes that encode these resistances. Even though there is still a need for further studies, the occurrence of resistant *E. coli* provides support for the assumption that stray cats may be fecal sources of resistance, so it is necessary to monitor these animals in antimicrobial resistance surveillance programs.

Citation: Gambino, D.; Galluzzo, F.G.; Cicero, L.; Cirincione, R.; Mannino, E.; Fiore, V.; Proverbio, D.; Spada, E.; Cassata, G.; Gargano, V. Antibiotic Resistance Genes Carried by Commensal *Escherichia coli* from Shelter Cats in Italy. *Vet. Sci.* **2023**, *10*, 680. https://doi.org/10.3390/vetsci10120680

Academic Editor: George Valiakos

Received: 17 October 2023
Revised: 22 November 2023
Accepted: 27 November 2023
Published: 28 November 2023

Abstract: Antimicrobial resistance is a widespread global health problem. The presence of resistant bacteria and antibiotic resistance genes has been demonstrated not only in humans but also in animals, including pets. Stray cats share the urban environment with people and pets. This may facilitate transmission of resistant bacteria and resistance genes between stray animals, people and domestic animals. Several studies have investigated the role of stray cats as a fecal carrier of ESBL-producing bacteria. However, there are many genes and resistance mechanisms that can be detected in commensal *E. coli*, which, because of its genetic plasticity, is considered an indicator for monitoring antibiotic resistance. In this study, rectal swabs were collected from stray cats from colonies and shelters in the city of Monza (Monza Brianza, Italy) to isolate commensal *E. coli*. Phenotypic tests, such as the minimum inhibitory concentration (MIC) and the double disc test (DDST), and molecular analyses to detect antimicrobial resistance genes (ARGs) were used to study the resistance of these isolates. The results obtained confirm that stray cats can carry ESBL-producing *E. coli* (6.7%) and genes conferring resistance to other important antibiotic classes such as tetracyclines and sulfonamides.

Keywords: *Escherichia coli*; antimicrobial resistance; stray cats

1. Introduction

Escherichia coli is a widely disseminated bacterium of the *Enterobacteriaceae* family. While pathogenic *E. coli* strains can cause infection in humans and animals, most strains colonize the intestine in a harmless way and only occasionally cause disease in healthy individuals [1]. However, non-pathogenic *E. coli* isolates show an increased ability to acquire virulence factors and resistance to various antimicrobials, as well as an efficient transmission and colonization capacity. The *E. coli* genome is characterized by the presence of mobile genetic elements such as insertion sequences (IS), integrons, plasmids and transposons, which promote horizontal gene transfer (HGT) between this species and other bacteria. Virulence genes are mostly associated with IS, whereas antimicrobial resistance genes (ARG) are mostly carried on plasmids, transposons and integrons (class 1, 2 and 3 integrons), and all these elements play a central role in rearrangement and gene transfer. Integrons, in particular, are able to extract ARGs from their environment and assemble them in their gene cassettes through specific recombination. This means that in the intestine, where bacterial population density and species diversity are high, commensal *E. coli* can acquire ARGs and transfer them to other commensal strains, as well as pathogens acting as an ARG reservoir [2]. Resistant *E. coli* strains are widely distributed, and the last decade has seen an increase in resistance, especially to certain classes of antibiotics such as beta-lactams and fluoroquinolones [3].

In Europe, the surveillance of antimicrobial resistance in commensal indicator *E. coli* isolated from the intestinal microbiota of healthy livestock serves as a valuable means to track the dissemination of resistant bacteria. Notably, both antimicrobial resistance bacteria (ARBs) and ARGs have the potential to be transferred between animal and human populations, thereby posing a risk for the transmission of resistance genes to pathogenic bacteria in both humans and animals [4]. The high levels of antimicrobial resistance detected among isolates from food-producing animals have highlighted the need for cross-sector collaboration between the human, veterinary and food production sectors as part of a "One-Health" strategy [5]. Although many previous publications have revealed the prevalence among pets of multi-resistant pathogens similar to those identified in humans, such as Gram-negative extended spectrum beta-lactamase producers (ESBL-producers), AmpC-type beta-lactamases or carbapenemases [6–9], demonstrating the existence of an emerging health problem, these animals are not included in systemic antimicrobial resistance monitoring programs [10]. Moreover, most data on antimicrobial resistance in pets refer to household pets, and focus on pathogenic bacteria and specific sites of infection (urinary tract, skin, ear, and gastrointestinal infections) [1,11–13]. Close contact between pets and owners, who share a home environment and have contact with the same surfaces and objects, can promote interspecies transmission of resistant bacteria [14–16]. Stray cats are synanthropic animals that meet and interact with humans in the urban environment and may also act as reservoirs of AMR. The urban environment is an extremely complex network in which factors, such as high human density, the presence of other animals and numerous microenvironments (buildings, open spaces, parks, sewers), provide opportunities for selection and transport of ARBs and ARGs [17]. Strays living in these environments can therefore acquire ARBs and ARGs from various urban sources (soil, garbage, feces of other animals, and sewers) and transmit them to humans, either indirectly, by sharing the urban environment, or directly, in the case of volunteers caring for them in shelters or colonies or owners adopting them.

In Italy, the count of stray cats is conducted based on the number of sterilizations performed by the national health system within trap, neuter and release (TNR) sterilization programs. These data are provided by each Italian region and published annually by the Ministry of Health, whose latest report for 2021 showed 11,228 stray cats sterilized in Lombardy. The Stray Animal Law requires that stray cats be trapped and transferred to shelters where they are sterilized and have all the necessary care before they are either adopted (if suitable) or returned to their colonies. Cat colonies are groups of cats living free in urban areas, usually at the same locations, and managed by animal welfare organizations or private volunteers [18]. In this context, the possibility of these animals coming into contact with humans, such as volunteers (who care for them in shelters or colonies) or new owners (in case of adoption), and transmitting ARBs or ARGs to them, should not be overlooked.

The aim of this work was to evaluate the role of stray cats as reservoirs and fecal vectors of ARBs or ARGs. In order to do this, phenotypic and molecular tests were carried out on commensal *E. coli* isolated from stray cats housed in animal shelters or belonging to cat colonies in the city of Monza (Monza Brianza, Italy).

2. Materials and Methods

2.1. *Escherichia coli* Strains Collection

A total of 60 rectal swabs from stray cats were analyzed in order to isolate *E. coli* strains and assess their phenotypic and genotypic antibiotic resistance. Sampling was performed in 2022 during activities related to the SARS-CoV-2 infection surveillance in stray cats in the city of Monza, Monza Brianza, in the Lombardy region, northern Italy.

Rectal swabs were collected from n = 35 cats from nine different colonies in the province of Monza Brianza (n = 7 from a colony of Monza, n = 6 from Brugherio, n = 5 from Muggiò, n = 4 from Agrate Brianza, n = 4 from Cornate D'Adda, n = 4 from Lissone, n = 3 from Cernusco sul Naviglio, n = 1 from Concorezzo and n = 1 from Caponago) and n = 25 from the ENPA shelter in Monza Brianza (Lombardy, Italy), which were captured and received general anesthesia to perform neutering surgery as part of trap, neuter and release (TNR) sterilization programs. TNR programs are carried out as part of a national program to control stray pet populations under Italian National Law (law no. 281/1991). Samples were collected with the informed consent of those legally responsible for the stray colonies or shelter cats, and in accordance with the study and animal welfare protocol, which was revised and authorized by the Animal Welfare Bioethical Committee of the University of Milan (approval number OPBA _91_2020, released on 15 January 2021 and OPBA_34_2021, released on 12 March 2021).

After collection, rectal swabs were stored at -4 °C and sent within 24–48 h to the Istituto Zooprofilattico Sperimentale (IZS) of Sicily (Palermo, Italy), where they were seeded on McConkey agar plates (Oxoid, Milan, Italy) incubated at 37 °C for 24 h. One or two colonies with morphology attributable to *E. coli* were selected from each plate, if available, and purified on brain heart infusion agar (Oxoid, Milan, Italy) incubated at 37 °C for 24 h. The isolated and purified strains were subjected to biochemical–enzymatic screening tests (citrate production, urea, H_2S, glucose and lactose fermentation) to discriminate *E. coli* from other enterobacteria. DNA of each strain was extracted with an automated King Fisher extractor (Thermo Fisher Scientific, Waltham, MA, USA) using the QIAamp One-For-All Nucleic Acid Kit as recommended by the manufacturer (QIAGEN Sciences, Germantown, Maryland, USA). Before performing the PCRs, DNA quality and concentrations were assessed using NanoDrop™ 8000 Microvolume UV-Vis spectrophotometer (Thermo Fisher Scientific, Inc., Wilmington, DE, USA).

PCR was carried out using a Platinum™ Taq DNA Polymerase High Fidelity (Thermo Fisher Scientific, MA USA) and following the factory instruction. In order to identify the strains, 16S ribosomal gene amplifications were then performed, as reported by Li et al. [19]. Briefly, the reaction mix was prepared with 0.4 mM of forward and reverse primer, 1X high fidelity PCR buffer, 2 mM $MgSO_4$, 0.2 mM of each dNTP, 0.2 of forward and reverse primer, 10 ng of genomic DNA and autoclaved distilled water to 50 µL. The amplification reaction involved an initial denaturation at 94 °C for 1 min and then 35 cycles consisting of denaturation for 1 min at 94° C, annealing for 1 min at 55 °C and extension for 1 min at 68 °C. In each PCR run, *E. coli* ATCC 25,922 (American Type Culture Collection, Rockville, MD, USA) was used as a positive control; nuclease-free water was used as the negative one. Subsequently, 5 µL of each amplification reaction was analyzed by electrophoresis analysis using E-Gel™ Go! Agarose Gels, 2% (Thermo Fisher Scientific, Waltham, MA, USA) to determine product size. Finally, the PCR products were purified and sequenced at BMR Genomics Srl (Padua, Italy).

2.2. Determination of Minimum Inhibitory Concentration (MIC)

The antibiotic susceptibility of *E. coli* isolates was assessed using the minimum inhibitory concentration (MIC) method. Using commercial plates (Thermo Scientific 96-well Sensititre™ Plate, Waltham, MA, USA), the MIC values (μg/mL) of 10 antibiotics were determined. The antibiotics and their dilutions tested were amoxicillin/clavulanic acid (0.25/0.12–32/16 μg/mL), ampicillin (0.25–32 μg/mL), cefazolin (0.5–8 μg/mL), cefotaxime (0.5–4 μg/mL), colistin (0.03–8 μg/mL), enrofloxacin (0.03–32 μg/mL), gentamicin (0.25–32 μg/mL), sulfamethoxazole/trimethoprim (0.06/1.19–16/304 μg/mL), sulfisoxazole (128–512 μg/mL), and tetracycline (0.5–16 μg/mL). After preparing a bacterial suspension with 0.5 McFarland turbidity in 5 mL of sterile water, 10 μL of this was mixed in 10 mL of Mueller–Hinton broth (Thermo Fisher Scientific, Waltham, MA, USA). Subsequently, 50 μL of inoculated broth was dispensed into each well of the MIC plate, which was incubated at 37 °C for 18–24 h. Manual plate reading was conducted with the Sensititre™ Manual Viewbox (Thermo Fisher Scientific, Waltham, MA, USA) and results were interpreted according to Clinical and Laboratory Standards Institute (CLSI) breakpoints [20].

2.3. Double Disk Sinergy Test

The double-disk synergy test (DDST) was performed as recommended by EUCAST [21]. Briefly, a 0.5 McFarland bacterial suspension was seeded on Muller—Hinton agar (Oxoid, Milan, Italy) and three discs containing cephalosporins (cefotaxime 30 μg, ceftazidime 30 μg, cefepime 30 μg) were positioned beside a disc containing clavulanic acid (amoxicillin–clavulanic acid 30 μg). The test was interpreted as positive if there were increased zones of inhibition around the cephalosporin discs or a "keyhole" towards the amoxicillin–clavulanic acid disc.

2.4. Multiplex Real-Time PCR for ESBLs Determination

A multiplex Real-Time PCR was performed to assess the presence of resistance genes to extended spectrum beta-lactamase antibiotics. Therefore, DNA obtained from each strain was subjected to a Real-Time PCR analysis specifically for the determination of the ESBL phenotype [22]. Real-time amplifications were performed in 25-μL reactions containing 12.5 μL of SsoAdvanced Universal Probes Supermix (Bio-Rad Laboratories S.r.l., California, USA), 0.4 μM of each forward and reverse primer (Table S1), 0.2 μM of TaqMan *blaTEM* probe, 0.4 μM of each of the other four TaqMan probes, 0.6 μL of sterile water, and 10 pg of DNA. Positive controls consisted of four ATCC® (Manassas, Virginia, USA) strains (*E. coli* BAA-3048™, *E. coli* BAA-3049™, *E. coli* BAA-3051™ and *Klebsiella pneumoniae* BAA-3060™, American Type Culture Collection, Rockville, MD, USA) harboring the searched genes, while DNAase- and RNAase-free water was used as a negative control.

2.5. Class 1 Integron and ARGs Detection for Tetracyclines, Sulfonamides and Fluoroquinolones

A reaction mix containing a concentration 1X of 5X Platinum II PCR Buffer, 10 mM dNTPs, 0.5 μM of each primer of the pairs shown in Table S2, 1.25 U of Platinum II Taq DNA Polymerase DNA polymerase (Thermo Fisher Scientific, MA USA), 10 ng of genomic DNA and nucleus-free water to obtain a volume of 50 μL was prepared. For each PCR reaction, DNA from two ATCC® (*E. coli* BAA-3048™ and *E. coli* BAA-3051™) strains harboring the researched genes was used as positive controls and DNAase- and RNAase-free water as a negative control. The size of all amplicons was verified by electrophoresis on E-Gel™ Go! agarose gel, 2% (Thermo Fisher Scientific, Waltham, MA, USA).

2.6. Statistical Analysis

Statistical analyses were conducted with R version 4.3.1 (16 June 2023). The phenotypic resistance variables were correlated with the genetic resistance to the antibiotics (resistance to beta-lactam, tetracyclines, sulfonamides, and fluoroquinolones). Fisher's exact probability test was used to evaluate if there is an association between phenotypic and genetic resistance. The p-value was considered as significant if $p < 0.05$.

3. Results

3.1. Phenotypic Profile

Phenotypic tests were carried out on the 60 *E. coli* isolates to determine the susceptibility to 10 antibiotics and the occurrence of ESBL-producing isolates. Table 1 shows the MIC values obtained for the 60 *E. coli* analyzed in this study.

Table 1. MIC values detected.

Antimicrobial Agent	Number of Isolates at the Indicated MIC Value (µg/mL)															S	R
	0.03	0.06	0.12	0.25	0.5	1	2	4	8	16	32	64	128	256	512	(%)	(%)
Amoxicillin/clavulanic acid				1	1	15	10	17		1	**5**					91.7	8.3
Ampicillin				3		12	16	15	2		**12**					80	20
Cefazolin					4	11	29	8	**8**							86.7	13.3
Cefotaxime					56			**4**								93.3	6.7
Colistin			1	10	37	8	1	**2**	**1**							95	5
Enrofloxacin	47	8	2	1	2											100	
Gentamicin				9	29	20	2									100	
Sulfamethoxazole/trimethoprim		53		2	1					**4**						93.3	6.7
Sulfisoxazole													49		**11**	81.7	18.3
Tetracycline					7	9	31	2		**11**						83.3	18.3

S = susceptible, R = resistant. Gray shaded areas indicate the antimicrobial concentration tested, while bold indicates the number of resistant strains according to the CLSI M-100 cut-off values [20].

Among the *E. coli* isolates, 40% (24/60) were resistant to at least one of the investigated antibiotics. Four isolates were multi-drug resistant (MDR), as one isolate showed resistance to four classes of antibiotics (beta-lactams, colistin, tetracyclines and sulfonamides) and three other isolates to three classes, i.e., beta-lactams, tetracyclines and sulfonamides. No strains showed resistance to enrofloxacin and gentamicin, but resistance was found to the other eight antibiotics tested (Table 1). The most common resistances were to ampicillin (20%), sulfisoxazole (18.3%), tetracycline (18.3%) and cefazolin (13.3%). Instead, less than 10% of the strains showed resistance to amoxicillin/clavulanic acid, sulfamethoxazole/trimethoprim, cefotaxime and colistin.

The DDST test was performed on all strains and allowed to identify 4/60 strains with an ESBL profile, the same strains that showed resistance to cefotaxime.

3.2. Genotypic Profile

The ARGs detection conducted on the 60 *E. coli* isolates showed that 31.6% (19/60) harbored at least one ARG among those screened for. The gene most frequently detected was *tet(A)*, present in 16.7% (10/60) of isolates, followed by *sul1* (13.3%, 8/60) and *tet(B)* (10%, 6/60). Also, although with low frequency, bla_{TEM} (8.3%, 5/60) and $bla_{CTX\text{-}M}$ (6.7%, 4/60) genes were also detected. No strains were found to harbor the bla_{SHV}, bla_{CMY}, *sul2*, *gyrA* and *parC* genes. Table 2 shows the phenotypic and corresponding genetic profiles detected in the tested isolates.

Table 2. Genetic and phenotypic profiles of 24 *E. coli* isolates resistant to tested antibiotics.

ARGs Detected	Phenotypic Resistance	Number of Isolates
sul1, *int1*	FIS	4
tet(A), *sul1*, *int1*	AMP-TET-SXT-FIS	3
tet(A), *tet*(B)	AMP-TET	3
bla_{TEM}, bla_{CTXM}, *tet*(A)	AMP-AUG2-FAZ-FOT-COL-TET	2
bla_{TEM}, tet(A), tet(B)	AMP- FAZ-TET	1
bla_{TEM}, *tet*(A), *tet*(B)	AMP-AUG2-TET-FIS	1

Table 2. *Cont.*

ARGs Detected	Phenotypic Resistance	Number of Isolates
bla$_{TEM}$	AMP-AUG2-FAZ	1
bla$_{CTXM}$	FAZ-FOT-FIS	1
sul1, *int1*	SXT	1
bla$_{CTXM}$	FOT	1
tet(B)	TET-FIS	1
*	FAZ-COL-FIS	1
*	AUG2-COL	1
*	AMP	1
*	FAZ	2

FIS: sulfisoxazole; AMP: ampicillin; TET: tetracycline; SXT: trimethoprim/sulfamethoxazole; AUG2: amoxicillin/clavulanic acid; FAZ: cefazolin; FOT: cefotaxime; COL: colistin; * No ARGs detected.

3.3. Data Analysis

Statistical analysis showed that beta-lactams, tetracyclines and sulfonamides had a significant correlation between the phenotypic resistance detected and the genes investigated (p-value < 0.05) (Figure 1). This correlation was also significant for fluoroquinolones, for which no discrepancies were found.

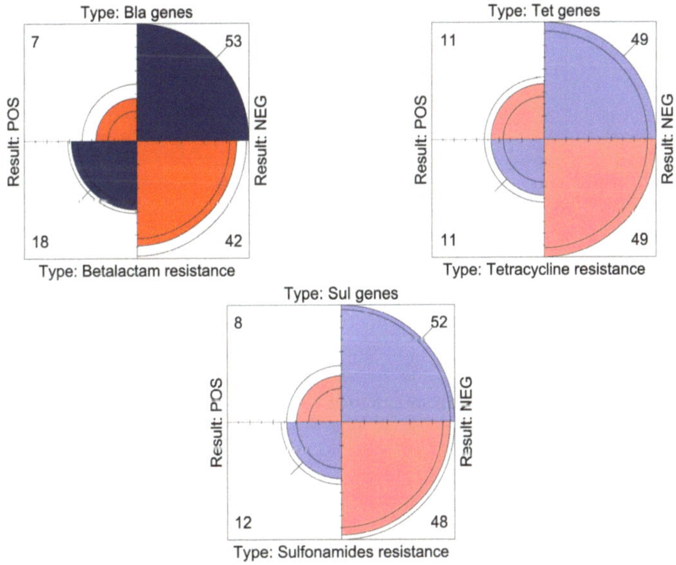

Figure 1. The graph displays, for each antibiotic class, a quadrant view of the relationship between the phenotypic resistance exhibited by the strains and the genes that were investigated. The area of each quadrant is proportional to the frequency of positives/negatives detected, indicated numerically in each corner.

4. Discussion

The transmission of antimicrobial-resistant bacteria or their resistance genes to humans can occur from both animals for food production and pets. Many studies have shown the attendance of resistant bacteria in pets and that several ARGs are spread between these animals and humans [6,7,23]. Stray cats can also act as reservoirs of AMR, as they can transmit ARBs and ARGs to humans and their pets but can also acquire them [6–9]. These animals live in urban environments where the conditions are favorable for the selection of resistance. Urban environments are characterized by the presence of selective pressures (antibiotics, heavy metals and biocides), high bacterial diversity and human density [17]. These factors, combined with the characteristics of bacteria, such as their ability to respond to stress, to implement mechanisms to minimize the fitness cost of ARGs, to form biofilms and, above all, the genetic plasticity of bacteria, are all factors that help favor the selection of resistant bacteria and resistance genes. As a consequence, stray cats living in this environment can acquire selected ARBs and ARGs and become reservoirs and disseminators of resistance. They can spread ARBs and ARGs in the environment through feces and, in countries such as Italy where stray animal management and adoption is encouraged, they can also spread them through direct contact with volunteers in shelters and colonies or, in the case of adoption, with owners.

In this study, we investigated the resistance of commensal *E. coli* isolated from rectal swabs of stray cats to assess the potential role of these animals as reservoirs and disseminators of AMR. Phenotypic and molecular tests were performed on the 60 isolates to determine their resistance, ESBL production and the presence of selected ARGs. The MICs showed that these isolates were resistant to three classes of antibiotic: beta-lactams (30%), sulfonamides (20%) and tetracyclines (18.3%). Looking at individual molecules, our results are consistent with those reported in other studies conducted on domestic and stray cats in Italy [15,23,24]. The highest resistance rates were found for ampicillin, sulfisoxazole and tetracyclines. These results are consistent with those reported in previous studies on stray and domestic cats [6,7,24]. Regarding sulfonamides, a higher resistance was found for sulfisoxazole (18.3%) and a lower for trimethoprim/sulfamethoxazole (6.7%), the latter also being lower than in the above-mentioned studies where the percentage of resistant strains was at least 33.6% [7,24]. The ESBL phenotype was detected in 6.7% of isolates by the DDST test, confirming that stray cats can harbor ESBL-producing bacteria even at low prevalence, which is in agreement with previous studies [25]. This result was confirmed by molecular analysis, which detected the presence of the $bla_{CTX\text{-}M}$ gene in our four ESBL-producing isolates. Although $bla_{CTX\text{-}M}$ gene variants were not tested in this study, it is known that all $bla_{CTX\text{-}M}$ variants encode extended spectrum beta-lactamases that confer resistance to most beta-lactam antibiotics, including third-generation cephalosporins. These genes are widely distributed internationally and have been detected in clinical ESBL isolates from humans and in isolates from several animal species, including dogs and cats [25,26]. Among the *bla* genes, the presence of bla_{TEM} (8.3%) has also been detected. There are now over 200 known variants of this gene, with different levels of resistance (narrow and extended spectrum) [27]. Although the variants were not determined in this study, the occurrence of the bla_{TEM} gene in isolates from dogs and cats, including strays, has been investigated and reported by other authors, confirming the possibility that these animals are not only carriers of resistant bacteria but also of the *bla* gene with its different spectrum of activity [7,9,25,28].

While nineteen *E. coli* isolates harbored one or more genes that could be involved in the phenotypic resistance they exhibited, this was not the case for five isolates in which no ARG was detected. Resistance to antimicrobials can be mediated by several major mechanisms encoded by numerous genes and variants. Therefore, it is possible that in these five *E. coli* isolates, the genes responsible for resistance are different from those we investigated [29]. In this study, the most frequently detected genes were the *tet* and *sul* genes. With regard to tetracycline resistance genes, the *tet(A)* gene was the predominant resistance determinant (16.6%), followed by the *tet(B)* gene (10%). Notably, the *tet(A)* gene was found in all tetracycline-resistant strains. The detection of the *tet(A)* gene suggests that active efflux is the main mechanism of tetracycline resistance in these *E. coli* isolates from stray cats, which is consistent with the epidemiological trend of tetracycline resistance genes in *E. coli* of animal origin [30,31]. The *sul1* gene was detected in 13.3% of *E. coli* isolates that also harbored the class 1 integrons. Sulfonamides resistance is widespread in Gram-negative bacteria from animals and humans worldwide [32]. The *sul1* is one of the genes coding for resistance to sulfonamides and was found almost exclusively on large conjugative plasmids and class 1 integrons. This class of integrons is the most abundant and clinically relevant and plays a crucial role in the spread of AMR genes [33].

The presence of ARGs in association with class 1 integrons in *E. coli* isolates from stray cats suggests that the discovered genes are organized in gene cassettes and could be transferred to other bacteria via HGT, which would contribute to the spread of resistance. Indeed, gene cassettes can be transferred from one bacterium to another via integrons. The acquisition, transfer and maintenance of class 1 integrons is thought to be one of the causes of the steady increase in the emergence of resistant *E. coli* over time [34].

5. Conclusions

In Europe, antimicrobial resistance surveillance programs only consider food-producing animals, and although the European Medicines Agency (EMA) monitors the commercial sales and consumption of antimicrobial products in animals, data on companion animals are not included in the annual reports of either the EU or the World Organization for Animal Health (WOAH). Antimicrobial resistance is now recognized as a global health issue that deserves a "One Health" approach, but the AMR surveillance programs implemented by many countries do not include domestic and stray animals, which can be a source of ARBs and ARGs. Our results, which are consistent with previous studies, confirm the carriage of not only ESBL-producing *E. coli* but also ARGs for tetracyclines and sulfonamides in the feces of stray cats, highlighting their potential role as reservoirs. Although this study has its limitations due to the lack of information on stray cats (age, permanence in shelter, clinical history, antimicrobial treatments, environmental contamination status, direct contact with humans and animals), the results obtained highlight the importance of including stray cats in surveillance programs, both to assess the prevalence of antimicrobial resistance and as possible sources for the spread of antimicrobial resistance. Stray cats are reservoirs and sentinels for the spread of AMR in city environments and therefore deserve to be controlled, but monitoring these animals is certainly not easy. Useful data on stray cats could be collected as part of TNR activities carried out in many European countries, including Italy, and would be useful for the implementation of AMR surveillance systems. Furthermore, since stray cats can be carriers of ARBs and ARGs, it would be appropriate to train and educate volunteers who care for them. For this purpose, the recommendations issued by the Italian Ministry of Health together with the National Plan to Combat Antimicrobial Resistance 2020–2025 are useful. In addition to emphasizing the need to prescribe and use antimicrobials correctly, these recommendations underline the need for appropriate hygiene and preventive measures to control infections and the spread of ARBs.

Supplementary Materials: The following supporting information can be downloaded at: https://www.mdpi.com/article/10.3390/vetsci10120680/s1; Table S1: Real-Time PCR primers for ESBL profile determination; Table S2: Target ARGs detected, primer pairs and annealing temperature used.

Author Contributions: Conceptualization, D.G., V.G., D.P. and E.S.; methodology, D.G. and V.G.; formal analysis, D.G., V.G. and D.P.; investigation, E.M., V.F., L.C., R.C. and G.C.; resources, E.S. and G.C.; data curation, D.G., F.G.G. and V.G.; writing—original draft preparation, D.G. and V.G.; writing—review and editing, D.G., V.G. and E.S.; visualization, D.G., V.G. and F.G.G.; supervision, E.S. and G.C.; project administration, E.S. and G.C. All authors have read and agreed to the published version of the manuscript.

Funding: This research was funded by the Italian Ministry of Health RC IZSSI 05/20.

Institutional Review Board Statement: Samples were collected following protocol for the study, and animal welfare was reviewed and approved by the Animal Welfare Bioethical Committee of the University of Milan (approval number OPBA _91_2020, released on 15 January 2021 and OPBA_34_2021, released on 12 March 2021). The study was conducted using samples collected during the trap, neuter and release (TNR) sterilization program activities prescribed by Italian law (law no. 281/1991) and submitted to the Istituto Zooprofilattico Sperimentale della Sicilia 'A. Mirri' for diagnostic purposes.

Informed Consent Statement: Samples were collected with the consent of legal representatives of the stray colonies or shelter cats.

Data Availability Statement: All data discussed are contained in the manuscript.

Conflicts of Interest: The authors declare no conflict of interest.

References

1. Cui, L.; Zhao, X.; Li, R.; Han, Y.; Hao, G.; Wang, G.; Sun, S. Companion Animals as Potential Reservoirs of Antibiotic Resistant Diarrheagenic Escherichia Coli in Shandong, China. *Antibiotics* **2022**, *11*, 828. [CrossRef]
2. Poirel, L.; Madec, J.-Y.; Lupo, A.; Schink, A.-K.; Kieffer, N.; Nordmann, P.; Schwarz, S. Antimicrobial Resistance in *Escherichia coli*. *Microbiol. Spectr.* **2018**, *6*, 27. [CrossRef] [PubMed]
3. Pormohammad, A.; Nasiri, M.J.; Azimi, T. Prevalence of Antibiotic Resistance in Escherichia Coli Strains Simultaneously Isolated from Humans, Animals, Food, and the Environment: A Systematic Review and Meta-Analysis. *Infect. Drug Resist.* **2019**, *12*, 1181–1197. [CrossRef] [PubMed]
4. European Food Safety; European Centre for Disease Prevention and Control (ECDC). The European Union Summary Report on Antimicrobial Resistance in Zoonotic and Indicator Bacteria from Humans, Animals and Food in 2020/2021. *EFSA J.* **2023**, *21*, e07867. [CrossRef]
5. European Centre for Disease Prevention and Control. *Antimicrobial Resistance Surveillance in Europe 2023–2021 Data*; World Health Organization: Stockholm, Sweden, 2023.
6. Gargano, V.; Gambino, D.; Orefice, T.; Cirincione, R.; Castelli, G.; Bruno, F.; Interrante, P.; Pizzo, M.; Spada, E.; Proverbio, D.; et al. Can Stray Cats Be Reservoirs of Antimicrobial Resistance? *Vet. Sci.* **2022**, *9*, 631. [CrossRef] [PubMed]
7. Ratti, G.; Facchin, A.; Stranieri, A.; Giordano, A.; Paltrinieri, S.; Scarpa, P.; Maragno, D.; Gazzonis, A.; Penati, M.; Luzzago, C.; et al. Fecal Carriage of Extended-Spectrum β-Lactamase-/AmpC-Producing Escherichia Coli in Pet and Stray Cats. *Antibiotics* **2023**, *12*, 1249. [CrossRef] [PubMed]
8. Jung, W.K.; Shin, S.; Park, Y.K.; Lim, S.-K.; Moon, D.-C.; Park, K.T.; Park, Y.H. Distribution and Antimicrobial Resistance Profiles of Bacterial Species in Stray Cats, Hospital-Admitted Cats, and Veterinary Staff in South Korea. *BMC Vet. Res.* **2020**, *16*, 109. [CrossRef] [PubMed]
9. Marchetti, L.; Buldain, D.; Gortari Castillo, L.; Buchamer, A.; Chirino-Trejo, M.; Mestorino, N. Pet and Stray Dogs as Reservoirs of Antimicrobial-Resistant *Escherichia coli*. *Int. J. Microbiol.* **2021**, *2021*, e6664557. [CrossRef] [PubMed]
10. Marco-Fuertes, A.; Marin, C.; Lorenzo-Rebenaque, L.; Vega, S.; Montoro-Dasi, L. Antimicrobial Resistance in Companion Animals: A New Challenge for the One Health Approach in the European Union. *Vet. Sci.* **2022**, *9*, 208. [CrossRef]
11. Gómez-Beltrán, D.A.; Villar, D.; López-Osorio, S.; Ferguson, D.; Monsalve, L.K.; Chaparro-Gutiérrez, J.J. Prevalence of Antimicrobial Resistance in Bacterial Isolates from Dogs and Cats in a Veterinary Diagnostic Laboratory in Colombia from 2016–2019. *Vet. Sci.* **2020**, *7*, 173. [CrossRef]
12. Hamame, A.; Davoust, B.; Cherak, Z.; Rolain, J.-M.; Diene, S.M. Mobile Colistin Resistance (Mcr) Genes in Cats and Dogs and Their Zoonotic Transmission Risks. *Pathogens* **2022**, *11*, 698. [CrossRef] [PubMed]
13. Salgado-Caxito, M.; Moreno-Switt, A.I.; Paes, A.C.; Shiva, C.; Munita, J.M.; Rivas, L.; Benavides, J.A. Higher Prevalence of Extended-Spectrum Cephalosporin-Resistant Enterobacterales in Dogs Attended for Enteric Viruses in Brazil Before and After Treatment with Cephalosporins. *Antibiotics* **2021**, *10*, 122. [CrossRef] [PubMed]
14. Zhao, R.; Hao, J.; Yang, J.; Tong, C.; Xie, L.; Xiao, D.; Zeng, Z.; Xiong, W. The Co-Occurrence of Antibiotic Resistance Genes between Dogs and Their Owners in Families. *iMeta* **2022**, *1*, e21. [CrossRef]
15. Naziri, Z.; Poormaleknia, M.; Ghaedi Oliyaei, A. Risk of Sharing Resistant Bacteria and/or Resistance Elements between Dogs and Their Owners. *BMC Vet. Res.* **2022**, *18*, 203. [CrossRef] [PubMed]
16. Carvalho, A.C.; Barbosa, A.V.; Arais, L.R.; Ribeiro, P.F.; Carneiro, V.C.; Cerqueira, A.M.F. Resistance Patterns, ESBL Genes, and Genetic Relatedness of *Escherichia coli* from Dogs and Owners. *Braz. J. Microbiol.* **2016**, *47*, 150–158. [CrossRef] [PubMed]
17. Vassallo, A.; Kett, S.; Purchase, D.; Marvasi, M. The Bacterial Urban Resistome: Recent Advances. *Antibiotics* **2022**, *11*, 512. [CrossRef] [PubMed]
18. Ministero della Salute Italiano. *Legge Quadro in Materia di Animali di Affezione e Prevenzione del Randagismo n 281. (GU Serie Generale n.203 del 30-08-1991)*; 1991.
19. Lu, J.-J.; Perng, C.-L.; Lee, S.-Y.; Wan, C.-C. Use of PCR with Universal Primers and Restriction Endonuclease Digestions for Detection and Identification of Common Bacterial Pathogens in Cerebrospinal Fluid. *J. Clin. Microbiol.* **2000**, *38*, 2076–2080. [CrossRef]
20. Clinical and Laboratory Standards Institute (CLSI). *Performance Standards for Antimicrobial Susceptibility Testing*, 33rd ed.; CLSI Supplement M100; Clinical and Laboratory Standards Institute: Wayne, PA, USA, 2023; ISBN 978-1-68440-171-0.
21. European Committee on Antimicrobial Susceptibility. *Testing Guideline for the Detection of Resistance Mechanisms and Specific Resistances of Clinical and/or Epidemiological Importance*; Version 2.0; EUCAST: Växjö, Sweden; European Committee on Antimicrobial Susceptibility: Växjö, Sweden, 2017.
22. Roschanski, N.; Fischer, J.; Guerra, B.; Roesler, U. Development of a Multiplex Real-Time PCR for the Rapid Detection of the Predominant Beta-Lactamase Genes CTX-M, SHV, TEM and CIT-Type AmpCs in *Enterobacteriaceae*. *PLoS ONE* **2014**, *9*, e100956. [CrossRef]

23. Salgado-Caxito, M.; Benavides, J.A.; Munita, J.M.; Rivas, L.; García, P.; Listoni, F.J.P.; Moreno-Switt, A.I.; Paes, A.C. Risk Factors Associated with Faecal Carriage of Extended-Spectrum Cephalosporin-Resistant Escherichia Coli among Dogs in Southeast Brazil. *Prev. Vet. Med.* **2021**, *190*, 105316. [CrossRef]
24. Piccolo, F.L.; Belas, A.; Foti, M.; Fisichella, V.; Marques, C.; Pomba, C. Detection of Multidrug Resistance and Extended-Spectrum/Plasmid-Mediated AmpC Beta-Lactamase Genes in Enterobacteriaceae Isolates from Diseased Cats in Italy. *J. Feline Med. Surg.* **2020**, *22*, 613–622. [CrossRef]
25. Salgado-Caxito, M.; Benavides, J.A.; Adell, A.D.; Paes, A.C.; Moreno-Switt, A.I. Global Prevalence and Molecular Characterization of Extended-Spectrum β-Lactamase Producing-Escherichia Coli in Dogs and Cats—A Scoping Review and Meta-Analysis. *One Health* **2021**, *12*, 100236. [CrossRef] [PubMed]
26. Castanheira, M.; Simner, P.J.; Bradford, P.A. Extended-Spectrum β-Lactamases: An Update on Their Characteristics, Epidemiology and Detection. *JAC Antimicrob. Resist.* **2021**, *3*, dlab092. [CrossRef] [PubMed]
27. Bush, K. Past and Present Perspectives on β-Lactamases. *Antimicrob. Agents Chemother.* **2018**, *62*, e01076-18. [CrossRef] [PubMed]
28. Weese, J.S.; O'Brien, T.; Bateman, S. Fecal Shedding of Extended-Spectrum Beta-Lactamase-Producing Enterobacterales in Cats Admitted to an Animal Shelter. *J. Feline Med. Surg.* **2022**, *24*, 1301–1304. [CrossRef] [PubMed]
29. Reygaert, W.C. An Overview of the Antimicrobial Resistance Mechanisms of Bacteria. *AIMS Microbiol.* **2018**, *4*, 482–501. [CrossRef] [PubMed]
30. Gargano, V.; Sciortino, S.; Gambino, D.; Costa, A.; Agozzino, V.; Reale, S.; Alduina, R.; Vicari, D. Antibiotic Susceptibility Profile and Tetracycline Resistance Genes Detection in *Salmonella* Spp. Strains Isolated from Animals and Food. *Antibiotics* **2021**, *10*, 809. [CrossRef]
31. Del Castillo, J.R.E. Tetracyclines. In *Antimicrobial Therapy in Veterinary Medicine*; Giguère, S., Prescott, J.F., Dowling, P.M., Eds.; Wiley: Hoboken, NJ, USA, 2013; pp. 257–268, ISBN 978-0-470-96302-9.
32. Wu, S.; Dalsgaard, A.; Hammerum, A.M.; Porsbo, L.J.; Jensen, L.B. Prevalence and Characterization of Plasmids Carrying Sulfonamide Resistance Genes among Escherichia Coli from Pigs, Pig Carcasses and Human. *Acta Vet. Scand.* **2010**, *52*, 47. [CrossRef]
33. Bhat, B.A.; Mir, R.A.; Qadri, H.; Dhiman, R.; Almilaibary, A.; Alkhanani, M.; Mir, M.A. Integrons in the Development of Antimicrobial Resistance: Critical Review and Perspectives. *Front. Microbiol.* **2023**, *14*, 1231938. [CrossRef]
34. Sun, W.; Wang, D.; Yan, S.; Xue, Y. Characterization of Escherichia Coli Strains Isolated from Geese by Detection of Integron-Mediated Antimicrobial Resistance. *J. Glob. Antimicrob. Resist.* **2022**, *31*, 10–14. [CrossRef]

veterinary sciences

Article

Characteristics of Extended-Spectrum β-Lactamase Producing Enterobacterales Isolated from Dogs and Cats, 2011–2021

Dennis J. Woerde [1], Krystle L. Reagan [2], Barbara A. Byrne [3], Bart C. Weimer [4], Steven E. Epstein [5], Cory Schlesener [4], Bihua C. Huang [4] and Jane E. Sykes [2,*]

1 William R Pritchard Veterinary Medical Teaching Hospital, University of California-Davis, 1 Garrod Drive, Davis, CA 95616, USA
2 Department of Veterinary Medicine and Epidemiology, University of California-Davis, 1 Garrod Drive, Davis, CA 95616, USA
3 Department of Veterinary Pathology, Microbiology & Immunology, University of California-Davis, 1 Garrod Drive, Davis, CA 95616, USA
4 Department of Population Health and Reproduction, 100K Pathogen Genome Project, School of Veterinary Medicine, University of California-Davis, 1 Garrod Drive, Davis, CA 95616, USA
5 Department of Veterinary Surgical and Radiological Sciences, University of California-Davis, 1 Garrod Drive, Davis, CA 95616, USA
* Correspondence: jesykes@ucdavis.edu; Tel.: +1-530-752-1393

Simple Summary: Emerging antimicrobial resistance is a major concern in both human and veterinary medicine. Of particular concern is the emergence of extended-spectrum beta-lactamase (ESBL)-producing bacteria. The ESBLs are a group of enzymes produced by bacteria that inactivate commonly used antimicrobials. Infections caused by ESBL-producing bacteria are increasingly being recognized in human medicine; however, information is lacking regarding the characteristics of ESBL-producing bacterial infections associated with clinical illness in dogs and cats. This study examined ESBL-producing bacterial infections in dogs and cats presenting to a veterinary teaching hospital from 2011–2021. *Escherichia coli* was the most commonly identified bacterial species, with urinary tract infection being the most common clinical presentation. Multi-drug resistance was present in 90% of ESBL-producing bacterial infections. Based on susceptibility patterns, antimicrobials such as piperacillin-tazobactam, amikacin, and cefoxitin may be alternative antibiotics to the current recommended regimen. Whole genome sequencing of bacteria was performed, which revealed $bla_{\text{CTX-M-15}}$ was the most common ESBL gene identified.

Abstract: The rising prevalence of extended-spectrum β-lactamase (ESBL)-producing Enterobacterales is a significant threat to animal and human health. This study aims to describe the clinical features, antimicrobial susceptibility patterns, and genotypic features of infections associated with ESBL-producing Enterobacterales in dogs and cats seen at a tertiary referral veterinary teaching hospital. Enterobacterales isolated from dogs and cats that underwent ESBL testing during the study period were identified using a search of the hospital antimicrobial susceptibility test software database. Medical records of confirmed ESBL isolates were reviewed, and the source of infection, clinical findings, and antimicrobial susceptibility were recorded. Genomic DNA from bacterial isolates was evaluated for antimicrobial resistance genes with whole genome sequencing. Thirty ESBL-producing isolates were identified based on phenotypic testing (twenty-nine from dogs, one from a cat); twenty-six were *Escherichia coli* and the remainder were *Klebsiella* spp. Bacterial cystitis was the most commonly identified (8/30, 27%) clinical problem associated with infection. Resistance to three or more antimicrobial classes was identified in 90% (27/30) of isolates, and all isolates were susceptible to imipenem. Over 70% of isolates were susceptible to piperacillin-tazobactam, amikacin, and cefoxitin. $Bla_{\text{CTX-M-15}}$ was the most common ESBL gene identified, present in 13/22 (59%) isolate genomes. A wide range of clinical infections were identified. Piperacillin-tazobactam and amikacin may be alternatives to carbapenem therapy. Further, larger-scale studies are needed.

Keywords: *Escherichia coli*; multidrug resistance; virulence; Enterobacterales

Citation: Woerde, D.J.; Reagan, K.L.; Byrne, B.A.; Weimer, B.C.; Epstein, S.E.; Schlesener, C.; Huang, B.C.; Sykes, J.E. Characteristics of Extended-Spectrum β-Lactamase Producing Enterobacterales Isolated from Dogs and Cats, 2011–2021. *Vet. Sci.* **2023**, *10*, 178. https://doi.org/10.3390/vetsci10030178

Academic Editor: George Valiakos

Received: 24 December 2022
Revised: 14 February 2023
Accepted: 17 February 2023
Published: 22 February 2023

1. Introduction

The increasing prevalence of antimicrobial-resistant bacterial infections is of significant concern in human and veterinary medicine. One of the most important causes of antimicrobial resistance within the family Enterobacterales (previously Enterobacteriaceae) is the acquisition of plasmid-mediated production of β-lactamase enzymes, resulting in resistance to β-lactam antimicrobials. More than 500 distinct β-lactamases have been identified [1] and several classification systems have been developed [2].

Extended-spectrum β-lactamases (ESBLs) are a group of β-lactamases capable of hydrolyzing third-generation cephalosporins yet are inhibited by clavulanic acid [3]. This inhibition by clavulanic acid differentiates ESBLs from AmpC type β-lactamases, another group of β-lactamases that confer resistance to third-generation cephalosporins and monobactams [4].

The major ESBL genes associated with bacteria isolated from human and animal patients belong to the groups $bla_{CTX\text{-}M}$, bla_{TEM}, and bla_{SHV} [2]. These resistance genes are most prevalent in *Escherichia coli* and *Klebsiella pneumoniae*, but can also be found in *Enterobacter* spp., *Salmonella* spp., *Proteus mirabilis*, and *Pseudomonas aeruginosa* [5]. The prevalence of infections associated with ESBL-producing bacteria is increasing worldwide [6], with multidrug resistance (MDR) being a common feature [5]. This feature often leads to treatment failure with empiric antimicrobial therapy, can limit clinical therapeutic options, and in human patients can lead to an increased mortality rate compared to infections with non-ESBL-producing Enterobacterales [7]. Hence, ESBL-producing bacteria pose a serious threat to human and animal health.

A paucity of information exists in veterinary medicine regarding the prevalence of infections with ESBL-producing bacteria, their clinical characteristics, bacterial species involved, and associated resistance mechanisms. Urinary tract infections, hepatobiliary infections, respiratory tract infections, bacteremia, and intra-abdominal infections have all been reported as secondary to ESBL-producing Enterobacterales in human medicine [8]. Several studies have identified ESBL-producing bacteria in the feces of healthy cats and dogs [9,10]; however, information is lacking regarding the phenotypic and genetic characteristics of isolates associated with clinical illness in dogs and cats.

The aims of this study were to describe the clinical features, antimicrobial susceptibility patterns, and genotypic features of ESBL-producing Enterobacterales infections in dogs and cats presenting to a veterinary teaching hospital.

2. Materials and Methods

2.1. Isolate Identification, Antimicrobial Susceptibility Test Results & Patient Data

Enterobacterales isolates that underwent ESBL testing from July 2011 to July 2021 were identified using a search of the database of the antimicrobial susceptibility testing software (Sensititre SWIN, V3,2,3 and V3.3, ThermoFisher Scientific, Waltham, MA, USA) used by the University of California-Davis Veterinary Medical Teaching Hospital Veterinary Microbiology laboratory. Briefly, a subset of the *Escherichia coli* and *Klebsiella* spp. isolates from dogs and cats that were identified as resistant to cefpodoxime (range of concentrations tested, 0.25–32 μg/mL) using broth microdilution (Sensititre, ThermoFisher Scientific COMP1F, COMPAN2F, or COMPGN1F panels) were further tested for ESBLs using the ESBLF panel (Sensititre, ThermoFisher Scientific).

Isolates that had at least a three, two-fold dilution difference in minimum inhibitory concentration (MIC) between one or both of cefotaxime (range of concentrations tested, 0.25–32 μg/mL) or ceftazidime (range of concentrations tested, 0.25–128 μg/mL) and when these antimicrobials were combined with clavulanic acid were classified as ESBL-producing isolates [11]. Quality control strains tested with all panels included *E. coli* American Type Cell Collection (ATCC), *Staphylococcus aureus* ATCC 25923, and *Pseudomonas aeruginosa* ATCC 27853. Susceptibility results were compiled and reported for each antimicrobial tested. Bacterial isolates were classified as susceptible, intermediate, or resistant using current veterinary-specific CLSI breakpoints [12] available at the time of writing. Urinary-specific breakpoints were used for isolates obtained from the urinary tract. Where veterinary-specific breakpoints were not available, human breakpoints were used. Multidrug resistance (MDR) was defined as resistance to three or more classes of antimicrobials [13]. All susceptibility testing was done at the time of isolation.

From 2014 to 2021, bacteria were identified to the species level using matrix-assisted laser-desorption ionization time-of-flight mass spectrometry (MALDI-TOF MS; Bruker, Billerica, MA, USA), and spot oxidase and indole testing. Prior to 2014, Enterobacterales were identified using conventional biochemical testing, including triple sugar iron, citrate, Christensen's urea agar (Hardy Diagnostics, Santa Maria, CA, USA), and sulfur-indole-motility agar (Biological Media Service, University of California, Davis, CA, USA), the aforementioned spot tests, or identification strips (API 20E, BioMerieux, Durham, NC, USA).

Identified bacterial isolates were then cross-referenced to individual patient medical records through a search of the University of California-Davis Veterinary Medical Teaching Hospital database (VMACS; Veterinary Medical and Administrative Computer System, University of California-Davis) to obtain data on signalment, clinical features, and specimen source.

2.2. Whole Genome Sequencing

Whole genome sequencing (WGS) of available banked isolates was performed. Banked isolates were stored as frozen stabilates at −80 °C using a bead-based system (Pro-Lab Microbank ™, ThermoFisher Scientific) until sequencing. Isolates were revived by subculture from the frozen stabilates on 5% sheep blood agar (Hardy Diagnostics, Santa Maria, CA, USA) and incubated overnight at 35 °C in 5% CO_2.

WGS was performed using the methods from the 100K Pathogen Genome Project [14,15]. Briefly, high-molecular-weight genomic DNA (gDNA) was extracted from banked bacterial isolate colonies using the Wizard Genomic DNA Jit (Promega, Madison, WI; cat#A1460) as described previously [16].

Approximately 600 ng of purified genomic DNA was used to construct a sequencing library using the KAPA HyperPlus library preparation kit (Roche Diagnostics). Library size distribution verification was performed on the Caliper LabChip GX (Perkin Elmer), and library quantification was performed with the KAPA Library Quantification Kit (Roche Diagnostics). Pooled libraries were sequenced on the Illumina HiSeq X Ten using the PE150 protocol. Reads were trimmed with Trimmomatic [17], assembled with SPAdes [18], and annotated with prokka [19], all with default settings. Antibiotic resistance genes were analyzed in every isolate genome using the Comprehensive Antibiotic Resistance Database (CARD) [20].

2.3. Genomic Similarity Comparison

Multi-Locus Sequence Typing (MLST) calls were made by scanning genome assemblies using "mlst" software version 2.23.0 [21] using PubMLST typing schemes [22]. Genome Assemblies were compared by Jaccard similarity index of k-mer (size = 31 bases) sequence-based profiles (sketch size = scaled 100 k/Mbp) using Sourmash version 3.2.3 [23]. Sequences were uploaded to the Sequence Read Archive database [24].

3. Results

3.1. Isolate Identification, Antimicrobial Susceptibility Test Results & Patient Data

A total of 30 ESBL-producing bacterial isolates were identified among approximately 5300 susceptibility panels performed over the 10-year study period on *Escherichia coli* and *Klebsiella* spp., 22 of which had been banked. Twenty-nine of the isolates were from dogs, and one was from a cat. A total of 299,187 dogs and 55,698 cats were examined at the hospital over the same 10-year period. Dogs were 5.3 times more commonly seen than cats at the hospital in the 10-year period of the study. The number of ESBL-producing Enterobacterales isolated varied per year (Figure 1); however, more than half (17/30) were obtained from the final two years of the study. Of the 30 isolates, 26 were *Escherichia coli*, 3 were *Klebsiella pneumoniae*, and 1 was *Klebsiella oxytoca*.

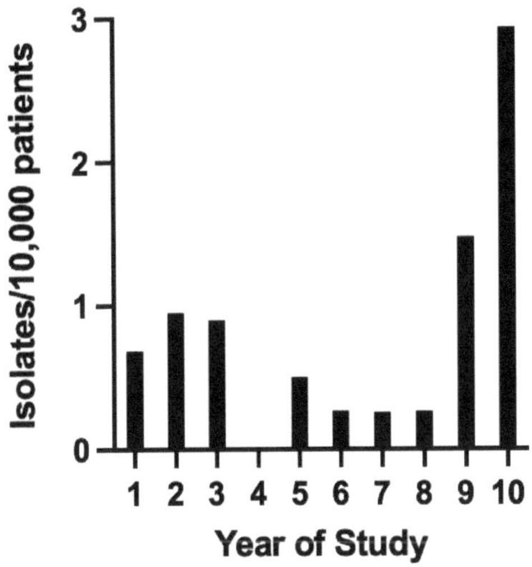

Figure 1. Prevalence of ESBL-producing Enterobacterales (isolates/10,000 patients for each year of the study, year 1 being July 2011–June 2012).

Antimicrobial susceptibility results of the isolates are summarized in Figure 2. All isolates were resistant to aztreonam, ampicillin/sulbactam, and piperacillin. The majority of isolates were resistant to the fluoroquinolones, including enrofloxacin (93%), marbofloxacin (90%), and orbifloxacin (90%). All isolates were susceptible to imipenem, meropenem, and cefotetan. Over 70% of isolates were susceptible to piperacillin-tazobactam, amikacin, and cefoxitin (Figure 2). MDR was identified in 27/30 (90%) of isolates. Resistance to cefepime, a 4th generation cephalosporin, was identified in 6/30 (20%) isolates.

Bacterial isolates were cultured from urine obtained via cystocentesis (11/30), skin swab specimens (7/30), pleural effusion collected via thoracocentesis (2/30), liver fine needle aspirate specimens (2/30), blood (2/30), peritoneal effusion collected by abdominocentesis (2/30), and one each of bronchoalveolar lavage (BAL) fluid, cholecystocentesis bile fluid, transtracheal lavage (TTL) fluid, and an ear swab specimen. Of the 11 bacterial isolates from urine, 8 were from dogs that had lower urinary tract signs consistent with bacterial cystitis (pollakiuria, hematuria, or malodorous urine). The remaining 3 isolates were from dogs without signs of lower urinary tract disease; 1 dog had clinical features consistent with pyelonephritis, and the other 2 dogs had subclinical bacteriuria but were being treated with glucocorticoids, and there was concern that this treatment may have been masking clinical signs of lower urinary tract disease. Two ESBLs were cultured at different time points from the same dog; an ESBL-producing *Klebsiella pneumoniae* was identified on the initial urine culture, and a urine culture performed 4 weeks later revealed the same ESBL-producing *K. pneumoniae* as well as an ESBL-producing *E. coli*.

Of the 4 isolates from the respiratory tract, 3 were from dogs that had clinical signs and radiographic abnormalities that were consistent with aspiration pneumonia. The remaining isolate was from a dog with pyothorax secondary to migrating grass awn. The cat with an ESBL-producing *E. coli* infection had bacterial peritonitis secondary to a dog bite wound.

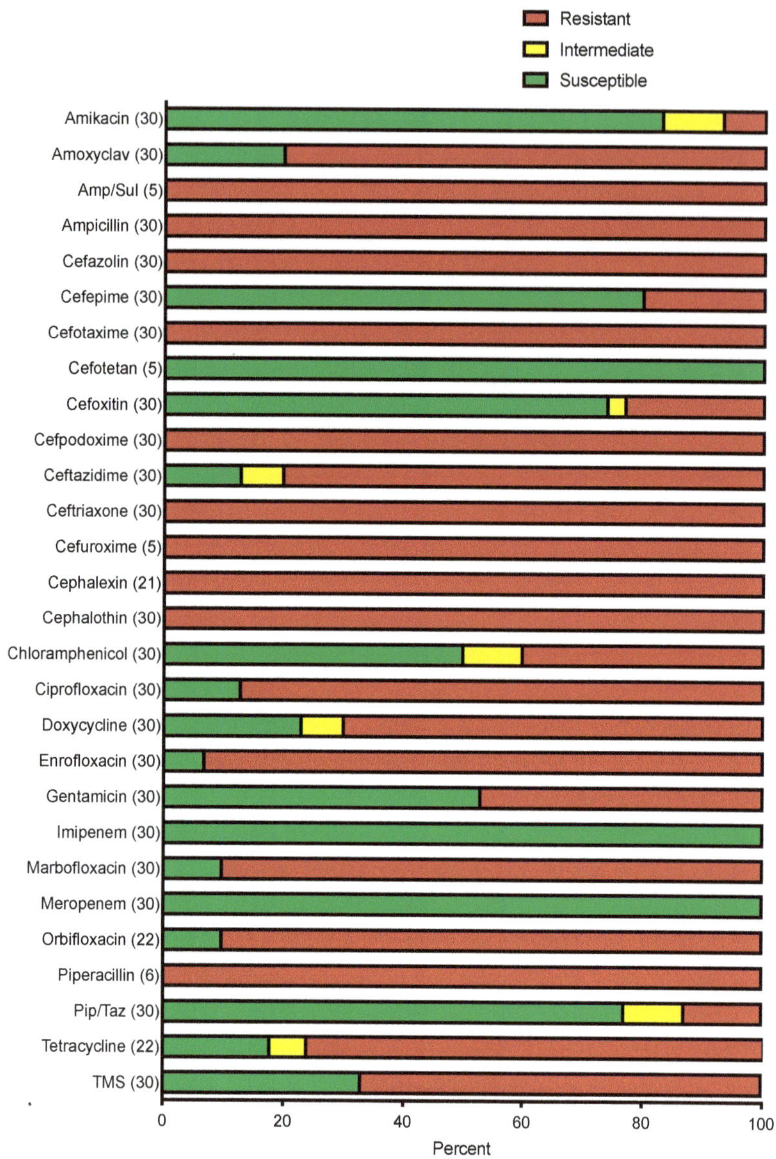

Figure 2. Susceptibility patterns for ESBL-producing bacterial isolates identified expressed as a percentage (susceptible, intermediate, and resistant, as defined by CLSI breakpoints; antimicrobials for which breakpoints were not available are not included). Number of isolates tested against each antimicrobial is listed on the vertical axis in parentheses. Amp/Sul, ampicillin/sulbactam; Amoxyclav, amoxicillin/clavulanic acid; Pip/Taz, piperacillin/tazobactam; TMS, trimethoprim sulfamethoxazole.

3.2. Whole Genome Sequencing

A total of 22/30 isolates were available for whole-genome sequencing, 4 *Klebsiella* spp. and 18 *E. coli* isolates, all from dogs. Genes associated with ESBL production were identified in each genome (Table 1). $Bla_{CTX\text{-}M\text{-}15}$ was the most common ESBL gene identified (13/22 isolate genomes). All isolates had multiple genes contributing to beta-lactam resistance. Either AmpC or AmpH β-lactamase genes or both were detected in all 22 isolates.

Table 1. Beta-lactamase genes identified in each isolate after whole-genome sequencing.

Isolate	Organism	Year Isolated	Source	SRA Name	MLST Number	ESBL Gene Product	Other β-Lactamase Gene Products
1	*Escherichia coli*	2012	BAL fluid	BCW_12610	410	CTX-M-15	AmpC, AmpH, OXA-1
2	*Klebsiella pneumoniae*	2020	Urine	BCW_12638	16	CTX-M-15, SHV-1, TEM-1	AmpH, CBP-1
3	*Escherichia coli*	2021	Urine	BCW_12625	224	CTX-M-1	AmpC, AmpH
4	*Escherichia coli*	2021	Urine	BCW_12626	162	CTX-M-14, TEM-1	AmpC, AmpH
5	*Escherichia coli*	2021	Skin	BCW_12627	46	CTX-M-15	AmpC, AmpH, CMY-136
6	*Klebsiella pneumoniae*	2021	Skin	BCW_12640	307	CTX-M-15, SHV-28	AmpH, CBP-1
7	*Escherichia coli*	2020	Pleural fluid	BCW_12621	10	CTX-M-15	AmpC, AmpH
8	*Escherichia coli*	2020	Skin	BCW_12622	450	CTX-M-15	AmpC, AmpH
9	*Escherichia coli*	2017	Urine	BCW_12615	162	CTX-M-14	AmpC, AmpH
10	*Escherichia coli*	2013	Bile	BCW_12611	410	CTX-M-15	AmpC, AmpH, OXA-1
11	*Escherichia coli*	2019	Bile	BCW_12616	90	CTX-M-15, TEM-1	AmpC, AmpH
12	*Escherichia coli*	2013	Skin	BCW_12614	410	CTX-M-15, TEM-1, FONA-6	AmpC, AmpH, OXA-1
13	*Escherichia coli*	2013	Ear swab	BCW_12613	44	CTX-M-15, FONA-6	AmpC, AmpH, OXA-1, DHA-1
14	*Escherichia coli*	2019	Urine	BCW_12617	1193	CTX-M-27	AmpC, AmpH, CMY-12
15	*Escherichia coli*	2020	Blood	BCW_12620	162	CTX-M-14	AmpC, AmpH
16	*Escherichia coli*	2019	Skin	BCW_12619	1148	FONA-6, TEM-1	AmpC, AmpH
17	*Escherichia coli*	2019	Tracheal wash	BCW_12618	162	CTX-M-14, TEM-1	AmpC, AmpH
18	*Klebsiella oxytoca*	2012	Urine	BCW_12637	101	TEM-1, SHV-66	AmpH, OXY-2-10
19	*Escherichia coli*	2021	Urine	BCW_12628	162	CTX-M-14, TEM-1	AmpC, AmpH
20	*Escherichia coli*	2021	Urine	BCW_12624	68	CTX-M-15	AmpC, AmpH, OXA-1
21	*Escherichia coli*	2021	Urine	BCW_12623	1431	CTX-M-15, TEM-1	AmpC, AmpH, CMY-2
22	*Klebsiella pneumoniae*	2021	Pleural fluid	BCW_12639	307	CTX-M-15, SHV-28, TEM-1	AmpH, OXA-1, CBP-1

BAL, bronchoalveolar lavage; SRA, sequence read archive; MLST, multi-locus sequence typing; ESBL, extended-spectrum β-Lactamase.

3.3. Genomic Similarity Comparison

Examination of sample identity by multi-locus sequence typing (MLST) [22] and comparison of genomic distance were carried out (Figure 3). Clustering of genomes was observed, but this did not correlate with the sample collection site (material, body location). Clusters with high genetic similarity were observed among isolates with the same MLST number, with the largest groups being *E. coli* (162) (n = 5, modest genetic diversity) and *E. coli* (410) (n = 3, modest genetic diversity). Two *E. coli* isolates (BCW_12624 and BCW_12617) showed greater divergence from each other and the remainder of the examined *E. coli* samples. Therefore, the *E. coli* isolates examined were diverse, and 12 MLST groups were identified.

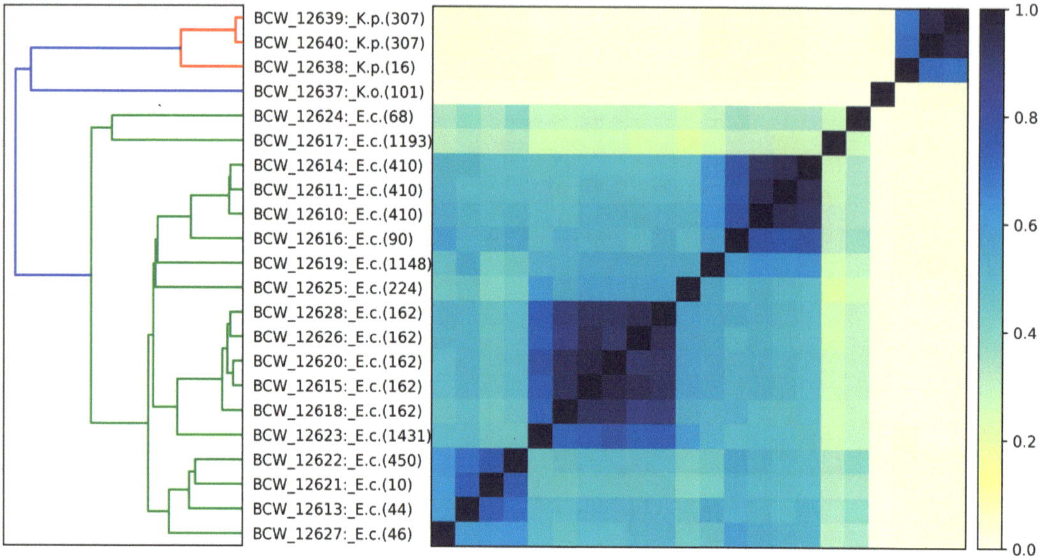

Figure 3. Sample genome assembly comparisons (DNA sequence k-mer (31 bp) matching) are shown in all against all matrix by Jaccard similarity index (color scale on right: 1.0 (dark blue) is near identical, 0.0 is no observed relationship). A hierarchical clustering dendrogram (left) shows structure of relatedness (branch lengths represent relative distance between clusters, colors correspond to species). Samples are labeled by Sequence Read Archive (SRA) ID number, species (E.c.: *Escherichia coli*, K.p.: *Klebsiella pneumoniae*, K.o.: *Klebsiella oxytoca*), and PubMLST schemes (*Escherichia coli* [Achtman], *Klebsiella pneumoniae* species complex [Pasteur], and *Klebsiella oxytoca*), respectively. The plot shows clustering corresponding to species and MLST.

4. Discussion

ESBL-producing Enterobacterales were associated with a wide range of clinical presentations in this study. Most of the specimens in this study in which ESBL-producing Enterobacterales were detected were from the urinary tract. Given that *E. coli* is the most common pathogen associated with urinary tract infections in dogs [25], this result was expected. The prevalence of ESBL-producing bacterial isolates in human patients with bacteriuria has been as high as 37% in some studies [26].

Almost all the ESBL-producing Enterobacterales in the study reported here were from dogs, with only one isolate from a cat. Furthermore, it is possible the isolate from the cat originated from the oral flora of a dog, given the source was dog-bite-associated peritonitis. A 2021 meta-analysis of studies pertaining to ESBL-producing *E. coli* in dogs and cats isolated from a variety of specimens however revealed no difference in the global prevalence of ESBL-producing Enterobacterales between dogs and cats [27]. Our hospital examined a total of 5.3 times more dogs than cats over the 10-year period, which may have contributed to the higher proportion of isolates from dogs. In addition, the prevalence of urinary tract infections (and bacterial infections in general) in cats is low compared to dogs [25], likely contributing to the lower number of isolates from cats in this study. The number of ESBL-producing Enterobacterales isolated varied per year (Figure 1), and more than half (17/30) were obtained from the final two years of the study. When examining the hospital data over the 10-year period, only 28% more patients were evaluated in 2020–2021 compared to 2011–2012, so the increase could not be explained by patient volume alone. However, conclusions about a change in the prevalence of ESBL-producing Enterobacterales cannot be drawn from this study because of inconsistent testing across the study period.

Multidrug resistance was a common feature of the identified isolates, present in 90% of isolates in this study. ESBLs typically confer resistance to third-generation cephalosporins, although, by definition, fourth (cefepime) and fifth (ceftaroline) generation cephalosporins remain active against ESBL-producing bacteria [28]. In this study, 6/30 (20%) isolates were resistant to cefepime. Carbapenems are considered the treatment of choice for ESBL-producing Enterobacterales infections in humans [29], and all isolates in our study were susceptible to carbapenems. Several clinical studies in human patients have revealed higher rates of mortality in patients with infections caused by ESBL-producing Enterobacterales treated with cefepime than those treated with carbapenems [30]. However, carbapenem-resistant Enterobacterales have been identified [31], and strategies to reduce carbapenem use should be employed when feasible. In the current study, 77% of isolates were susceptible to piperacillin/tazobactam, an extended-spectrum penicillin and beta-lactamase inhibitor combination antimicrobial with a broad spectrum of activity against Gram-positive and Gram-negative bacteria. Whilst some conjecture in the literature exists between in vitro activity and clinical responses, a study in human patients with urinary tract infections caused by ESBL-producing *E. coli* revealed that carbapenem and piperacillin/tazobactam had similar clinical efficacy and mortality rates were similar between groups [32]. Similarly, 83% of isolates in this study were susceptible to amikacin, so this represents an additional alternative antimicrobial for the treatment of ESBL-producing Enterobacterales infections, at least for the population of animals seen at the authors' institution. Isolates were largely susceptible to cefoxitin and cefotetan (second-generation cephamycins). ESBLs typically do not hydrolyze cephamycins, although isolates resistant to cephamycins have been reported [28]. Several studies of human patients have reported the efficacy of cephamycins in the treatment of urinary tract infections caused by ESBL-producing Enterobacterales [33], although data are limited on their use for the treatment of infections outside the urinary tract or in small animals.

Bla$_{\text{CTX-M-15}}$ was the most prevalent ESBL gene and is the most common ESBL gene reported in *E. coli* isolates from dogs and cats worldwide [27]. The prevalence of various ESBL genes in bacteria from companion animals may vary geographically, although the number of isolates in these studies is low. The *bla*$_{\text{TEM}}$ gene family was the most common in a Brazilian study (47 isolates) [34], *bla*$_{\text{CTX-M-1}}$ in a French study (10 isolates) [35], and *bla*$_{\text{CTX-M-14}}$ in New Zealand (36 isolates) [36]. A study performed in 2011 that characterized ESBL genes from 54 *E. coli* isolates from companion animals in the USA revealed 78% of isolates carried the *bla*$_{\text{CTX-M-15}}$ gene [37]. Concerningly, all isolates in our study possessed multiple genes that could contribute to beta-lactam resistance.

While isolates in this study were obtained from patients with clinical disease, ESBL-producing Enterobacterales have been detected in the feces and saliva of healthy dogs and cats [9,10]. Future studies are needed to determine whether there are differences in the phenotypic and genotypic characteristics of ESBL-producing bacteria from diseased versus healthy companion animals. Additional studies are also needed to identify risk factors for infection and the outcomes of infection. In human medicine, significant previous antimicrobial use [38,39], urinary catheter placement [38], and overcrowded households [39] have all been implicated as risk factors for ESBL colonization. In one study that examined the presence of ESBL-producing Enterobacterales in fecal specimens from healthy dogs in the Netherlands [10], consumption of raw meat was the main risk factor identified. A similar study performed in the UK on the fecal carriage of ESBL-producing Enterobacterales [40] found that dogs with a history of antimicrobial therapy in the past year and dogs obtained from a shelter or breeder were at increased risk for colonization. Recently, the presence of ESBL-producing Enterobacterales in the feces of dogs and cats was investigated in animals admitted to a veterinary hospital in Brazil on admission and at discharge [34]. A total of 11/47 patients had ESBL-producing Enterobacterales in their feces at hospital discharge and not at admission, suggesting veterinary hospitals as a major source of acquisition. Evidence of household transfer of ESBL/AmpC-producing Enterobacterales among humans and dogs has been reported [41], although other studies have suggested exposure to a common environmental source or clonal transmission between pets and humans is more likely [10].

There were several limitations to this study. Only 22 of the 30 isolates were banked and available for WGS. The results reflect the population of bacteria and their resistance patterns seen at the University of California-Davis, which is a tertiary referral hospital, and may not reflect bacterial populations seen at other hospitals. Given the high prevalence of AmpC and AmpH β-Lactamase genes identified in this study, some ESBL-producing Enterobacterales may have been overlooked using our search process. This is because ESBL detection is often masked by high-level production of AmpC [42]. Susceptibility testing and ESBL confirmation were not performed on all *E. coli* and *Klebsiella* isolates obtained during the study period. As a result, meaningful temporal information on prevalence may be lacking. Furthermore, routine susceptibility panels did not include all antimicrobials, such as nitrofurantoin.

5. Conclusions

This study lays the groundwork for future large-scale epidemiological studies on dogs and cats caused by clinical infections with ESBL-producing Enterobacterales. Improved understanding of prevalence, risk factors, outcomes, and mechanisms of resistance gene acquisition is needed for companion animals. Information on clinical cases will ideally help clinicians identify animals that are at high risk for ESBL infections and aid in antimicrobial selection. Surveillance for the emergence of ESBL-producing Enterobacterales is important in both veterinary medicine and human healthcare, given the potential for zoonotic and healthcare-associated transmission.

Author Contributions: Conceptualization, D.J.W., K.L.R. and J.E.S. investigation, D.J.W.; data curation, D.J.W., B.A.B. and S.E.E.; formal analysis, B.A.B., B.C.W., B.C.H. and C.S.; writing—original draft preparation, D.J.W., K.L.R. and J.E.S.; writing—review and editing, D.J.W., K.L.R., B.A.B., B.C.W., S.E.E. and J.E.S. All authors have read and agreed to the published version of the manuscript.

Funding: This research was funded by the Center for Companion Animal Health (CCAH), School of Veterinary Medicine, University of California, Davis.

Institutional Review Board Statement: Not applicable.

Informed Consent Statement: Not applicable.

Data Availability Statement: Further inquiries can be directed to the corresponding authors.

Conflicts of Interest: Sykes is a member of the board of directors for LexaGene, Inc., which markets tests for detection of resistance markers. The remaining authors declare no conflict of interest.

References

1. Ur Rahman, S.; Ali, T.; Ali, I.; Khan, N.A.; Han, B.; Gao, J. The Growing Genetic and Functional Diversity of Extended Spectrum Beta-Lactamases. *BioMed Res. Int.* **2018**, *2018*, 9519718. [CrossRef]
2. Bush, K.; Jacoby, G.A.; Medeiros, A.A. A functional classification scheme for beta-lactamases and its correlation with molecular-structure. *Antimicrob. Agents Chemother.* **1995**, *39*, 1211–1233. [CrossRef]
3. Paterson, D.L.; Bonomo, R.A. Extended-spectrum beta-lactamases: A clinical update. *Clin. Microbiol. Rev.* **2005**, *18*, 657–686. [CrossRef]
4. Jacoby, G.A. AmpC beta-lactamases. *Clin. Microbiol. Rev.* **2009**, *22*, 161–182. [CrossRef]
5. Rawat, D.; Nair, D. Extended-spectrum β-lactamases in Gram Negative Bacteria. *J. Glob. Infect. Dis.* **2010**, *2*, 263–274. [CrossRef]
6. Jernigan, J.A.; Hatfield, K.M.; Wolford, H.; Nelson, R.E.; Olubajo, B.; Reddy, S.C.; McCarthy, N.; Paul, P.; McDonald, L.C.; Kallen, A.; et al. Multidrug-Resistant Bacterial Infections in U.S. Hospitalized Patients, 2012–2017. *N. Engl. J. Med.* **2020**, *382*, 1309–1319. [CrossRef]
7. Melzer, M.; Petersen, I. Mortality following bacteraemic infection caused by extended spectrum beta-lactamase (ESBL) producing *E. coli* compared to non-ESBL producing *E. coli*. *J. Infect.* **2007**, *55*, 254–259. [CrossRef]
8. Pitout, J.D.; Nordmann, P.; Kevin, B.; Laupland, K.B.; Poirel, L. Emergence of Enterobacteriaceae producing extended-spectrum β-lactamases (ESBLs) in the community. *J. Antimicrob. Chemother.* **2005**, *56*, 52–59. [CrossRef]
9. Abdel-Moein, K.A.; Samir, A. Occurrence of extended spectrum β-lactamase-producing Enterobacteriaceae among pet dogs and cats: An emerging public health threat outside health care facilities. *Am. J. Infect. Control.* **2014**, *42*, 796–798. [CrossRef]

10. Van den Bunt, G.; Fluit, A.C.; Spaninks, M.P.; Timmerman, A.J.; Geurts, Y.; Kant, A.; Scharringa, J.; Mevius, D.; Wagenaar, J.A.; Bonten, M.; et al. Faecal carriage, risk factors, acquisition and persistence of ESBL-producing Enterobacteriaceae in dogs and cats and co-carriage with humans belonging to the same household. *J. Antimicrob. Chemother.* **2020**, *75*, 342–350. [CrossRef]
11. Clinical and Laboratory Standards Institute. *Performance Standards for Antimicrobial Susceptibility Testing*, 30th ed.; CLSI supplement M100; Clinical and Laboratory Standards Institute: Wayne, PA, USA, 2020.
12. Clinical and Laboratory Standards Institute. *Performance Standards for Antimicrobial Disk and Dilution Susceptibility Tests for Bacteria Isolated From Animals*, 5th ed.; CLSI supplement VET01S; Clinical and Laboratory Standards Institute: Wayne, PA, USA, 2020.
13. Magiorakos, A.P.; Srinivasan, A.; Carey, R.B.; Carmeli, Y.; Falagas, M.E.; Giske, C.G.; Harbarth, S.; Hindler, J.F.; Kahlmeter, G.; Olsson-Liljequist, B.; et al. Multidrug-resistant, extensively drug-resistant and pandrug-resistant bacteria: An international expert proposal for interim standard definitions for acquired resistance. *Clin. Microbiol. Infect.* **2012**, *18*, 268–281. [CrossRef]
14. Kong, N.; Davis, M.; Arabyan, N.; Huang, B.C.; Weis, A.M.; Chen, P.; Thao, K.; Ng, W.; Chin, N.; Foutouhi, S.; et al. Draft Genome Sequences of 1183 *Salmonella* Strains from the 100K Pathogen Genome Project. *Genome Announc.* **2017**, *5*, e00518-17. [CrossRef]
15. Weis, A.M.; Clothier, K.A.; Huang, B.C.; Kong, N.; Weimer, B.C. Draft Genome Sequence of Multidrug-Resistant Abortive *Campylobacter jejuni* from Northern California. *Genome Announc.* **2017**, *5*, e00171-17. [CrossRef]
16. Hernández-Juárez, L.E.; Camorlinga, M.; Méndez-Tenorio, A.; Calderón, J.F.; Huang, B.C.; Bandoy, D.R.; Weimer, B.C.; Torres, J. Analyses of publicly available *Hungatella hathewayi* genomes revealed genetic distances indicating they belong to more than one species. *Virulence* **2021**, *12*, 1950–1964. [CrossRef]
17. Bolger, A.M.; Lohse, M.; Usadel, B. Trimmomatic: A flexible trimmer for Illumina sequence data. *Bioinformatics* **2014**, *30*, 2114–2120. [CrossRef]
18. Bankevich, A.; Nurk, S.; Antipov, D.; Gurevich, A.A.; Dvorkin, M.; Kulikov, A.S.; Lesin, V.M.; Nikolenko, S.I.; Pham, S.; Prjibelski, A.D.; et al. SPAdes: A new genome assembly algorithm and its applications to single-cell sequencing. *J. Comput. Biol.* **2012**, *19*, 455–477. [CrossRef]
19. Seemann, T. Prokka: Rapid prokaryotic genome annotation. *Bioinformatics* **2014**, *30*, 2068–2069. [CrossRef]
20. The Comprehensive Antibiotic Resistance Database. Available online: https://card.mcmaster.ca/ (accessed on 1 May 2022).
21. Seemann, T.A.; Goncalves da Silva, A.; Bulach, D.M.; Schultz, M.B.; Kwong, J.C.; Howden, B.P. "Nullarbor" Github. Available online: https://github.com/tseemann/nullarbor (accessed on 25 January 2023).
22. Jolley, K.A.; Maiden, M.C. BIGSdb: Scalable analysis of bacterial genome variation at the population level. *BMC Bioinform.* **2010**, *11*, 595. [CrossRef]
23. Brown, C.T.; Irber, L. Sourmash: A library for MinHash sketching of DNA. *J. Open Source Softw.* **2016**, *1*, 27. [CrossRef]
24. Sequence Read Archive Database. Available online: https://www.ncbi.nlm.nih.gov/sra (accessed on 20 January 2023).
25. Byron, J.K. Urinary tract infection. *Vet. Clin. North Am. Small Anim. Pract.* **2019**, *49*, 211–221. [CrossRef]
26. Quan, J.; Dai, H.; Liao, W.; Zhao, D.; Shi, Q.; Zhang, L.; Shi, K.; Akova, M.; Yu, Y. Etiology and prevalence of ESBLs in adult community-onset urinary tract infections in East China: A prospective multicenter study. *J. Infect.* **2021**, *83*, 175–181. [CrossRef]
27. Salgado-Caxito, M.; Benavides, J.A.; Adell, A.D.; Paes, A.C.; Moreno-Switt, A.I. Global prevalence and molecular characterization of extended-spectrum β-lactamase producing-*Escherichia coli* in dogs and cats—A scoping review and meta-analysis. *One Health* **2021**, *12*, 100236. [CrossRef]
28. Bradford, P.A. Extended-spectrum beta-lactamases in the 21st century: Characterization, epidemiology, and detection of this important resistance threat. *Clin. Microbiol. Rev.* **2001**, *14*, 933–951. [CrossRef]
29. Bassetti, M.; Vena, A.; Roberto-Giacobbe, D.; Castaldo, N. Management of Infections Caused by Multidrug-resistant Gram-negative Pathogens: Recent Advances and Future Directions. *Arch. Med. Res.* **2021**, *52*, 817–827. [CrossRef]
30. Patel, H.B.; Lusk, K.A.; Cota, J.M. The Role of Cefepime in the Treatment of Extended-Spectrum Beta-Lactamase Infections. *J. Pharm. Pract.* **2019**, *32*, 458–463. [CrossRef]
31. Yousfi, M.; Touati, A.; Mairi, A.; Brasme, L.; Gharout-Sait, A.; Guillard, T.; De Champs, C. Emergence of Carbapenemase-Producing *Escherichia coli* Isolated from Companion Animals in Algeria. *Microb. Drug Resist.* **2016**, *22*, 342–346. [CrossRef]
32. Seo, Y.B.; Lee, J.; Kim, Y.K.; Lee, S.S.; Lee, J.A.; Kim, H.Y.; Uh, Y.; Kim, H.S.; Song, W. Randomized controlled trial of piperacillin-tazobactam, cefepime and ertapenem for the treatment of urinary tract infection caused by extended-spectrum beta-lactamase-producing *Escherichia coli*. *BMC Infect. Dis.* **2017**, *17*, 404. [CrossRef]
33. Senard, O.; Bouchand, F.; Deconinck, L.; Matt, M.; Fellous, L.; Rottman, M.; Perronne, C.; Dinh, A.; Davido, B. Efficacy of cefoxitin for the treatment of urinary tract infection due to extended-spectrum-beta-lactamase-producing *Escherichia coli* and *Klebsiella pneumoniae* isolates. *Ther. Adv. Infect. Dis.* **2019**, *6*, 2049936118811053. [CrossRef]
34. Sfaciotte, R.; Parussolo, L.; Melo, F.D.; Wildemann, P.; Bordignon, G.; Israel, N.D.; Leitzke, M.; Wosiacki, S.R.; Salbego, F.Z.; da Costa, U.M.; et al. Identification and Characterization of Multidrug-Resistant Extended-Spectrum Beta-Lactamase-Producing Bacteria from Healthy and Diseased Dogs and Cats Admitted to a Veterinary Hospital in Brazil. *Microb. Drug Resist.* **2021**, *27*, 855–864. [CrossRef]
35. Dupouy, V.; Abdelli, M.; Moyano, G.; Arpaillange, N.; Bibbal, D.; Cadiergues, M.C.; Lopez-Pulin, D.; Sayah-Jeanne, S.; de Gunzburg, J.; Saint-Lu, N.; et al. Prevalence of Beta-Lactam and Quinolone/Fluoroquinolone Resistance in Enterobacteriaceae From Dogs in France and Spain-Characterization of ESBL/pAmpC Isolates, Genes, and Conjugative Plasmids. *Front. Vet. Sci.* **2019**, *6*, 279. [CrossRef]

36. Karkaba, A.; Hill, K.; Benschop, J.; Pleydell, E.; Grinberg, A. Carriage and population genetics of extended spectrum β-lactamase-producing *Escherichia coli* in cats and dogs in New Zealand. *Vet. Microbiol.* **2019**, *233*, 61–67. [CrossRef]
37. Shaheen, B.W.; Nayak, R.; Foley, S.L.; Kweon, O.; Deck, J.; Park, M.; Rafii, F.; Boothe, D.M. Molecular characterization of resistance to extended-spectrum cephalosporins in clinical *Escherichia coli* isolates from companion animals in the United States. *Antimicrob. Agents Chemother.* **2011**, *55*, 5666–5675. [CrossRef]
38. Hsieh, C.J.; Shen, Y.H.; Hwang, K.P. Clinical implications, risk factors and mortality following community-onset bacteremia caused by extended-spectrum β-lactamase (ESBL) and non-ESBL producing *Escherichia coli*. *J. Microbiol. Immunol. Infect.* **2010**, *43*, 240–248. [CrossRef]
39. Otter, J.A.; Natale, A.; Batra, R.; Tosas Auguet, O.; Dyakova, E.; Goldenberg, S.D.; Edgeworth, J.D. Individual- and community-level risk factors for ESBL Enterobacteriaceae colonization identified by universal admission screening in London. *Clin. Microbiol. Infect.* **2019**, *25*, 1259–1265. [CrossRef]
40. Belas, A.; Salazar, A.S.; Gama, L.T.; Couto, N.; Pomba, C. Risk factors for faecal colonisation with *Escherichia coli* producing extended-spectrum and plasmid-mediated AmpC β-lactamases in dogs. *Vet. Rec.* **2014**, *175*, 202. [CrossRef]
41. Ljungquist, O.; Ljungquist, D.; Myrenås, M.; Rydén, C.; Finn, M.; Bengtsson, B. Evidence of household transfer of ESBL-/pAmpC-producing Enterobacteriaceae between humans and dogs—A pilot study. *Infect. Ecol. Epidemiol.* **2016**, *6*, 31514. [CrossRef]
42. Kaur, J.; Mahajan, G.; Chand, K.; Sheevani; Chopra, S. Enhancing Phenotypic Detection of ESBL in AmpC co-producers by using Cefepime and Tazobactam. *J. Clin. Diagn. Res.* **2016**, *10*, DC05–DC8. [CrossRef]

veterinary sciences

MDPI

Article

Occurrence of Multidrug-Resistant Bacteria Resulting from the Selective Pressure of Antibiotics: A Comprehensive Analysis of ESBL *K. pneumoniae* and MRSP Isolated in a Dog with Rhinorrhea

Inês C. Rodrigues [1,2,†], Marisa Ribeiro-Almeida [1,3,†], Jorge Ribeiro [1], Leonor Silveira [4], Joana C. Prata [1,5], Angela Pista [4] and Paulo Martins da Costa [1,2,*]

1 School of Medicine and Biomedical Sciences, University of Porto (ICBAS-UP), Rua de Jorge Viterbo Ferreira, 228, 4050-313 Porto, Portugal; icrodrigues@icbas.up.pt (I.C.R.); up201008465@edu.icbas.up.pt (M.R.-A.); jribeiro@icbas.up.pt (J.R.); joana.prata@iucs.cespu.pt (J.C.P.)
2 Interdisciplinary Centre of Marine and Environmental Research (CIIMAR), Terminal de Cruzeiros do Porto, de Leixões, Av. General Norton de Matos s/n, 4450-208 Matosinhos, Portugal
3 UCIBIO—Applied Molecular Biosciences Unit, Rede de Química e Tecnologia (REQUIMTE), Laboratory of Microbiology, Department of Biological Sciences, Faculty of Pharmacy, University of Porto, 4050-313 Porto, Portugal
4 Department of Infectious Diseases, National Institute of Health, Av. Padre Cruz, 1600-609 Lisbon, Portugal; leonor.silveira@insa.min-saude.pt (L.S.); angela.pista@insa.min-saude.pt (A.P.)
5 TOXRUN—Toxicology Research Unit, University Institute of Health Sciences, Cooperativa de Ensino Superior Politécnico e Universitário (CESPU), 4585-116 Gandra, Portugal
* Correspondence: pmcosta@icbas.up.pt
† These authors contributed equally to this work.

Citation: Rodrigues, I.C.; Ribeiro-Almeida, M.; Ribeiro, J.; Silveira, L.; Prata, J.C.; Pista, A.; Martins da Costa, P. Occurrence of Multidrug-Resistant Bacteria Resulting from the Selective Pressure of Antibiotics: A Comprehensive Analysis of ESBL *K. pneumoniae* and MRSP Isolated in a Dog with Rhinorrhea. *Vet. Sci.* **2023**, *10*, 326. https://doi.org/10.3390/vetsci10050326

Academic Editors: George Valiakos and Karl Pedersen

Received: 31 March 2023
Revised: 28 April 2023
Accepted: 29 April 2023
Published: 2 May 2023

Simple Summary: Antimicrobial resistance (AMR) poses a major threat to human and animal health. One of the causes underlying the emergence of increasingly resistant strains is antibiotic selective pressure. This study aimed to evaluate the impact of treatment with amikacin on an extended spectrum β-lactamase (ESBL)-producing *Klebsiella pneumoniae* isolated in a dog with rhinorrhea. In the middle of the treatment, methicillin-resistant *Staphylococcus pseudintermedius* (MRSP) was isolated from the left nasal cavity of the dog. At the end of the treatment, *K. pneumoniae* was not recovered from nasal swab samples, while MRSP displayed phenotypical and genotypical changes. Six weeks after the end of the treatment, only commensal flora was observed in both nasal cavities. These results warn of the effects of antibiotic pressure, which can lead to the emergence of multidrug-resistant strains either by directly promoting the enrichment of bacteria with resistance to multiple antimicrobial agents or via the subsequent acquisition of resistance genes. Therefore, adapting clinical practice to this new reality is crucial to limit the selection and spread of multi-resistant bacteria among pets, humans and the environment.

Abstract: Because of public health concerns, much greater scrutiny is now placed on antibiotic use in pets, especially for antimicrobial agents that have human analogs. Therefore, this study aimed to characterize the phenotypic and genotypic profiles of multidrug-resistant bacteria isolated from nasal swabs samples taken from a one-year-old male Serra da Estrela dog with rhinorrhea that was treated with amikacin. An extended-spectrum β-lactamases (ESBL) *Klebsiella pneumoniae* was isolated in the first sample taken from the left nasal cavity of the dog. Seven days later, methicillin-resistant (MRSP) *Staphylococcus pseudintermedius* was also isolated. Nevertheless, no alterations to the therapeutic protocol were performed. Once the inhibitory action of the antibiotic disappeared, the competitive advantage of the amikacin-resistant MRSP was lost, and only commensal flora was observed on both nasal cavities. The genotypic profile of extended-spectrum β-lactamase (ESBL)-producing *Klebsiella pneumoniae* revealed the same characteristics and close relation to other strains, mainly from Estonia, Slovakia and Romania. Regarding MRSP isolates, although resistance to aminoglycosides was present in the first MRSP, the second isolate carried *aac(6′)-aph(2″)*, which enhanced its resistance to amikacin. However, the veterinary action was focused on the treatment of the primary agent (ESBL

K. pneumoniae), and the antibiotic applied was according to its phenotypic profile, which may have led to the resolution of the infectious process. Therefore, this study highlights the importance of targeted therapy, proper clinical practice and laboratory-hospital communication to safeguard animal, human and environmental health.

Keywords: antibiotic pressure; dog; multidrug-resistant bacteria; ESBL *Klebsiella pneumoniae*; MRSP *Staphylococcus pseudintermedius*

1. Introduction

Antimicrobial resistance (AMR) poses a major threat to human and animal health worldwide [1,2]. The indiscriminate use and overuse of antimicrobials are some of the most important causes underlying the emergence of increasingly resistant strains through selective pressure [3–5]. Empirical antibiotic therapy frequently uses broad-spectrum antimicrobials or combinations of antimicrobials, which may also be required in polymicrobial infections or life-threatening conditions [6]. Indeed, 76% of veterinary clinicians support antimicrobial selection based on personal experience [4]. In addition to veterinary and human medicines sharing antibiotics, the transmission of resistant bacteria may occur due to the proximity between humans and companion animals [7,8].

Antibiotic therapy, either empirical or pathogen-directed, exerts selective pressure triggering multiple survival strategies resulting in vertical (mutations) or horizontal (of mobile genetic elements) transmission [9]. Bacterial populations present a notorious adaptative potential and high plasticity when facing different types of stress, including antibiotic therapy [9,10]. Indeed, multidrug-resistant strains can be selected under antibiotic pressure as a result of antibiotic treatment [11,12]. Therefore, physicians and veterinarians are important actors in the control of antimicrobial resistance as part of "One Health," especially for critical pathogens (WHO, ECDC) [13–16].

Klebsiella pneumoniae (*K. pneumoniae*) is one of the leading pathogens associated with the emergence of antibiotic resistance and a clinically significant nosocomial pathogen, also associated with high morbidity and mortality rates in companion animals [17,18]. Another emerging zoonotic pathogen of canine origin is methicillin-resistant *Staphylococcus pseudintermedius* (*S. pseudintermedius*; MRSP), which is transmitted by direct contact with or bites to pet owners or veterinary staff [19].

This study aimed at characterizing the phenotypic and genotypic profiles of multidrug-resistant strains isolated during a single infectious process, evaluating the impact of antibiotic selective pressure. A single case study of a dog presenting with mucopurulent rhinorrhea caused by an extended spectrum β-lactamases (ESBL) producing *K. pneumoniae* was investigated, followed by the appearance of MRSP. Antimicrobial resistance profiling and Whole Genome Sequencing (WGS) were performed on the strains isolated from this case. The impact of the applied antibiotic, amikacin, was evaluated during the therapeutic protocol on the nasal flora of the animal.

2. Materials and Methods

2.1. Case Selection

Cases admitted to the Veterinary Hospital (UPVET) of the Institute of Biomedical Sciences Abel Salazar, University of Oporto (ICBAS/UP) from the 1st of January 2022 to the 31st of December 2022 were analyzed (n = 8338). Eligibility criteria for case enrolment were: (i) admission to the UPVET for bacteriological infection, (ii) follow-up performed by UPVET, (iii) sending of more than one consecutive sample to the microbiology laboratory of ICBAS-UP during the same infectious process, (iv) isolation of a pure bacterial culture with a multidrug-resistant profile and clinically relevant under the One Health approach. A single case was selected based on these criteria. Informed consent was obtained from all of the UPVET clients for the use of data of patients for scientific study and teaching purposes.

Data was safely stored and anonymized according to Data Protection laws (Regulation (EU) 2016/679).

The selected case pertains to a 1-year-old, unneutered, giant breed (Serra da Estrela) dog with up-to-date vaccination and deworming, followed at UPVET. The owner of this animal had requested an emergency appointment since his dog showed signs of vomiting and mucopurulent rhinorrhea from the left nostril (Figure 1). Clinical examination revealed pain on abdominal palpation and mucopurulent rhinorrhea from the left nostril with no other abnormalities. The dog had been submitted for the surgical correction of a gastric dilatation volvulus a week earlier. Therefore, the clinicians decided to hospitalize the animal until the vomiting stopped. During hospitalization, supportive medication was administered, keeping the antibiotic treatment instituted by the hospital where the surgery was performed, consisting of cephazolin and metronidazole. After 2 days, the dog returned home with gastric support medication, along with a cephalexin prescription. After the first microbiological result of the left nasal cavity, the dog was again hospitalized, and a computed tomography (CT) scan was performed to understand the severity of the infectious process in the nasal cavities. The CT scan showed rhinosinusitis in the left nasal cavity, decreased nasal turbinates in the middle cavity and homogeneous material partially occupying the left nasal cavity (Appendix A). In order to discard fungal involvement, a rhinoscopy was performed, in which no signs of fungal plaques were detected. Besides, the biochemical analysis of blood to monitor urea and creatinine was also performed, and no alterations were recorded. In parallel, 4 samples were collected in order to follow the microbiological evolution of the clinical case. Samples were collected at 3 different times during antibiotic treatment: 8 and 18 days after the antibiotics' start and 6 weeks after the end of antibiotic treatment.

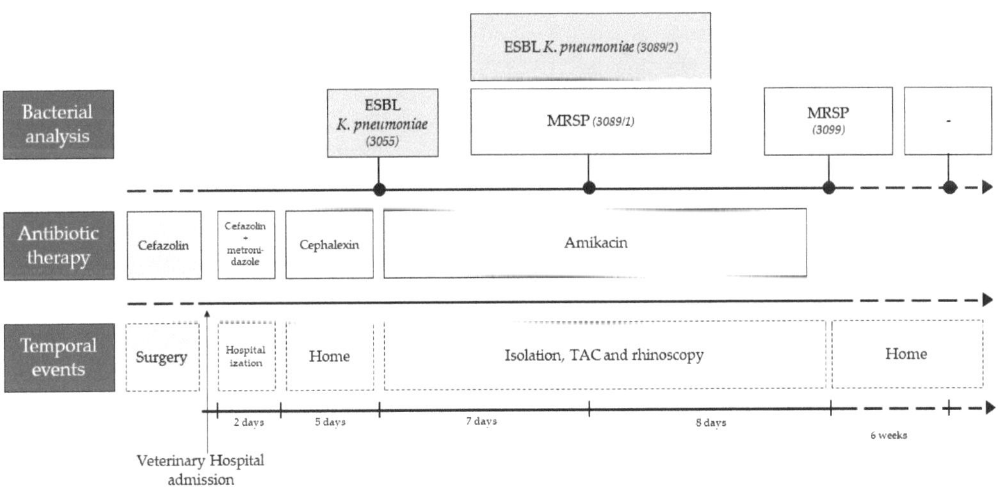

Figure 1. Timeline of the clinical case highlighting the main temporal events, antibiotic therapy and bacterial analysis.

2.2. Sample Collection and Bacterial Isolation

Samples were collected using a sterile swab and vigorously rubbing the most caudal side of the nasal cavity, as previously described by the Centers for Disease Control and Prevention (CDC) [20]. The collected samples were immediately transported to the microbiology laboratory and processed within 2 h of collection.

The first analysis was processed according to UK Standards for Microbiology Investigations [21]. As ESBL *K. pneumoniae* was isolated in the first sample, Simmons Citrate Agar (SCA) containing 1% (*w*/*v*) of myo-inositol (SCAi) was used along with blood agar media (BA, Tryptone Soy Agar containing 5% of laked horse blood Agar). In the third sample,

CHROMagar™ *Staphylococcus aureus* (CSA) was also used since, in the second sample, *K. pneumoniae* and MRSP were isolated. Finally, in the fourth sample, these 3 culture mediums were used: BA, CSA and SCAi. The schematic representation of the culture media used throughout the five samples is shown in Figure 2.

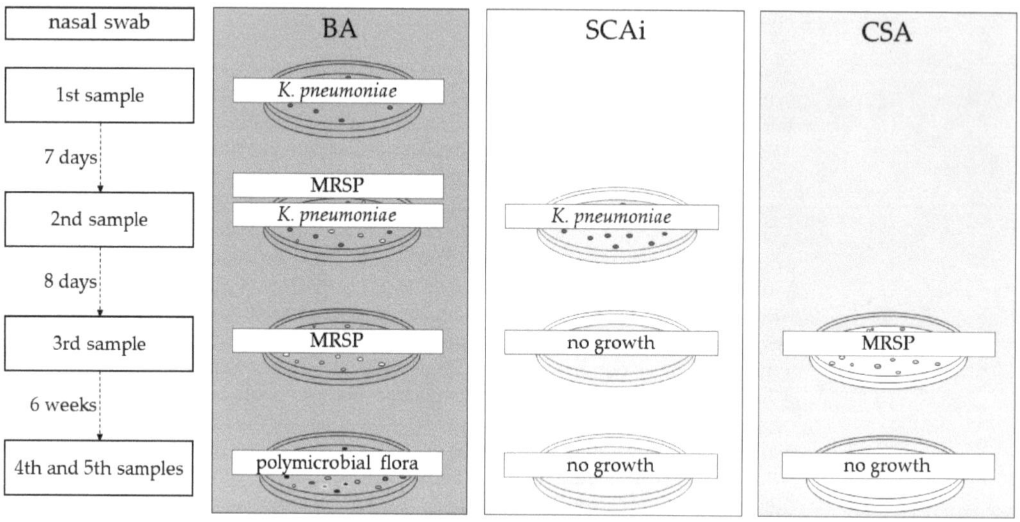

Figure 2. Culture media used in the microbiological analysis of the five samples. Additional culture media were added to subsequent samples to specifically culture the previously identified bacteria. BA: Blood agar, SCAi: Simmons Citrate agar contain 1% of inositol, CSA: CHROMagar™ *S. aureus* (CSA).

The plates with BA and CSA were incubated for 24 h, while the SCAi plates were left for 48 h at 37 °C. Bacterial isolates obtained from BA were Gram-stained and identified with conventional biochemical tests. Mauve to purple colonies growing on CSA were suspected to be coagulase-positive Staphylococcus. Moreover, yellow, dome-shaped, often mucoid colonies growing on SCAi were suspicious for *Klebsiella* spp. To confirm at the species level of *Klebsiella* spp. isolates, the RapID™ ONE System (Thermo Fischer Scientifics, Waltham, MA, USA) was used.

All isolated colonies were frozen in buffered peptone water (BPW) containing 1.5% (v/v) glycerol at −20 °C.

2.3. Antimicrobial Susceptibility Testing

Antimicrobial susceptibility testing was performed and interpreted according to the Clinical and Laboratory Standards Institute guidelines [22] using the Kirby–Bauer method. Antimicrobials were selected to represent a wide range of classes, and selection between different antimicrobial agents of the same class was based on the availability of clinical CLSI breakpoints [23].

A panel of 19 antimicrobials was used for *K. pneumoniae*: amikacin (AMK, 30 μg), amoxicillin/clavulanic acid (AMC, 30 μg), ampicillin (AMP, 10 μg), azithromycin (AZM, 15 μg), aztreonam (ATM; 30 μg), cefotaxime (CTX; 30 μg), cefoxitin (FOX; 30 μg), ceftazidime (CAZ; 30 μg), cephazolin (CFZ; 30 μg), chloramphenicol (CHL; 30 μg), ciprofloxacin (CIP; 5 μg), doxycycline (DOX; 30 μg), gentamycin (GEN; 120 μg), imipenem (IMP; 10 μg), levofloxacin (LEV; 5 μg), nitrofurantoin (NIT; 300 μg), streptomycin (STR; 10 μg), sulfamethoxazole/trimethoprim (SXT; 25 μg), tetracycline (TET; 30 μg), and tobramycin (TOB; 10 μg). For *S. pseudintermedius*, 17 antimicrobials were tested: azithromycin (AZM; 15 μg), cefoxitin (FOX; 30 μg), chloramphenicol (CHL; 30 μg), ciprofloxacin (CIP; 5 μg), clindamycin (CLI; 2 μg), doxycycline (DOX; 30 μg), erythromycin (ERY; 15 μg), gentamycin (GEN; 120 μg), levofloxacin (LEV; 5 µg), linezolid (LZD; 30 µg), nitrofurantoin (NIT; 300 µg), oxacillin (OXA; 1 µg), penicillin (PEN; 10 UI), quinupristin-dalfopristin (QDA; 15 µg), rifampicin (RIF; 5 µg), sulfamethoxazole/trimethoprim (SXT; 25 µg), tetracycline (TET; 30 µg), and tobramycin (TOB; 10 µg). All antimicrobial disks were from Oxoid (Basingstoke, UK).

Bacterial isolates were classified as susceptible, intermediate or resistant using current CLSI breakpoints [23]. Isolates resistant to 3 or more antibiotics classes were defined as multidrug-resistant (MDR) bacteria [24].

2.4. DNA Extraction and WGS Tecnhique

Genomic DNA was extracted from fresh cultures of each isolate using the Isolate II Genomic DNA Kit (Bioline, London, UK), followed by quantification in the Qubit fluorometer (Invitrogen, Waltham, MA, USA) with the dsDNA HS Assay Kit (Thermo Fisher Scientific, Waltham, MA, USA), according to the manufacturer's instructions. The DNA was subjected to the NexteraXT library preparation protocol (Illumina, San Diego, CA, USA) prior to cluster generation and paired-end sequencing (2 × 150 bp) on a NextSeq 550 instrument (Illumina), according to the manufacturer's instructions. FastQC v0.11.5 (https://www.bioinformatics.babraham.ac.uk/projects/fastqc/ (accessed on 16 February 2023)) was used for quality control and Trimmomatic v0.38 [25] for trimming low-quality bases.

2.4.1. Bioinformatic Analysis of *K. pneumoniae*

Online bioinformatic tools from PathogenWatch v20.0.13 (https://pathogen.watch/; accessed on 8 February 2023), specifically, Kleborate v2.2.0, were used to evaluate *K. pneumoniae* antibiotic resistance genes or known mutations, virulence genes, plasmid typing, Multilocus Sequence Typing (MLST) [26], core genome Multilocus Sequence Typing (cgMLST), capsular polysaccharide (K) and lipopolysaccharide O locus types and serotypes [27]. The phylogenetic analysis inferred by the neighbor-joining tree was based on the Pathogenwatch pairwise-distance matrix, based on the single nucleotide polymorphism (SNP) distances of a core gene library (1972 genes) [28]. Closely related genomes and the associated metadata (country, source and date) were collected from all public genomes available from Pathogenwatch after cgMLST single-linkage clustering and the selection of those with less than 5 allele differences. The neighbor-joining tree was edited using iToL [29].

2.4.2. Bioinformatic Analysis of *S. pseudintermedius*

For bioinformatic analysis for *S. pseudintermedius* strains, DNA was assembled using the Bacterial and Viral Bioinformatics Resource Center (BV-BRC) platform (https://www.bv-brc.org/app/Assembly2; accessed on 8 February 2023). Moreover, tools from Centre for Genomic and Epidemiology (http://www.genomicepidemiology.org; accessed on 26 January 2023) were used to assess antibiotic resistance genes or known mutations (ResFinder 4.1; https://cge.food.dtu.dk/services/ResFinder/; accessed on 9 February 2023), virulence genes (VirulenceFinder 2.0; https://cge.food.dtu.dk/services/VirulenceFinder/; accessed on 9 February 2023), plasmid replicons (PlasmidFinder 2.1; https://cge.food.dtu.dk/services/PlasmidFinder/; accessed on 16 February 2023), SCC*mec* elements (SCCmecFinder 1.2; https://cge.food.dtu.dk/services/SCCmecFinder/; accessed on 16 February 2023) and Multilocus Sequence Typing (MLST 2.0; https://cge.food.dtu.dk/services/MLST/; accessed on 16 February 2023).

2.4.3. Data Availability

Sequence data were submitted to the European Nucleotide Archive (ENA) under BioProject accession number PRJEB61067. Each strain was stored with the accession numbers ERS14859644-ERS14859647, and the genomics sequences can be accessed with the accession numbers ERR11179010-ERR11179013.

3. Results

3.1. Dog Hospital Procedures and Bacterial Analysis

A sample of the mucopurulent discharge of the left nostril was collected during the emergency appointment and immediately transported to the microbiology laboratory. Microbiological analysis of this first sample detected growth of a pure culture on BA medium, being identified as a *K. pneumoniae* (strain 3055). Antimicrobial susceptibility results of this isolate revealed the expression of extended-spectrum β-lactamases (ESBL). Therefore, the clinicians decided to proceed with a second hospitalization in the isolation ward and administer injectable amikacin (by slow intravenous infusion). During the administration of antibiotics, the dog was kept in the isolation ward, where the following care was performed: (i) cleaning both nostrils every eight hours with saline and nostril aspiration; (ii) total restriction to public space (no access to the street); (iii) proper disposal of all organic (feces, urine) and non-organic materials (e.g., gloves); (iv) no external visits; and (v) the biochemical analysis of blood to monitor urea and creatinine every 5 days, in order to control renal function.

Eight days after the start of injectable antibiotic therapy, a new sample was taken from the left nasal cavity of the dog. In this sample, growth on BA and SCAi media was observed, as it involved a recovered *K. pneumoniae* isolate (strain 3089/2, BA and SCAi) and an *S. pseudintermedius* isolate (strain 3089/1, BA). Both strains revealed a multidrug-resistant profile (ESBL Klebsiella pneumoniae and MRSP). Nevertheless, no alteration was made to the clinical protocol: antibiotic therapy with amikacin in the isolation ward was maintained, which lasted 14 days. After 4 days without medication, the third sample was collected from the left nostril, and only MRSP could be found in microbiological samples. Therefore, the dog was discharged without any antimicrobial therapy. New samples of both nostrils (fourth and fifth samples) were taken six weeks after ending antimicrobial therapy, and no growth on SCAi nor on MAC mediums was observed. The polymicrobial flora presented on these samples were compatible with commensal flora, and none of the previous isolates were identified.

3.2. Antimicrobial Susceptibility Testing

The antimicrobial profile of the four isolated strains (two *K. pneumoniae* and two *S. pseudintermedius*) is presented in Table 1. All the strains were resistant to more than three antibiotic classes, being classified as MDR.

The *K. pneumoniae* isolates recovered in this study were classified as ESBL since they were resistant to aztreonam, cefotaxime and ceftazidime (Table 1) [23]. The antimicrobial profile of both *K. pneumoniae* strains (3055 and 3089/2) displayed the same antibiotic susceptibility pattern with resistance to penicillin, cephalosporins, monobactam, macrolides, tetracyclines, fluoroquinolones, folate inhibitor, phenicol, nitrofuran and aminoglycosides antibiotic class. Regarding the aminoglycoside class, both *K. pneumoniae* strains only showed resistance to tobramycin and streptomycin (Table 1).

The two *S. pseudintermedius* isolates recovered were resistant to oxacillin, being classified as MRSP. The antimicrobial profile of *S. pseudintermedius* isolates (3089/1 and 3099) revealed antibiotic resistance to penicillin, cephalosporins, aminoglycosides, tetracyclines, fluoroquinolones, lincosamides and folate inhibitor classes. Although both strains demonstrated susceptibility to doxycycline, the diameter of inhibition was near the lower limit of the breakpoint.

Table 1. Antimicrobial susceptibility profile of *K. pneumoniae* and *S. pseudintermedius* isolated from the dog in different sampling times.

Antibiotic Class	Antibiotic	*Klebsiella pneumoniae*		*Staphylococcus pseudintermedius*	
		3055	**3089/2**	**3089/1**	**3099**
Penicillin	AMP	R	R	-	-
	AMC	R	R	-	-
	OXA	-	-	R	R
	PEN	-	-	R	R
Cephalosporins	CAZ	R	R	-	-
	CFZ	R	R	-	-
	CTX	R	R	-	-
	FOX	R	R	R	R
Monobactam	ATM	R	R	-	-
Macrolides	AZM	R	R	R	R
	ERY	-	-	R	R
Aminoglycosides	AMK	S	S	-	-
	GEN	S	S	I	I
	STR	R	R	-	-
	TOB	R	R	R	R
Tetracyclines	DOX	I	I	S	S
	TET	I	I	R	R
Fluoroquinolones	CIP	R	R	R	R
	LEV	R	R	R	R
Ansamycin	RIF	-	-	S	S
Lincosamide	CLI	-	-	R	R
Folate inhibitor	SXT	R	R	R	R
Phenicol	CHL	R	R	S	S
Nitrofuran	NIT	R	R	S	S
Carbapenems	IMP	S	S	-	-
Streptogramins	QDA	-	-	S	S
Oxazolidinones	LZD	-	-	S	S
Sampling order		1st	2nd	2nd	3rd

R, resistant; I, intermediate; S, susceptible; AMK, amikacin, AMC, amoxicillin/clavulanic acid; AMP, ampicillin; ATM, aztreonam; AZM, azithromycin; CAZ, ceftazidime; CHL, chloramphenicol; CIP, ciprofloxacin, CTX, cefotaxime; CLI, clindamycin; DOX, doxycycline; ERY, erythromycin; FOX, cefoxitin; GEN, gentamycin, IMP, imipenem; CFZ: cephazolin; LEV, levofloxacin; LZD, linezolid; NIT, nitrofurantoin; OXA, oxacillin; PEN, penicillin; QDA, quinupristin-dalfopristin; RIF, rifampicin, STR, streptomycin; SXT, sulfamethoxazole/trimethoprim; TET, tetracycline; TOB, tobramycin.

3.3. *WGS and In Silico Genomic Characterization*

3.3.1. ESBL *K. pneumoniae* Strains Characterization

Both ESBL *K. pneumoniae* strains presented the same seventeen acquired genes related to aminoglycoside (*aac(6′)-Ib-cr, aph3-Ia, strA* and *strB*), 3rd generation cephalosporins ($bla_{\text{CTX-M-15}}$), fluoroquinolones (*qnrB1*, *qnrB4*, *gyrA*-83I and *parC*-80I), penicillins ($bla_{\text{DHA-1}}$, $bla_{\text{OXA-1}}$, $bla_{\text{TEM-1D}}$ and $bla_{\text{SHV-11}}$), phenicols (*catB3*), sulfonamides (*sul1* and *sul2*) and trimethoprim resistances (*dfrA14*) (Table 2). No genes mediating resistance to carbapenems, 3rd generation cefalosporins or penicillins combined with beta-lactamase inhibitors, colistin, fosfomycin, tetracyclines, monobactams, nitrofurans and tigecycline were found. Interestingly, the resistance of *K. pneumoniae* to monobactam, tetracycline and nitrofuran was observed, but no associated genes were identified (Appendix B).

Table 2. Whole genome characterization of *Klebsiella pneumoniae*.

Sample ID	Antimicrobial Resistance Genes	Virulence Genes	Plasmid Typing	MLST	Closest cgMLST	Capsule (K) Locus	O Serotype Locus
3055	*aac(6′)-Ib-cr, aph3-Ia, strA, strB,* $bla_{CTX-M-15}$, *qnrB1, qnrB4, gyrA*-83I, *parC*-80I, bla_{DHA-1}, bla_{OXA-1}, bla_{TEM-1D}, bla_{SHV-11}, *catB3, sul1, sul2, dfrA14*	*ybt 1*	IncFII(K), IncFIB(K), IncR	11	1509	KL105	O1/O2v2
3089/2	*aac(6′)-Ib-cr, aph3-Ia, strA, strB,* $bla_{CTX-M-15}$, *qnrB1, qnrB4, gyrA*-83I, *parC*-80I, bla_{DHA-1}, bla_{OXA-1}, bla_{TEM-1D}, bla_{SHV-11}, *catB3, sul1, sul2, dfrA14*	*ybt 1*	IncFII(K), IncFIB(K), IncR	11	1509	KL105	O1/O2v2

In terms of virulence-associated genes, only the siderophore yersiniabactin gene was found on both ESBL *K. pneumoniae* strains (Table 2).

Both ESBL *K. pneumoniae* strains presented the same three types of plasmids: IncFII(K), IncFIB(K) and IncR. These strains also possessed identical MLST, closest cgMLST, capsule locus and serotype O (11, 1509, KL105 and O1/O2v2, respectively; Table 2).

The neighbor-joining tree generated by comparing the cgMLST of *K. pneumoniae* genomes isolated in this study with those available in PathogenWatch revealed an association with isolates from human infections and a cat (Figure 3). Geographically, this group was identified mainly in Romania and Slovakia, followed by Estonia, France and Croatia (Figure 3).

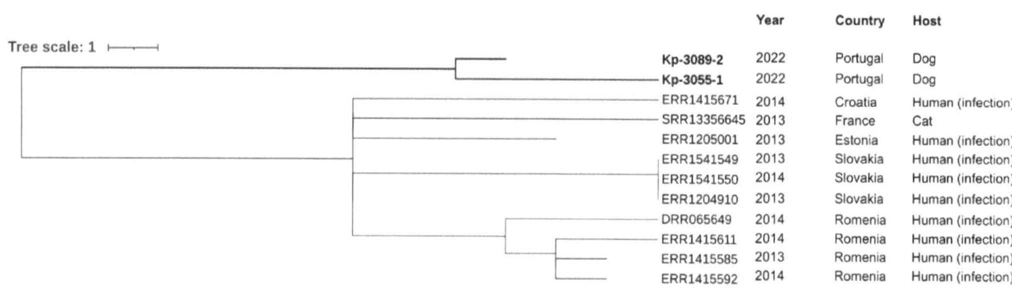

Figure 3. Neighbor-joining tree representing the phylogenetic relationships among *K. pneumoniae* genomes obtained in this study and those available in PathogenWatch with less than 22 SNPs. The cgMLST single linkage clustering was used for genome selection with a threshold of 5 allele differences, and the neighbor-joining tree was inferred from the PathogenWatch pairwise-distance matrix. The metadata of each isolate (country, source and date) was added using iTOL (https://itol.embl.de/; accessed on 10 March 2023).

3.3.2. MRSP Strains Characterization

Genes mediating resistance to penicillins (*blaZ* and *mecA*), macrolides (*erm(B)*), aminoglycosides (*aph(3′)-III* and *ant(6)-Ia*), tetracyclines (*tet(K)* and *tet(M)*), clindamycin (*erm(B)*) and trimethoprim (*drfG*) were found on both strains (Table 3). The *aac(6′)-aph(2″)* gene was also detected in the 3099 strain. However, fluoroquinolone, streptogramin or phenicol genes mediating resistance were not found. Resistance of *S. pseudintermedius* to cephalosporin and fluoroquinolones was observed, but no associated genes were identified (Appendix B).

Table 3. Whole genome characterization of Staphylococcus pseudintermedius.

Sample ID	Antimicrobial Resistance Genes	Plasmid	MLST	SCC*mec* Type
3089/1	*aph(3′)-III, ant(6)-Ia, erm(B), drfG, blaZ, mecA, tet(K), tet(M)*	rep7a, repUS43	551	Vc(5C2&5)
3099	*aac(6′)-aph(2″), aph(3′)-III, ant(6)-Ia, erm(B), dfrG, blaZ, mecA, tet(K), tet(M)*	rep7a, repUS43	551	Vc(5C2&5)

Both isolates have the rep7a and repUS43 plasmid replicons and the Vc(5C2&5) SCC*mec* element. No virulence-associated genes were identified in the MRSP strains (Table 3).

4. Discussion

The frequent occurrence of multidrug-resistant bacteria has become a global threat to public health [2]. Overuse of antibiotics has been identified as the leading driver of AMR [30,31]. A case of a dog with rhinorrhea caused by an ESBL *K. pneumoniae* was investigated. Due to the resistance profile presented by this isolate, injectable amikacin was administered, and the animal was hospitalized in the isolation ward. Other causes were ruled out using CT and rhinoscopy. While under antibiotic treatment, ESBL *K. pneumoniae* was again isolated along with MRSP. This last MDR bacterial strain was considered opportunistic [32], derived from the selective pressure and depletion of the natural nasal microbiome caused by the antibiotics [33]. Therefore, no additional treatment was prescribed, and 6 weeks after the antibiotic treatment had ceased, only commensal flora was found in samples from both nostrils.

The antimicrobial resistance profiles of the two strains of *K. pneumoniae* (3055 and 3089/2) showed identical antibiotic susceptibility patterns, both being considered ESBL. The high level of resistance was remarkable, especially for the antibiotic classes of penicillin, cephalosporin and fluoroquinolone, which can be explained by two reasons: these are the most frequently prescribed antibiotics in veterinary medicine [34], and the dog underwent an emergency gastric surgery one week before the first sample collection, in which cephazolin antibiotherapy was prescribed. For instance, in Portugal, fluoroquinolones and cephalosporins represented the second and fourth most often prescribed antibiotic classes in both human and animal medicines [35]. Nonetheless, *K. pneumoniae* is commonly resistant to aminopenicillins [23].

Considering the resistance genes found by PathogenWatch in *K. pneumoniae* isolates, the ESBL phenotype was held by the detection of β-lactamase resistance genes (*bla*$_{\text{CTX-M-15}}$; Table 2). Also, fluoroquinolone, phenicol, aminoglycoside, sulfonamide and trimethoprim resistance genes were detected, in accordance with previous studies, which demonstrated that at least 80% of ESBL producers were also resistant to sulfonamides, quinolones and aminoglycosides [36].

Nevertheless, a few discrepancies were found between phenotypic and genotypic resistance profiles. Although *aac(6′)-Ib-cr*, *aph3-Ia*, *strA* and *strB* were detected, ESBL *K. pneumoniae* strains were phenotypically susceptible to gentamycin and amikacin. Likewise, genes mediating resistance to penicillin combined with β-lactamase inhibitors, monobactams, macrolides, tetracyclines and nitrofuran were not found, despite ESBL *K. pneumoniae* strains showing intermediate resistance to tetracycline, doxycycline and resistant to amoxicillin/clavulanic acid, aztreonam, azithromycin and nitrofurantoin.

The virulence gene *Ybt 1* was found in both strains of *K. pneumoniae*, which encode the iron-scavenging siderophore yersiniabactin, promoting systemic survival and dissemination [37]. Also, previous studies demonstrated that this virulence gene favored the maximum growth and lethality of *K. pneumoniae* in respiratory tract infection [38,39], which may have largely contributed to the pathogenicity in this case.

Plasmids often transport resistance genes and virulence genes that can disseminate by horizontal gene transfer mechanisms [40]. IncFII(K), IncFIB(K) and IncR plasmids were detected for both strains of *K. pneumoniae*, being this type of plasmids associated with epidemic *K. pneumoniae* and implicated in the worldwide spread of multidrug resistance [41]. The same authors found an association between the *bla*$_{CTX-M-15}$ gene and the IncR plasmid in ESBL *K. pneumoniae* isolates from Portuguese hospitals [41].

The two strains of *K. pneumoniae* recovered in this study presented the capsular-type KL105 and Sequence Type 11, previously associated with MDR and virulence determinants (yersiniabactin and colibactin) [42]. The ST11 KL105 clade has been successfully disseminated in Europe, even circulating in Portuguese hospitals for years [43,44]. Therefore, it can be hypothesized that a human previously hospitalized or working in a hospital may have had contact with this dog.

Moreover, lipopolysaccharide O locus serotypes were identified as O1/O2v2 for both strains of *K. pneumoniae*, which was associated with hypervirulent strains and was found more frequently in clinical genomes, including in a Portuguese clinical genome [45].

Although strains circulating in Portugal with the same MLST, lipopolysaccharide O locus and capsular type have been described, these Portuguese isolates of *K. pneumoniae* were not available on PathogenWatch, and it was impossible to establish a phylogenetic correlation between them. Therefore, the results of the phylogenetic tree supported kinship (<22 SNPs) to strains mainly from eastern countries (Estonia, Slovakia and Romania) and isolated especially from human infections (Figure 3). These data can be explained by the migration (of both people and animals) from eastern countries to Portugal and the consumption of imported food and feed. In addition, *K. pneumoniae* ST11 was first reported in France (in 1997) and has since been reported all over the world, including in America, Asia and most countries in Europe, such as The Netherlands, Norway, Poland, Slovakia and Portugal [46–48]. ST11-*K. pneumoniae* lineage has only been reported in humans, and no data was available in dogs. Moreover, in the 2000s, there was a wave of immigration from Eastern European countries, namely from Ukraine, which is the third country with the largest group of immigrants to Portugal [49]. Also, both strains of *K. pneumoniae* isolated were closely related (5 allele differences).

Since ESBL *K. pneumonia* isolated in this study possessed a myriad of genetic determinants, previously characterized with high pathogenicity and antimicrobial resistance, it was assumed that the isolated strains were at the origin of the mucopurulent rhinorrhea and that veterinary medical action was correctly adjusted to the microbiological findings.

Regarding the results of *S. pseudintermedius* strains (3089/1 and 3099), the isolation of these strains during amikacin treatment may have been caused by antibiotic selection pressure [32]. Hence, the susceptibility profiles were identical. Besides showing resistance to cefoxitin and oxacillin, detection of the *mecA* resistance gene enabled both strains to be classified as MRSP. Similar to ESBL *K. pneumoniae* strains, *S. pseudintermedius* presented resistance to penicillin, aminoglycosides, macrolides, tetracyclines and trimethoprim classes. Moreover, the *aac(6′)-aph(2″)* gene was only detected on the 3099 strain, and it has been described that it confers resistance to a broad spectrum of aminoglycosides [50,51]. The acquisition of another gene to reinforce the resistance to aminoglycosides might have been caused by the selective pressure of amikacin, allowing the bacteria to gain a competitive advantage over other bacteria [52].

However, a few disparities were also found between the phenotypic and genotypic resistance profiles. Indeed, no genes for resistance to fluoroquinolones were identified, and phenotypically, both MRSP strains showed resistance to ciprofloxacin and levofloxacin. These discrepancies observed in the two bacterial species (*K. pneumoniae* and *S. pseudintermedius*) were not pursued. However, potential antimicrobial mechanisms without resistance gene expression include activation of multidrug efflux pumps or decreased outer membrane permeability [53], which should be further explored for fluoroquinolone resistance. Moreover, databases can differ essentially in the number and type of genes and resistance determinants they comprise [54], so there is the possibility that quinolone-resistant determinants were not found due to the database data used. Hence, phenotypic and genomic evaluation are complementary, both being required for a complete account of resistance.

In both strains of MRSP, rep7a and repUS43 were identified. These plasmids frequently carried the *tet(M)* and *tet(K)* resistance genes [55], which is in agreement with antimicrobial resistance genes results.

In addition, the two MRSP strains were identified as ST551, being recorded between 2015 and 2018, 12 *S. pseudintermedius* ST551 strains in the PubMLST database, from different geographical locations and animal hosts (https://pubmlst.org/, last accessed 28 February 2023). Among the 12 records, six samples were isolated from dogs (50%), four from cats (33%) and two from humans (17%). Indeed, *S. pseudintermedius* has been correlated to infections in dogs, being considered an important pathogen in canine pyodermas [56,57]. As for localization, these isolates were from Poland (50%), Switzerland (25%), Sweden (17%) and the USA (8%), evidencing the spread of ST551 throughout Europe since 2015.

Also, both MRSP isolates harbored the SCC*mec* type Vc (5C2&5) element. Since these strains possess the *tet(K)* gene, this is in agreement with previous studies, which showed that isolates carrying the Vc (5C2&5) element co-harbor *tet(K)* in a higher proportion than isolates with other SCC*mec* elements [58].

In this study, the right choice of antibiotic in combination with inpatient hospitalization in the isolation ward might have contributed to the clinical success of the case. While the choice based on the antibiogram allowed the elimination of the primary agent of infection, the isolation of the animal possibly prevented the dissemination and spread of multidrug-resistant bacteria. Although antibiotic therapy may have been at the origin of MRSP recovery, once the inhibitory action of the prescribed aminoglycoside disappeared, the competitive advantage of MRSP on nasal flora dissipated. Thereby, the decision of clinicians to focus only on eliminating the ESBL *K. pneumoniae* strains may have been the correct one.

Some limitations should be considered in the present study. Firstly, data on antimicrobial prescriptions before the emergency appointment were not available. Secondly, since only one clinical case was investigated, some bias in the interpretation of results may be present. Despite these limitations, the results of this study provide valuable information on the dynamics established between the antibiotic and the bacteria during a therapeutic protocol of an infectious process.

Hence, pets can act as reservoirs of AMR genes that may transfer to other inhabitants of the house, both humans and animals. Therefore, veterinary practices, along with microbiology laboratory guidance, must adapt to this new reality, ensuring effective treatment of infections and protection of animal, human and environmental health.

5. Conclusions

The present study intended to investigate the effect of antibiotic pressure on the isolation of multidrug-resistant bacteria. Our results showed that antibiotic therapy may have been the cause of antimicrobial resistance and MRSP recovery. The isolation of MRSP followed by its elimination may have been the result of antibiotic pressure for a long period, combined with the competitive action of the commensal flora. The discrepancies observed in this study between phenotypic and genotypic determinants of antimicrobial resistance demonstrated their complementarity. Moreover, the geographical distribution of isolates with similar characteristics to the isolates in this study showed the wide dispersion of the bacteria. Thus, this study highlights the importance of readapting veterinary practices to safeguard the effective treatment of infection and the protection of human, animal and environmental health.

Author Contributions: Conceptualization, I.C.R. and M.R.-A.; methodology, I.C.R. and M.R.-A.; software, M.R.-A. and A.P.; formal analysis, P.M.d.C. and A.P.; investigation, I.C.R.; resources, P.M.d.C. and A.P.; data curation, M.R.-A. and J.C.P.; writing—original draft preparation, I.C.R.; writing—review and editing, M.R.-A., L.S., J.R., J.C.P., A.P. and P.M.d.C.; supervision, P.M.d.C. and A.P. All authors have read and agreed to the published version of the manuscript.

Funding: This research received no external funding.

Institutional Review Board Statement: Ethical review and approval were waived for this study. The management of the animal included in this study was conducted on the practitioner's experience and routine therapy plan and was not influenced in any way by this study.

Informed Consent Statement: Written informed consent was obtained from all subjects involved in this study.

Data Availability Statement: Not applicable.

Acknowledgments: The authors express their gratitude to the Technology and Innovation Unit and to Alexandra Nunes from the Genomics and Bioinformatics Unit of the National Institute of Health Doutor Ricardo Jorge for performing the Next Generation Sequencing runs and for helping in the ENA submission process, respectively. The authors thank Joana Santos for performing the CT exam and providing the respective results. The authors are also grateful to all veterinary staff of UPVET for providing samples and their technical support. The authors would also like to thank Elizabete Lopes for her technical assistance.

Conflicts of Interest: The authors declare no conflict of interest.

Appendix A

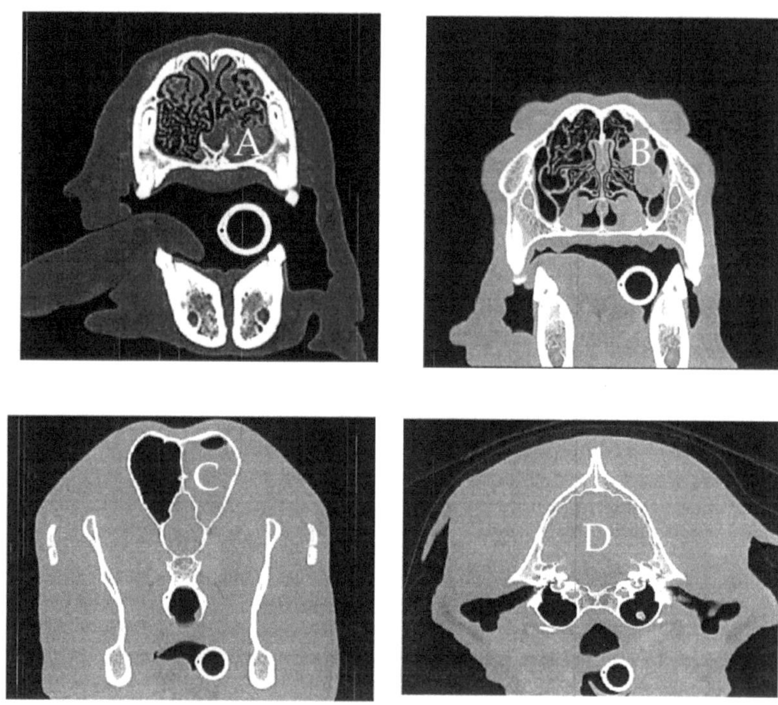

Figure A1. Cranioencephalic computed tomography of a one-year-old Serra da Estrela dog. Left rhinosinusitis with no evidence of a foreign body with the left nasal cavity occupied with a homogeneous material with preservation of the nasal turbinates (**A**). This material extends caudally until the etmoturbinates and left frontal sinus (**B**,**C**), with no signs of osteolysis. The tympanic bullae are regular with normal aerial content, with the exception of the presence of an otolith in the left bulla (**D**). The cerebral parenchyma was normal (**D**).

Appendix B

Table A1. Summary of phenotypic resistance and resistance genes by the antibiotic class of the four isolates.

		Klebsiella pneumoniae			*Staphylococcus pseudintermedius*		
Antibiotic Class		**3055**	**3089/2**	**Genome**	**3089/1**	**3099**	**Genome**
Penicillin	R S	2/2 0/2	2/2 0/2	*bla*$_{DHA-1}$, *bla*$_{OXA-1}$, *bla*$_{TEM-1D}$, *bla*$_{SHV-11}$	2/2 0/2	2/2 0/2	*blaZ*, *mecA*
Cephalosporin	R S	4/4 0/4	4/4 0/4	*bla*$_{CTX-M-15}$	1/1 0/1	1/1 0/1	Missing
Monobactam	R S	1/1 0/1	1/1 0/1	Missing	na na	na na	na
Macrolides	R S	1/1 0/1	1/1 0/1	Missing	2/2 0/2	2/2 0/2	*erm(B)*
Aminoglycosides	R S	2/4 2/4	2/4 2/4	*aac(6′)-Ib-cr*, *aph3-Ia*, *strA*, *strB*	2/2 0/2	2/2 0/2	*aph(3′)-III*, *ant(6)-Ia)*, *aac(6′)-aph(2″)* *
Tetracyclines	R S	2/2 0/2	2/2 0/2	Missing	1/2 1/2	1/2 1/2	*tet(K)*, *tet(M)*
Fluoroquinolones	R S	2/2 0/2	2/2 0/2	*qnrB1*, *qnrB4*, *gyrA*-83I, *parC*-80I	2/2 0/2	2/2 0/2	Missing
Ansamycin	R S	na na	na na	na	0/1 1/1	0/1 1/1	na
Lincosamide	R S	na na	na na	na	1/1 0/1	1/1 0/1	*erm(B)*
Folate inhibitor	R S	1/1 0/1	1/1 0/1	*sul1*, *sul2*, *dfrA14*	1/1 0/1	1/1 0/1	*dfrG*
Phenicol	R S	1/1 0/1	1/1 0/1	*catB3*	0/1 1/1	0/1 1/1	na
Nitrofuran	R S	1/1 0/1	1/1 0/1	Missing	0/1 1/1	0/1 1/1	na
Carbapenems	R S	0/1 1/1	0/1 1/1	na	na na	na na	na
Streptogramins	R S	na na	na na	na	0/1 1/1	0/1 1/1	na
Oxazolidinones	R S	na na	na na	na	0/1 1/1	0/1 1/1	na

* *aac(6′)-aph(2″)* was only identified in 3099 strain. All results with intermediate susceptibility were classified as resistant. na, not available.

References

1. Murray, C.J.; Ikuta, K.S.; Sharara, F.; Swetschinski, L.; Robles Aguilar, G.; Gray, A.; Han, C.; Bisignano, C.; Rao, P.; Wool, E.; et al. Global burden of bacterial antimicrobial resistance in 2019: A systematic analysis. *Lancet* **2022**, *399*, 629–655. [CrossRef]
2. Palma, E.; Tilocca, B.; Roncada, P. Antimicrobial resistance in veterinary medicine: An overview. *Int. J. Mol. Sci.* **2020**, *21*, 1914. [CrossRef] [PubMed]
3. Mengistu, T.S.; Garcias, B.; Castellanos, G.; Seminati, C.; Molina-López, R.A.; Darwich, L. Occurrence of multidrug resistant Gram-negative bacteria and resistance genes in semi-aquatic wildlife—*Trachemys scripta*, *Neovison vison* and *Lutra lutra*—As sentinels of environmental health. *Sci. Total Environ.* **2022**, *830*, 154814. [CrossRef]
4. Grakh, K.; Mittal, D.; Kumar, T.; Thakur, S.; Panwar, D.; Singh, L.; Kumar, M.; Jindal, N. Attitude, opinions, and working preferences survey among pet practitioners relating to antimicrobials in India. *Antibiotics* **2022**, *11*, 1289. [CrossRef]
5. Skalet, A.H.; Cevallos, V.; Ayele, B.; Gebre, T.; Zhou, Z.; Jorgensen, J.H.; Zerihun, M.; Habte, D.; Assefa, Y.; Emerson, P.M.; et al. Antibiotic selection pressure and macrolide resistance in nasopharyngeal *Streptococcus pneumoniae*: A cluster-randomized clinical trial. *PLoS Med.* **2010**, *7*, e1000377. [CrossRef] [PubMed]

6. Redding, L.; Grunwald, H.; Cole, S.; Rankin, S.; Nolen-Walston, R. Modification of empirical antimicrobial regimens in large animal medicine. *Vet. Rec.* **2020**, *187*, e78. [CrossRef] [PubMed]
7. Moulin, G.; Cavalié, P.; Pellanne, I.; Chevance, A.; Laval, A.; Millemann, Y.; Colin, P.; Chauvin, C. A comparison of antimicrobial usage in human and veterinary medicine in France from 1999 to 2005. *J. Antimicrob. Chemother.* **2008**, *62*, 617–625. [CrossRef]
8. Tompson, A.C.; Mateus, A.L.P.; Brodbelt, D.C.; Chandler, C.I.R. Understanding antibiotic use in companion animals: A literature review identifying avenues for future efforts. *Front. Vet. Sci.* **2021**, *8*, 1125. [CrossRef]
9. Windels, E.M.; Van Den Bergh, B.; Michiels, J. Bacteria under antibiotic attack: Different strategies for evolutionary adaptation. *PLoS Pathog.* **2020**, *16*, e1008431. [CrossRef]
10. Goossens, H. Antibiotic consumption and link to resistance. *Clin. Microbiol. Infect.* **2009**, *15*, 12–15. [CrossRef]
11. van Opijnen, T.; Dedrick, S.; Bento, J. Strain dependent genetic networks for antibiotic-sensitivity in a bacterial pathogen with a large pan-genome. *PLoS Pathog.* **2016**, *12*, e1005869. [CrossRef]
12. Chevin, L.M.; Hoffmann, A.A. Evolution of phenotypic plasticity in extreme environments. *Philos. Trans. R. Soc. B Biol. Sci.* **2017**, *372*, 20160138. [CrossRef]
13. European Commission. EU Guidelines for the prudent use of antimicrobials in human health (2017/C 212/01). *EN Off. J. Eur. Union* **2007**, *44*, 159–177.
14. The Council of the European Union. Council conclusions on the next steps towards making the EU a best practice region in combatting antimicrobial resistance (2019/C 214/01). *Off. J. Eur. Union* **2019**, *19*, 1–7.
15. WHO. *Global Action Plan on Antimicrobial Resistance*; World Health Organization: Geneva, Switzerland, 2015; pp. 1–28. Available online: https://www.amcra.be/swfiles/files/WHO%20actieplan_90.pdf (accessed on 2 February 2023).
16. European Centre for Disease Prevention and Control. *Survey of Healthcare Workers' Knowledge, Attitudes and Behaviours on Antibiotics, Antibiotic Use and Antibiotic Resistance in the EU/EEA*; European Centre for Disease Prevention and Control: Solna, Sweden, 2019; 108p. Available online: www.ecdc.europa.eu (accessed on 10 February 2023).
17. Alvarez-Uria, G.; Gandra, S.; Mandal, S.; Laxminarayan, R. Global forecast of antimicrobial resistance in invasive isolates of *Escherichia coli* and *Klebsiella pneumoniae*. *Int. J. Infect. Dis.* **2018**, *68*, 50–53. [CrossRef]
18. Fursova, N.K.; Astashkin, E.I.; Ershova, O.N.; Aleksandrova, I.A.; Savin, I.A.; Novikova, T.S.; Fedyukina, G.N.; Kislichkina, A.A.; Fursov, M.V.; Kuzina, E.S.; et al. Multidrug-resistant *Klebsiella pneumoniae* causing severe infections in the neuro-icu. *Antibiotics* **2021**, *10*, 979. [CrossRef] [PubMed]
19. Bünsow, D.; Tantawy, E.; Ostermeier, T.; Bähre, H.; Garbe, A.; Larsen, J.; Winstel, V. Methicillin-resistant *Staphylococcus pseudintermedius* synthesizes deoxyadenosine to cause persistent infection. *Virulence* **2021**, *12*, 989–1002. [CrossRef] [PubMed]
20. Centers for Disease Control and Prevention. How to Collect an Anterior Nasal Swab Specimen for COVID-19 Testing Set-Up. 2020. Available online: https://stacks.cdc.gov/view/cdc/90444/cdc_90444_DS1.pdf (accessed on 3 December 2022).
21. Public Health England. *UK Standards for Microbiology Investigations*; Public Health England: London, UK, 2014.
22. CLSI. *Performance Standards for Antimicrobial Disk Susceptibility Tests: Approved Standard, M02*, 11th ed.; CLSI: St. Louis, MO, USA, 2012; Volume 32, pp. 1–58.
23. Clinical and Laboratory Standards Institute. *M100. Performance Standards for Antimicrobial Susceptibility Testing*, 28th ed.; Clinical and Laboratory Standards Institute: St. Louis, MO, USA, 2018; pp. 30–220.
24. Magiorakos, A.-P.; Srinivasan, A.; Carey, R.B.; Carmeli, Y.; Falagas, M.E.; Giske, C.G.; Harbarth, S.; Hindler, J.F.; Kahlmeter, G.; Olsson-Liljequist, B.; et al. Multidrug-resistant, extensively drug-resistant and pandrug-resistant bacteria: An international expert proposal for interim standard definitions for acquired resistance. *Clin. Microbiol. Infect.* **2012**, *18*, 268–281. [CrossRef] [PubMed]
25. Bolger, A.M.; Lohse, M.; Usadel, B. Trimmomatic: A flexible trimmer for Illumina sequence data. *Bioinformatics* **2014**, *30*, 2114–2120. [CrossRef]
26. Diancourt, L.; Passet, V.; Verhoef, J.; Grimont, P.A.D.; Brisse, S. Multilocus sequence typing of *Klebsiella pneumoniae* nosocomial isolates. *J. Clin. Microbiol.* **2005**, *43*, 4178–4182. [CrossRef]
27. Wyres, K.L.; Holt, K.E. *Klebsiella pneumoniae* Population genomics and antimicrobial-resistant clones. *Trends Microbiol.* **2016**, *24*, 944–956. [CrossRef] [PubMed]
28. Argimón, S.; Yeats, C.A.; Goater, R.J.; Abudahab, K.; Taylor, B.; Underwood, A.; Sánchez-Busó, L.; Wong, V.K.; Dyson, Z.A.; Nair, S.; et al. A global resource for genomic predictions of antimicrobial resistance and surveillance of *Salmonella typhi* at pathogenwatch. *Nat. Commun.* **2021**, *12*, 2879. [CrossRef] [PubMed]
29. Letunic, I.; Bork, P. Interactive tree of life (iTOL) v5: An online tool for phylogenetic tree display and annotation. *Nucleic Acids Res.* **2021**, *49*, W293–W296. [CrossRef]
30. Irfan, M.; Almotiri, A.; AlZeyadi, Z.A. Antimicrobial resistance and its drivers—A review. *Antibiotics* **2022**, *11*, 1362. [CrossRef] [PubMed]
31. Tan, S.Y.; Khan, R.A.; Khalid, K.E.; Chong, C.W.; Bakhtiar, A. Correlation between antibiotic consumption and the occurrence of multidrug-resistant organisms in a Malaysian tertiary hospital: A 3-year observational study. *Sci. Rep.* **2022**, *12*, 3106. [CrossRef] [PubMed]
32. Walther, B.; Tedin, K.; Lübke-Becker, A. Multidrug-resistant opportunistic pathogens challenging veterinary infection control. *Vet. Microbiol.* **2017**, *200*, 71–78. [CrossRef]
33. Khan, R.; Petersen, F.C.; Shekhar, S. Commensal bacteria: An emerging player in defense against respiratory pathogens. *Front. Immunol.* **2019**, *10*, 1203. [CrossRef] [PubMed]

34. Shea, A.; Mccarthy, R.; Lindenmayer, J. Therapeutic antibiotic use patterns in dogs: Observations from a veterinary teaching hospital. *J. Small Anim. Pract.* **2011**, *52*, 310–318.
35. Almeida, A.; Duarte, S.; Nunes, R.; Rocha, H.; Pena, A.; Meisel, L. Human and veterinary antibiotics used in portugal—A ranking for ecosurveillance. *Toxics* **2014**, *2*, 188–225. [CrossRef]
36. Müller-Schulte, E.; Tuo, M.N.; Akoua-Koffi, C.; Schaumburg, F.; Becker, S.L. High prevalence of ESBL-producing *Klebsiella pneumoniae* in clinical samples from central Côte d'Ivoire. *Int. J. Infect. Dis.* **2020**, *91*, 207–209. [CrossRef]
37. Wyres, K.L.; Nguyen, T.N.T.; Lam, M.M.C.; Judd, L.M.; Van Vinh Chau, N.; Dance, D.A.B.; Ip, M.; Karkey, A.; Ling, C.L.; Miliya, T.; et al. Genomic surveillance for hypervirulence and multi-drug resistance in invasive *Klebsiella pneumoniae* from south and southeast Asia. *Genome Med.* **2020**, *12*, 11. [CrossRef] [PubMed]
38. Bachman, M.A.; Oyler, J.E.; Burns, S.H.; Caza, M.; Lépine, F.; Dozois, C.M.; Weiser, J.N. *Klebsiella pneumoniae* yersiniabactin promotes respiratory tract infection through evasion of lipocalin. *Infect. Immun.* **2011**, *79*, 3309–3316.
39. Farzand, R.; Rajakumar, K.; Barer, M.R.; Freestone, P.P.E.; Mukamolova, G.V.; Oggioni, M.R.; O'Hare, H.M. A virulence associated siderophore importer reduces antimicrobial susceptibility of *Klebsiella pneumoniae*. *Front. Microbiol.* **2021**, *12*, 607512. [CrossRef] [PubMed]
40. McCarthy, A.J.; Lindsay, J.A. The distribution of plasmids that carry virulence and resistance genes in *Staphylococcus aureus* is lineage associated. *BMC Microbiol.* **2012**, *12*, 104. [CrossRef]
41. Rodrigues, C.; Machado, E.; Ramos, H.; Peixe, L.; Novais, Â. Expansion of ESBL-producing *Klebsiella pneumoniae* in hospitalized patients: A successful story of international clones (ST15, ST147, ST336) and epidemic plasmids (IncR, IncFIIK). *Int. J. Med. Microbiol.* **2014**, *304*, 1100–1108. [CrossRef]
42. Morgado, S.; Fonseca, E.; Vicente, A.C. Genomics of *Klebsiella pneumoniae* species complex reveals the circulation of high-risk multidrug-resistant pandemic clones in human, animal, and environmental sources. *Microorganisms* **2022**, *10*, 2281. [CrossRef]
43. Novais, Â.; Ferraz, R.V.; Viana, M.; da Costa, P.M.; Peixe, L. NDM-1 Introduction in Portugal through a ST11 KL105 *Klebsiella pneumoniae* widespread in Europe. *Antibiotics* **2022**, *11*, 92. [CrossRef]
44. Mendes, G.; Ramalho, J.F.; Duarte, A.; Pedrosa, A.; Silva, A.C.; Méndez, L.; Caneiras, C. First outbreak of NDM-1-producing *Klebsiella pneumoniae* ST11 in a portuguese hospital centre during the COVID-19 pandemic. *Microorganisms* **2022**, *10*, 251. [CrossRef]
45. Rocha, J.; Henriques, I.; Gomila, M.; Manaia, C.M. Common and distinctive genomic features of *Klebsiella pneumoniae* thriving in the natural environment or in clinical settings. *Sci. Rep.* **2022**, *12*, 10441. [CrossRef]
46. Andrade, L.N.; Curiao, T.; Ferreira, J.C.; Longo, J.M.; Clímaco, E.C.; Martinez, R.; Bellissimo-Rodrigues, F.; Basile-Filho, A.; Evaristo, M.A.; Del Peloso, P.F.; et al. Dissemination of blaKPC-2 by the spread of *Klebsiella pneumoniae* clonal complex 258 clones (ST258, ST11, ST437) and plasmids (IncFII, IncN, IncL/M) among Enterobacteriaceae species in Brazil. *Antimicrob. Agents Chemother.* **2011**, *55*, 3579–3583. [CrossRef]
47. Tan, D.; Zhang, Y.; Cheng, M.; Le, S.; Gu, J.; Bao, J.; Qin, J.; Guo, X.; Zhu, T. Characterization of *Klebsiella pneumoniae* ST11 isolates and their interactions with lytic phages. *Viruses* **2019**, *11*, 1080. [CrossRef]
48. Lam, M.M.C.; Wick, R.R.; Wyres, K.L.; Holt, K.E. Genomic surveillance framework and global population structure for *Klebsiella pneumoniae*. *bioRxiv* **2020**. [CrossRef]
49. Borrego, B. *Demographics of Immigration in Portugal*; European Commission: Brussels, Belgium, 2016.
50. Chow, J.W.; Kak, V.; You, I.; Kao, S.J.; Petrin, J.; Clewell, D.B.; Lerner, S.A.; Miller, G.H.; Shaw, K.J. Aminoglycoside resistance genes *aph(2′)-Ib* and *aac(6′)-Im* detected together in strains of both *Escherichia coli* and *Enterococcus faecium*. *Antimicrob. Agents Chemother.* **2001**, *45*, 2691–2694. [CrossRef] [PubMed]
51. Souza-Silva, T.; Rossi, C.C.; Andrade-Oliveira, A.L.; Vilar, L.C.; Pereira, M.F.; de Araújo Penna, B.; Giambiagi-deMarval, M. Interspecies transfer of plasmid-borne gentamicin resistance between *Staphylococcus* isolated from domestic dogs to *Staphylococcus aureus*. *Infect. Genet. Evol.* **2022**, *98*, 105230. [CrossRef] [PubMed]
52. Simner, P.J.; Antar, A.A.R.; Hao, S.; Gurtowski, J.; Tamma, P.D.; Rock, C.; Opene, B.N.A.; Tekle, T.; Carroll, K.C.; Schatz, M.C.; et al. Antibiotic pressure on the acquisition and loss of antibiotic resistance genes in *Klebsiella pneumoniae*. *J. Antimicrob. Chemother.* **2018**, *73*, 1796–1803. [CrossRef]
53. Yamba, K.; Kapesa, C.; Mpabalwani, E.; Hachaambwa, L.; Smith, A.M.; Young, A.L.; Gally, D.; Mainda, G.; Mukuma, M.; Samutela, M.T.; et al. Antimicrobial susceptibility and genomic profiling of *Salmonella enterica* from bloodstream infections at a tertiary referral hospital in Lusaka, Zambia, 2018–2019. *IJID Reg.* **2022**, *3*, 248–255. [CrossRef]
54. Papp, M.; Solymosi, N. Review and comparison of antimicrobial resistance gene databases. *Antibiotics* **2022**, *11*, 339. [CrossRef]
55. Asante, J.; Hetsa, B.A.; Amoako, D.G.; Abia, A.L.K.; Bester, L.A.; Essack, S.Y. Genomic analysis of antibiotic-resistant *Staphylococcus epidermidis* isolates from clinical sources in the Kwazulu-Natal Province, South Africa. *Front. Microbiol.* **2021**, *12*, 656306. [CrossRef]
56. Carroll, K.C.; Burnham, C.A.D.; Westblade, L.F. From canines to humans: Clinical importance of *Staphylococcus pseudintermedius*. *PLoS Pathog.* **2021**, *17*, e1009961. [CrossRef]

57. Wegener, A.; Broens, E.M.; van der Graaf-Van Bloois, L.; Zomer, A.L.; Visser, C.E.; van Zeijl, J.; van der Meer, C.; Kusters, J.G.; Friedrich, A.W.; Kampinga, G.A.; et al. Absence of host-specific genes in canine and human *Staphylococcus pseudintermedius* as inferred from comparative genomics. *Antibiotics* **2021**, *10*, 854. [CrossRef]
58. Larsen, J.; Clasen, J.; Hansen, J.E.; Paulander, W.; Petersen, A.; Larsen, A.R.; Frees, D. Copresence of *tet(K)* and *tet(M)* in livestock-associated methicillin-resistant *Staphylococcus aureus* clonal complex 398 is associated with increased fitness during exposure to sublethal concentrations of tetracycline. *Antimicrob. Agents Chemother.* **2016**, *60*, 4401–4403. [CrossRef] [PubMed]

veterinary sciences

MDPI

Article

Multidrug-Resistant Commensal and Infection-Causing *Staphylococcus* spp. Isolated from Companion Animals in the Valencia Region

Ana Marco-Fuertes [1], Clara Marin [1], Concepción Gimeno-Cardona [2], Violeta Artal-Muñoz [3], Santiago Vega [1,*] and Laura Montoro-Dasi [1,*]

1 Departamento de Producción y Sanidad Animal, Salud Pública Veterinaria y Ciencia y Tecnología de los Alimentos, Facultad de Veterinaria, Instituto de Ciencias Biomédicas, Universidad Cardenal Herrera-CEU, CEU Universities, Calle Santiago Ramón y Cajal 20, Alfara del Patriarca, 46115 Valencia, Spain; ana.marcofuertes@uchceu.es (A.M.-F.); clara.marin@uchceu.es (C.M.)

2 Servicio de Microbiología, Consorcio Hospital General Universitario de Valencia, Facultad de Medicina, Universidad de Valencia, 46014 Valencia, Spain; concepcion.gimeno@uv.es

3 Servicio de Microbiología, Consorcio Hospital General Universitario de Valencia, 46014 Valencia, Spain; artal_vio@gva.es

* Correspondence: svega@uchceu.es (S.V.); laura.montoro@uchceu.es (L.M.-D.)

Simple Summary: The increase in microorganisms resistant to antimicrobials poses a growing risk to the effectiveness of medical treatments, both in humans and animals. This surveillance is essential to understand and address the magnitude of the problem and its impact on public health. Therefore, it is crucial to monitor antimicrobial resistance not only in human medicine but also in veterinary medicine. Companion animals, in particular, play a significant role as they live in close contact with their owners, potentially facilitating the transmission of these antimicrobial resistance between people and animals. Thus, this study aimed to evaluate the epidemiological situation of antimicrobial resistance in dogs and cats to the opportunistic pathogen *Staphylococcus* spp. The main results showed a high prevalence of antimicrobial resistance in the study population (healthy and diseased dogs and cats), even to the last resort of antibiotics in human medicine, which poses a threat to global public health.

Abstract: The emergence of antimicrobial resistance (AMR) and multidrug resistance (MDR) among microorganisms to commonly used antibiotics is a growing concern in both human and veterinary medicine. Companion animals play a significant role in the epidemiology of AMR, as their population is continuously increasing, posing a risk of disseminating AMR, particularly to strains of public health importance, such as methicillin-resistant *Staphylococcus* strains. Thus, this study aimed to investigate the prevalence of AMR and MDR in commensal and infection-causing *Staphylococcus* spp. in dogs and cats in Valencia region. For this purpose, 271 samples were taken from veterinary centers to assess antimicrobial susceptibility against 20 antibiotics, including some of the most important antibiotics for the treatment of *Staphylococcus* infections, including the five last resort antibiotics in this list. Of all the samples, 187 *Staphylococcus* spp. strains were recovered from asymptomatic and skin-diseased dogs and cats, of which *S. pseudintermedius* (≈60%) was more prevalent in dogs, while *S. felis* (≈50%) was more prevalent in cats. In the overall analysis of the isolates, AMR was observed for all antibiotics tested, including those crucial in human medicine. Furthermore, over 70% and 30% of the strains in dogs and cats, respectively, exhibited MDR. This study highlights the significance of monitoring the trends in AMR and MDR among companion animals. The potential contribution of these animals to the dissemination of AMR and its resistance genes to humans, other animals, and their shared environment underscores the necessity for adopting a One Health approach.

Keywords: companion animals; antimicrobial resistance; *Staphylococcus*; methicillin-resistant *Staphylococcus aureus*; methicillin-resistant *Staphylococcus pseudintermedius*

check for updates

Citation: Marco-Fuertes, A.; Marin, C.; Gimeno-Cardona, C.; Artal-Muñoz, V.; Vega, S.; Montoro-Dasi, L. Multidrug-Resistant Commensal and Infection-Causing *Staphylococcus* spp. Isolated from Companion Animals in the Valencia Region. *Vet. Sci.* **2024**, *11*, 54. https://doi.org/10.3390/vetsci11020054

Academic Editor: George Valiakos

Received: 12 December 2023
Revised: 10 January 2024
Accepted: 24 January 2024
Published: 26 January 2024

1. Introduction

In an ever-changing society, companion animals are increasingly living in close contact with their owners in their homes, but they also share public spaces, such as parks or beaches, with animals (domestic or wild) and other people, including the elderly, children, or immunosuppressed patients [1]. In fact, the population of companion animals continues to grow in importance and members in European households (230 million dogs and cats) [2,3].

In this context, new challenges have arisen because zoonotic pathogens, multi-resistant bacteria, and their resistance genes can be spread and acquired through the environment they share [4,5]. Among these hazards, AMR and the emergence of multidrug resistance (MDR) are one of the most important problems facing public health, according to the World Health Organisation (WHO) [6]. In fact, the study conducted in 2019 revealed that there were 1.27 million deaths directly caused by AMR per year [7].

Over time, the trends in AMR have evolved, conditioned by the implementation of new regulations focused on controlling past overuse of antibiotics in both human and animal health [8,9]. As a result of these efforts, including surveillance and monitoring programmes and the European Medicines Agency (EMA) categorisation for the responsible use of antibiotics in veterinary medicine [10], the use of antibiotics in animal production has decreased. However, companion animals have not been included in all of these control measures. Within this framework, the European Union (EU) is developing the European Antimicrobial Resistance Surveillance network in veterinary medicine (EARS-Vet), which aims to monitor AMR in the main pathogens affecting companion animals (dogs and cats) together with food-producing animals [11,12], to complement the existing European AMR Surveillance Network (EARS-Net) in human medicine [13] in order to achieve a global view of this problem under the "One Health" strategy. Nevertheless, there is a lack of studies evaluating the epidemiological situation of AMR in companion animals, although it is necessary to establish a starting point [5].

To study the epidemiology of AMR, *Escherichia coli* has been the main sentinel bacterium used, due to its ability to acquire and transfer AMR genes, as it is a commensal bacterium that is part of the microbiota of animals and humans [14,15]. However, it is necessary to research the AMR problem from more perspectives. For this reason, Gram-positive bacteria belonging to the family *Staphylococcaceae*, which are considered part of the commensal microbiota of the skin and mucous membranes of animals and humans, are also used as an indicator of resistance [16]. Within this family, two groups are distinguished: coagulase-positive *Staphylococcus* (CoPS) [17] and coagulase-negative *Staphylococcus* (CoNS) [18]. Most CoPS are opportunistic pathogens and cause the majority of infections at the dermal level in humans and animals [19]. They are known to acquire resistant genes to a large extent, so treatment options against these bacteria are limited, making infections difficult to treat, especially those caused by methicillin-resistant *Staphylococcus aureus* (MRSA) and methicillin-resistant *Staphylococcus pseudintermedius* (MRSP) [20–22]. Regarding CoNS, although they are not as common in causing infections, they are widely recognised as commensal organisms of the skin microbiota and opportunistic pathogens of humans and animals [18,23]. Therefore, the aim of this study was to establish the presence of commensal vs. infection-causing *Staphylococcus* spp. and the epidemiological situation of their AMR and MDR in companion animals (dogs and cats) in the Valencia region.

2. Materials and Methods

2.1. Experimental Design

The animal sampling procedure was evaluated and authorised by the Animal Ethics Committees of UCH-CEU University (permit N°. CEEA 22/04).

Veterinary hospitals (VHs) and clinics (VCs) located across the Valencia region were invited to take part voluntarily in this study. Out of these, eight veterinary centers voluntarily consented to collaborate: three large reference VHs, handling cases from the entire Valencia region, and five VCs, spread throughout the Valencia region.

2.2. Epidemiological Data Collection

First, an epidemiological questionnaire for each animal was completed, together with the informed consent signed by the owners (Supplementary Materials, Part A), in order to classify the animals depending on their epidemiologic characteristics and be able to evaluate their effect on the appearance of AMR and MDR. The information collected was related to the origin of the animals (VH vs. VC) and general information such as sex and age. Regarding the age of the animals, a general classification described by Marco-Fuertes et al. (2023) was used to group dogs and cats [24]. Moreover, whether they cohabit with other animals and the clinical data of each animal were included (chronic diseases, daily medication, and antibiotic treatment received). Finally, the data regarding dogs and cats were analysed individually.

2.3. Sample Collection

Dogs and cats were sampled between October 2022 and June 2023 in order to isolate *Staphylococcus* spp. To isolate commensal *Staphylococcus*, a single swab (Cary–Blair sterile transport swabs, DELTALAB, Barcelona, Spain) was first introduced in the nasal cavity and then in the auricular cavity, approximately 3 cm [25,26], from healthy asymptomatic dogs and cats. Before taking the samples, the veterinarians performed a clinical examination in which they assessed the animals' vital signs to confirm that they were within normal ranges, thus classifying them as asymptomatic healthy animals. For the isolation of infection-causing *Staphylococcus*, animals with active skin infections were sampled by taking a Cary–Blair sterile transport swabs, which were then introduced into skin-infected wounds. After collecting the samples, all of them were preserved in Cary–Blair transport medium and transported under refrigeration at ≤4 °C to the microbiology laboratory within 24 h of sampling to the Faculty of Veterinary Sciences of the University CEU Cardenal Herrera.

2.4. Staphylococcus Isolation

A pre-enrichment in buffered peptone water (BPW; Scharlau, Barcelona, Spain), at a ratio of 1:10 v/v, of the sample swabs collected were carried out, followed by an incubation at 37 ± 1 °C for 24 h. After that, the suspension was streaked onto the non-specific agar Columbia CNA agar with 5% sheep blood, Improved II (BD, Becton Dickinson, Madrid, Spain), and incubated at 37 ± 1 °C for 24–48 h. The plates were examined at 24 and 48 h, and the suspected colonies, matching the typical morphology of *Staphylococcus* spp. in blood agar and the positive result of the catalase test, were identified using a MALDI-TOF MS Biotyper System (Bruker Daltonics, Madrid, Spain) at the Microbiology Service of the *Consorcio Hospital General Universitario de Valencia*.

2.5. Antimicrobial Susceptibility Testing

Antimicrobial susceptibility was evaluated using the minimum inhibition concentration (MIC) assay (Thermo Scientific™ Sensititre™ Plates, Madrid, Spain) using a panel of 20 antibiotics applied in human medicine and of importance in public health (Table 1) [11]. In addition, the plate presented two D-test wells. D-test wells combine two antibiotics (clindamycin (CLI) and erythromycin (ERY)), indicating whether the strain tested has inducible resistance to CLI in the presence of ERY and may therefore lead to therapeutic failure. The interpretation was carried according to the Spanish Society of Infectious Diseases and Clinical Microbiology (*SEIMC*, from its Spanish acronym *Sociedad Española de Enfermedades Infecciones y Microbiología Clínica*) [27].

Table 1. Antibiotics, World Health Organisation classification, and their concentrations included in the Sensititre plate for Gram-positive bacteria GPALL1F (Thermo Scientific™ Sensititre™, Madrid, Spain).

Antibiotic Group	Antibiotic	Abbreviation	WHO	Concentration
Aminoglycosides	Gentamicin	GEN	CIA	2–16 μg/mL
Amphenicols	Chloramphenicol	CHL	HIA	2–16 μg/mL
Cephalosporins	Cefoxitin [1]	CXI	HIA	6 μg/mL
Folate inhibitor pathway	Trimethoprim/ sulfamethoxazole	TRS	HIA	1/19–8/152 μg/mL
Glycopeptides	Vancomycin	VAN	NA	0.25–32 μg/mL
Glycylcyclines	Tigecycline	TIG	NA	0.03–0.5 μg/mL
Lincosamides	Clindamycin	CLI	HIA	0.5–2 μg/mL
Lipopeptides	Daptomycin	DAP	NA	0.5–4 μg/mL
Macrolides	Erythromycin	ERY	CIA	0.25–4 μg/mL
Nitrofurans	Nitrofurantoin	NIT	NA	32–64 μg/mL
Oxazolidinones	Linezolid	LIN	NA	1–8 μg/mL
Penicillins	Ampicillin	AMP	HIA	0.25–8 μg/mL
	Oxacillin + 2% NaCl [1]	OXA+	HIA	0.25–4 μg/mL
	Penicillin	PEN	HIA	0.06–8 μg/mL
Quinolones	Levofloxacin (FQ)	LEV	HPCIA	0,25–4 μg/mL
	Ciprofloxacin (FQ)	CIP	HPCIA	1–2 μg/mL
	Moxifloxacin (FQ)	MOX	HPCIA	0.25–4 μg/mL
Tetracyclines	Tetracycline	TET	HIA	2–16 μg/mL
Ansamycins	Rifampicin	RIF	CIA	0.5–4 μg/mL
Streptogramins	Quinupristin/ dalfopristin	QUD	HIA	0.5–4 μg/mL
D-test	Erythromycin (E) + clindamycin (C)	DT		4 μg/mL (E) + 0.5 μg/mL (C)

FQ: fluoroquinolone. [1]: cefoxitin and oxacillin + 2% NaCl are two antibiotics used to screen methicillin-resistant *Staphylococcus* strains. WHO: World Health Organisation (this column indicates the last update of the classification of medically important antimicrobials authorised by the WHO for human and animal use in order to protect public health, updated in 2023 [28]). HIA: highly important antimicrobial. CIA: critically important antimicrobial. HPCIA: highest priority critical important antimicrobial. NA: not authorised for animal use.

Each bacterial strain was cultured and revived on nutrient agar and then incubated at 37 ± 1 °C for 24 h. After the incubation period, the colonies were transferred into 5 mL of sterile demineralised water (T3339; Thermo Fisher Scientific™, Madrid, Spain). The suspension of each bacterium was mixed and adjusted to achieve a 0.5 McFarland using a nephelometer (Sensititre™ Nephelometer, Thermo Fisher Scientific™, Madrid, Spain). Subsequently, 10 μL of the suspension were introduced into a vial containing 11 mL of Mueller–Hinton broth (T3462; Thermo Fisher Scientific™, Madrid, Spain) and mixed. From this suspension, 50 μL of the vial contents were transferred into each Sensititre plate well (GPALL1F, Thermo Fisher Scientific™, Madrid, Spain). Then, the plates were incubated at 37 ± 1 °C for 24 h and manually examined using a Sensititre Vizion (Thermo Scientific™ Sensititre™ Vizion™ Digital MIC Viewing System, Thermo Fisher Scientific, Madrid, Spain).

Finally, the results were interpreted following the guidelines established by the European Committee on Antimicrobial Susceptibility Testing (EUCAST) in its last report (14th ed., 2024) [29]. Methicillin-resistant *Staphylococcus* (MRS) strains were studied by monitoring the AMR observed against oxacillin + 2% NaCl for *S. pseudintermedius* (the antibiotic used for screening MRSP strains) and against cefoxitin for *S. aureus* and CoNS (the antibiotic used for screening MRSA and MR-CoNS strains). However, some MIC values for these two antibiotics for screening MRSP and methicillin-resistant coagulase-negative *Staphylococcus* (MR-CoNS) are not currently available in the EUCAST, so the Clinical and Laboratory Standards Institute (CLSI) recommendations specified in M100 [30] and VET01 [31] were followed. Moreover, MDR was characterized as the acquired resistance to at least one agent in three or more antimicrobial classes [32]. Finally, according to the EARS-Vet [11], the following detailed results are those obtained for *S. aureus* and *S. pseudintermedius*, while the rest of the information detailed on the AMR observed for each of the isolated species can be found in the Supplementary Materials, Table S1.

2.6. Statistical Analysis

A generalised linear model (GLM) with a probit link function, assuming a binomial distribution, was applied to the data to examine the influence of external factors on AMR and MDR patterns. This analysis aimed to determine associations with categorical variables such as animal origin, sex, cohabitation with other animals (and number of animals, if applicable), relationship with animals outside the household, and clinical information regarding chronic diseases, daily medication, and previous antibiotic treatments. In addition, a probit link function GLM was performed, assuming a binomial distribution for AMR patterns in *Staphylococcus* spp. from dogs and cats, for the microbiological results. A *p*-value of ≤ 0.05 was considered indicative of a statistically significant difference. Data were presented as the least squares means $\pm$ standard error of least squares means. Statistical analyses were performed using the R software (version 4.3.1) packages EMMs [33], car [34], and multicompView [35].

3. Results

3.1. Epidemiological Data

Among the sampled population (n = 271), there were 152 dogs and 119 cats. Regarding the samples' origin, 43.9% of the samples were taken in VHs (79/152 and 40/119, dogs and cats, respectively), and 56.1% of the samples were collected in VCs (73/152 and 70/119, dogs and cats, respectively).

As reported in the Materials and Methods section, an epidemiological survey for each animal was collected. Figure 1 compiles all the information collected in the questionnaire regarding the dog samples, while Figure 2 compiles all the information related to the cat samples.

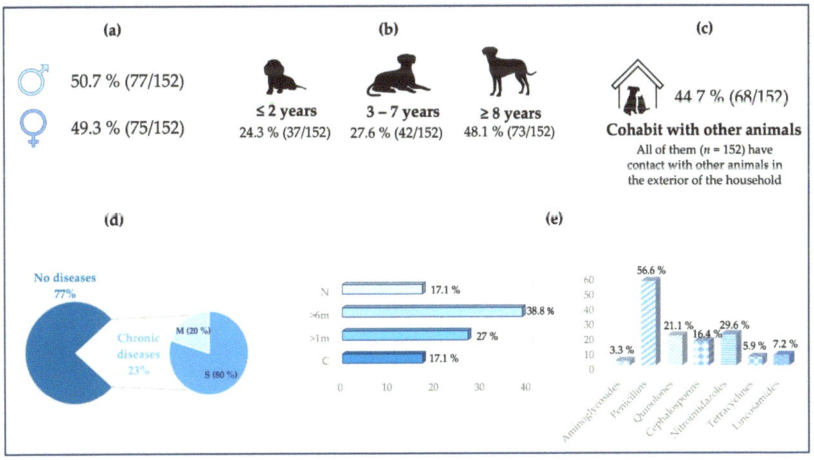

Figure 1. Epidemiological data for all the dogs sampled. (**a**) Distribution of the study population by sex. (**b**) Distribution of the study population by age. (**c**) The relationship of the animals in the study population with other animals. n: total number of animals. (**d**) Whether the animals of the study population present any disease and of which type. M: musculoskeletal. S: systemic. (**e**) Previous antibiotic therapy (**left** graph) and antibiotics administered at some point in their lives (**right** graph). N: never. >6 m: in the last six months. >1 m: in the last month. C: currently.

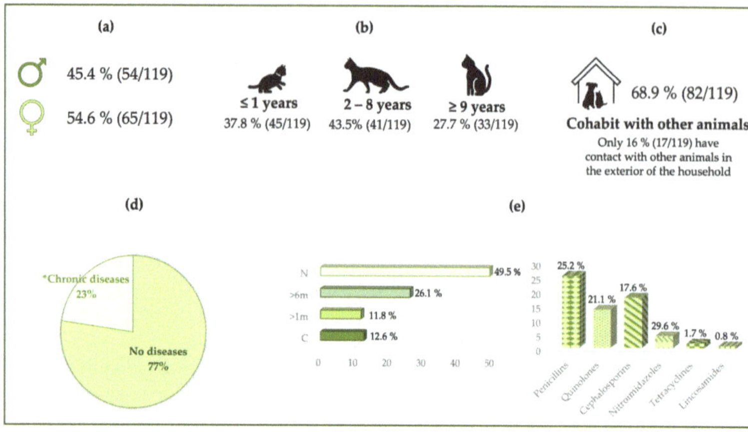

Figure 2. Epidemiological data for all the cats sampled. (**a**) Distribution of the study population by sex. (**b**) Distribution of the study population by age. (**c**) The relationship of the animals in the study population with other animals. (**d**) Whether the animals of the study population present any disease. *Chronic diseases: all of them are classified as systemic diseases. (**e**) Previous antibiotic therapy (**left** graph) and antibiotics administered at some point in their lives (**right** graph). N: never. >6 m: in the last six months. >1 m: in the last month. C: currently.

3.2. *Staphylococcus Prevalence*

The prevalence of *Staphylococcus* spp. from all the samples taken, including dogs and cats, was 69% (187/271).

From all the canine samples collected, the prevalence of *Staphylococcus* was 74.3% (113/152), of which 74.3% (84/113) and 25.7% (29/113) were commensal and infection-causing *Staphylococcus*, respectively.

Regarding the samples collected from cats, the prevalence of this bacterium was 62.2% (74/119). About the prevalence of this bacterium according to the type of sample, 87.8% (65/74) were commensal *Staphylococcus*, and 12.2 % (9/74) were infection-causing *Staphylococcus*. All *Staphylococcus* species isolated from dogs and cats and the type of sample from which they are derived are detailed in Table 2.

Table 2. Prevalence of *Staphylococcus* species isolated from commensal mucosa and active skin infection samples from dogs and cats identified using a MALDI-TOF MS Biotyper System (Bruker Daltonics, Madrid, Spain).

	Type of Sample	Prevalence of *Staphylococcus* by Class	*Staphylococcus* Species	*n* and (%) Prevalence of Each Species
Dog	Commensal mucosa	CoPS—79.8%	*S. aureus*	6 (7.1)
			S. pseudintermedius	54 (64.2)
			S. schleiferi	7 (8.3)
		CoNS—20.2%	*S. cohnii*	1 (1.2)
			S. epidermidis	4 (4.8)
			S. haemolyticus	1 (1.2)
			S. hominis	3 (3.6)
			S. sciuri [1]	2 (2.4)
			S. simulans	2 (2.4)
			S. warneri	2 (2.4)
			S. xylosus	2 (2.4)
	Active skin infection	CoPS—82.8%	*S. aureus*	4 (13.8)
			S. pseudintermedius	18 (62.2)
			S. schleiferi	2 (6.9)
		CoNS—17.2%	*S. canis*	1 (3.4)
			S. chromogenes	1 (3.4)
			S. epidermidis	2 (6.9)
			S. felis	1 (3.4)

Table 2. *Cont.*

	Type of Sample	Prevalence of *Staphylococcus* by Class	*Staphylococcus* Species	*n* and (%) Prevalence of Each Species
Cat	Commensal mucosa	CoPS—16.9%	*S. aureus*	6 (9.2)
			S. pseudintermedius	3 (4.6)
			S. schleiferi	2 (3.1)
		CoNS—83.1%	*S. capitis*	2 (3.1)
			S. epidermidis	2 (3.1)
			S. felis	32 (49.2)
			S. hominis	1 (1.5)
			S. pettenkoferi	2 (3.1)
			S. saprophyticus	1 (1.5)
			S. sciuri [1]	4 (6.2)
			S. simulans	6 (9.2)
			S. xylosus	4 (6.2)
	Active skin infection	CoPS—11.1%	*S. aureus*	1 (11.1)
		CoNS—88.9%	*S. epidermidis*	1 (11.1)
			S. felis	5 (55.6)
			S. hominis	1 (11.1)
			S. pasteuri	1 (11.1)

CoPS: coagulase-positive *Staphylococcus*. CoNS: coagulase-negative *Staphylococcus*. *n*: number of isolated strains. [1]: *S. sciuri* is still identified as such in all identification databases but now belongs to a new genus due to new phylogenomic studies named *Mammaliicoccus sciuri*.

3.3. Antimicrobial Susceptibility in Staphylococcus Strains

3.3.1. Methicillin Resistance

In all the *Staphylococcus* strains, 30.5% (57/187) were MRS, of which 71.9% (41/57) belonged to dogs and 28.1% (16/57) belonged to cats. All the results regarding the sampled animals (dogs or cats), the strain species, and the strain's origin (commensal or infection) are represented in Figure 3.

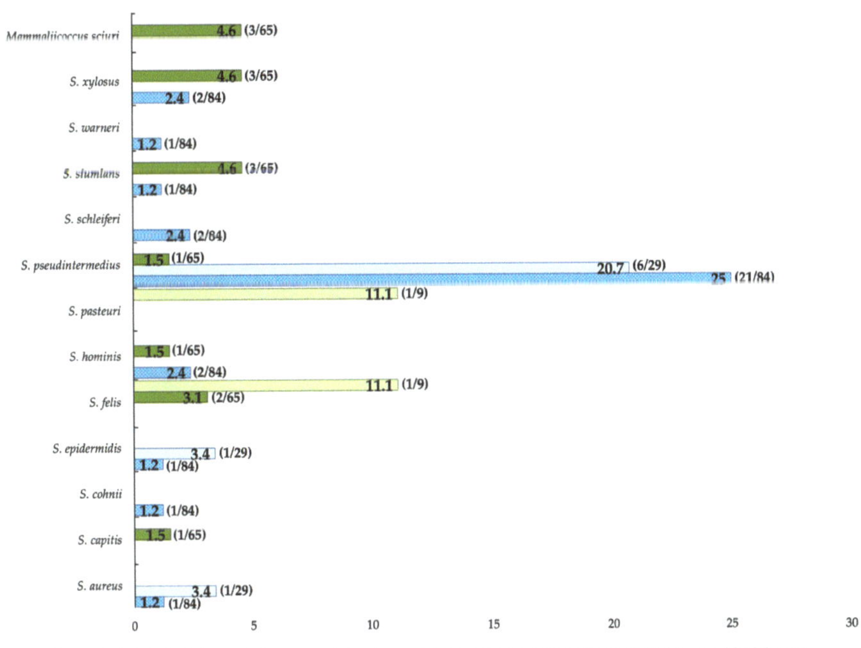

Figure 3. Percentage and number of methicillin-resistant *Staphylococcus* strains classified by *Staphylococcus* species according to their observed phenotypic resistance to oxacillin + 2% NaCl and cefoxitin (the two antibiotics used to screen methicillin-resistant *Staphylococcus* strains).

3.3.2. Dogs

Of all the commensal *Staphylococcus* isolated from healthy asymptomatic dogs, 95.2% (80/84) showed AMR to at least 1 of the 20 antibiotics studied, and 72.6% (61/84) were considered MDR, while only 4.8% (4/84) of the strains were sensitive to all the antibiotics studied. Regarding the AMR of these strains for the different antibiotic groups evaluated, they are ordered from the highest to the lowest percentage: 59.5% for amphenicols, 56% for macrolides, 54.4% for penicillins, 47.6% for glycylcyclines, 46.4% for tetracyclines, 42.9% for lincosamides, 40.5% for cephalosporins, 28.6% for oxazolidinones, 28.6% for quinolones, 11.9% for aminoglycosides, 10.7% for folate inhibitor pathway, 9.5% for streptogramins, 7.1% for ansamycins, 3.6% for glycopeptides, 3.6% for nitrofurans, and 2.4% for lipopeptides. In the D-test we performed, 35.7% (30/84) of the strains were positive.

On the other hand, only 10.3% (3/29) of the infection-causing *Staphylococcus* isolated from animals with active skin infections were sensitive to all the antibiotics tested, while 89.7% (26/29) of the strains presented AMR and 55.2% (16/29) were MDR. The AMR in each antibiotic group were (from the highest percentage to the lowest percentage) as follows: 51.7% for macrolides, 49.4% for penicillins, 44.8% for amphenicols, 44.8% for lincosamides, 41.4% for tetracyclines, 26.4% for quinolones, 24.1% for cephalosporins, 20.7% for aminoglycosides, 17.2% for folate inhibitor pathway, 10.3% for oxazolidinones, 6.9% for glycylcyclines, 6.9% for streptogramins, and 3.4% for glycopeptides, and no resistance was found for the lipopeptides, nitrofurans, and ansamycins. Concerning the D-test results, 41.4% (12/29) of the strains tested positive.

In addition, no correlation was observed between the clinical data collected in the questionnaire and the appearance of AMR and MDR (p-value > 0.05).

Of all the strains isolated, *S. aureus* and *S. pseudintermedius* were the main strains with importance in public health. Therefore, their AMR levels are detailed in Tables 3 and 4.

Table 3. Antimicrobial resistance in commensal and infection-causing *Staphylococcus aureus* isolated from healthy dogs and dogs with an active skin infection.

AB Group	AB	% AMR/AB in Commensal *S. aureus*	% AMR/AB in Infection-Causing *S. aureus*
Aminoglycosides	GEN	16.7 ^{a,b} (1/6) ± 15.2	0 ^{a} (0/4) ± 0
Amphenicols	CHL	83.3 ^{c} (5/6) ± 15.2	75 ^{b} (3/4) ± 21.7
Cephalosporins	CXI	16.7 ^{a,b} (1/6) ± 15.2	25 ^{a,b} (1/4) ± 21.7
Folate inhibitor pathway	TRS	0 ^{b} (0/6) ± 0	0 ^{a} (0/4) ± 0
Glycopeptides	VAN	50 ^{a,c} (3/6) ± 20.4	0 ^{a} (0/4) ± 0
Glycylcyclines	TIG	16.7 ^{a,b} (1/6) ± 15.2	25 ^{a,b} (1/4) ± 21.7
Lincosamides	CLI	50 ^{a,c} (3/6) ± 20.4	75 ^{b} (3/4) ± 21.7
Lipopeptides	DAP	0 ^{b} (0/6) ± 0	0 ^{a} (0/4) ± 0
Macrolides	ERY	83.3 ^{c} (5/6) ± 15.2	25 ^{a,b} (1/4) ± 25
Nitrofurans	NIT	0 ^{b} (0/6) ± 0	0 ^{a} (0/4) ± 0
Oxazolidinones	LIN	83.3 ^{c} (5/6) ± 15.2	25 ^{a,b} (1/4) ± 21.7
Penicillins	AMP	50 ^{a,c} (3/6) ± 20.4	50 ^{a,b} (2/4) ± 25
	PEN	83.3 ^{c} (5/6) ± 15.2	75 ^{b} (3/4) ± 21.7
Quinolones	LEV	16.7 ^{a,b} (1/6) ± 15.2	0 ^{a} (0/4) ± 0
	CIP	0 ^{b} (0/6) ± 0	0 ^{a} (0/4) ± 0
	MOX	16.7 ^{a,b} (1/6) ± 15.2	0 ^{a} (0/4) ± 0
Tetracyclines	TET	33.3 ^{a,b} (2/6) ± 19.2	25 ^{a,b} (1/4) ± 21.7
Ansamycins	RIF	16.7 ^{a,b} (1/6) ± 15.2	0 ^{a} (0/4) ± 0
Streptogramins	QUD	0 ^{b} (0/6) ± 0	25 ^{a,b} (1/4) ± 21.7

AB: antibiotic. AMR: antimicrobial resistance. GEN: gentamicin. CHL: chloramphenicol. CXI: cefoxitin. TRA: trimethoprim–sulfamethoxazole. TIG: tigecycline. CLI: clindamycin. DAP: daptomycin. ERY: erythromycin. NIT: nitrofurantoin. LIN: linezolid. AMP: ampicillin. penicillin. LEV: levofloxacin. CIP: ciprofloxacin. MOX: marbofloxacin. TET: tetracycline. RIF: rifampicin. QUD: quinupristin/dalfopristin. $^{a–c}$: the different superscripts in each column denote statistically significant variations (p-value ≤ 0.05) in the observed resistance to the antibiotics examined. ±: standard error.

Table 4. Antimicrobial resistance in commensal and infection-causing *Staphylococcus pseudintermedius* isolated from healthy dogs and dogs with an active skin infection.

AB Group	AB	% AMR/AB in Commensal *S. pseudintermedius*	% AMR/AB in Infection-Causing *S. pseudintermedius*
Aminoglycosides	GEN	11.1 [a] (10/54) ± 4.3	27.8 [a,b,c,d] (5/18) ± 10.6
Amphenicols	CHL	68.5 [e,f] (37/54) ± 6.3	50 [a,c,e] (9/18) ± 11.8
Folate inhibitor pathway	TRS	13 [a,i] (7/54) ± 4.6	22.2 [b,c,d] (4/18) ± 9.8
Glycopeptides	VAN	0 [g] (0/54) ± 0	5.6 [b,g] (1/18) ± 5.4
Glycylcyclines	TIG	46.3 [c,d] (25/54) ± 6.8	5.6 [b,g] (1/18) ± 5.4
Lincosamides	CLI	46.3 [c,d] (25/54) ± 6.8	50 [a,c,e] (9/18) ± 11.8
Lipopeptides	DAP	1.9 [g,h] (1/54) ± 1.8	0 [g] (0/18) ± 0
Macrolides	ERY	57.4 [c,e] (31/54) ± 6.7	55.6 [a,e,f] (10/18) ± 11.7
Nitrofurans	NIT	3.7 [a,g,h] (2/54) ± 2.6	0 [g] (0/18) ± 0
Oxazolidinones	LIN	25.9 [i,j] (14/54) ± 6	11.1 [b,d,g] (2/18) ± 7.4
Penicillins	AMP	44.4 [b,c,d] (27/54) ± 6.8	66.6 [e,f] (12/18) ± 11.1
	OXA+	37 [b,d,j] (21/54) ± 6.6	33.3 [a,c,d] (6/18) ± 11.1
	PEN	77.8 [f] (41/54) ± 5.7	83.3 [f] (15/18) ± 8.8
Quinolones	LEV	42.6 [b,c,d,j] (23/54) ± 6.7	44.4 [a,c,e] (8/18) ± 11.7
	CIP	0 [g] (0/54) ± 0	38.9 [a,c,e] (7/18) ± 11.5
	MOX	42.6 [b,c,d,j] (23/54) ± 6.7	33.3 [a,c,d] (6/18) ± 11.1
Tetracyclines	TET	51.9 [c,d,e] (28/54) ± 6.8	50 [a,c,e] (9/18) ± 11.8
Ansamycins	RIF	3.7 [a,g,h] (2/54) ± 2.6	0 [g] (0/18) ± 0
Streptogramins	QUD	7.4 [a,h] (3/54) ± 3.6	5.6 [b,g] (1/18) ± 5.4

AB: antibiotic. AMR: antimicrobial resistance. GEN: gentamicin. CHL: chloramphenicol. TRA: trimethoprim–sulfamethoxazole. TIG: tigecycline. CLI: clindamycin. DAP: daptomycin. ERY: erythromycin. NIT: nitrofurantoin. LIN: linezolid. AMP: ampicillin. OXA+: oxacillin + 2% NaCl. PEN: penicillin. LEV: levofloxacin. CIP: ciprofloxacin. MOX: marbofloxacin. TET: tetracycline. RIF: rifampicin. QUD: quinupristin/dalfopristin. [a–j]: each superscript in each column signify statistically significant differences (p-value $\leq$ 0.05) in the resistance observed against the various antibiotics investigated. ±: standard error.

3.3.3. Cats

Regarding all the commensal *Staphylococcus* strains isolated from healthy asymptomatic cats, 21.5% (14/65) were susceptible to the 20 antibiotics studied, while 75.4% (49/65) showed AMR to at least one of the antibiotics studied, and 32.3% (21/65) were MDR. In addition, the AMR observed of all cat strains studied against each group of antibiotics, ordered from the highest to the lowest percentage, was: 32.3% for macrolides, 27.7% for lincosamides, 21.5% for amphenicols, 21.5% for tetracyclines, 25.1% for penicillins, 21.5% for cephalosporins, 15.4% for ansamycins, 13.8% for streptogramins, 11.8% for quinolones, 9.2% for aminoglycosides, 7.7% for lipopeptides, 6.2% for nitrofurans, and 4.6% for the folate inhibitor pathway, glycopeptides, glycylcyclines, and oxazolidinones. In the results observed from the D-test, 13.8% (9/65) tested positive.

For all the infection-causing *Staphylococcus* isolated from cats with active skin infections, 11.1% (1/9) were sensitive to all the antibiotics studied, while 88.9% (8/9) were AMR, and 55.6% (5/9) were MDR. Ordered from the highest to the lowest percentage, the AMR of all the strains in each antibiotic group was: 51.9% for penicillins, 44.4% for amphenicols, 44.4% for lincosamides, 44.4% for macrolides, 44.4% for tetracyclines, 37% for quinolones, 22.2% for cephalosporins, and 11.1% for the glycopeptides, lipopeptides, nitrofurans, ansamycins, and streptogramins. None of the isolated strains showed resistance to aminoglycosides, the folate inhibitor pathway, glycylcyclines, or oxazolidinones. Regarding the D-test performed in the infection-causing *Staphylococcus* strains, 22.2% (2/9) were positive. Furthermore, no relationship was observed between the clinical data collected in the questionnaire and the manifestation of AMR and MDR (p-value > 0.05).

As mentioned for dogs, *S. aureus* and *S. pseudintermedius* were the main strains with importance in public health. Therefore, their AMR levels are shown in Tables 5 and 6. However, no infection-causing *S. pseudintermedius* was isolated from cats with active skin infections.

Table 5. AMR in commensal and infection-causing *Staphylococcus aureus* isolated from healthy cats and cats with an active skin infection.

AB Group	AB	% AMR/AB in Commensal *S. aureus*	% AMR/AB in Infection-Causing *S. aureus*
Aminoglycosides	GEN	0 ^{a} (0/6) ± 0	0 ^{a} (0/1) ± 0
Amphenicols	CHL	16.7 ^{a,b} (1/6) ± 15.2	0 ^{a} (0/1) ± 0
Cephalosporins	CXI	0 ^{a} (0/6) ± 0	0 ^{a} (0/1) ± 0
Folate Inhibitor Pathway	TRS	0 ^{a} (0/6) ± 0	0 ^{a} (0/1) ± 0
Glycopeptides	VAN	0 ^{a} (0/6) ± 0	100 ^{b} (1/1) ± 0
Glycylcyclines	TIG	0 ^{a} (0/6) ± 0	0 ^{a} (0/1) ± 0
Lincosamides	CLI	0 ^{a} (0/6) ± 0	0 ^{a} (0/1) ± 0
Lipopeptides	DAP	0 ^{a} (0/6) ± 0	0 ^{a} (0/1) ± 0
Macrolides	ERY	50 ^{b} (3/6) ± 20.4	100 ^{b} (1/1) ± 0
Nitrofurans	NIT	0 ^{a} (0/6) ± 0	100 ^{b} (1/1) ± 0
Oxazolidinones	LIN	0 ^{a} (0/6) ± 0	0 ^{a} (0/1) ± 0
Penicillins	AMP	16.7 ^{a,b} (1/6) ± 15.2	100 ^{b} (1/1) ± 0
	PEN	16.7 ^{a,b} (1/6) ± 15.2	100 ^{b} (1/1) ± 0
Quinolones	LEV	0 ^{a} (0/6) ± 0	100 ^{b} (1/1) ± 0
	CIP	0 ^{a} (0/6) ± 0	100 ^{b} (1/1) ± 0
	MOX	0 ^{a} (0/6) ± 0	100 ^{b} (1/1) ± 0
Tetracyclines	TET	0 ^{a} (0/6) ± 0	100 ^{b} (1/1) ± 0
Ansamycins	RIF	0 ^{a} (0/6) ± 0	100 ^{b} (1/1) ± 0
Streptogramins	QUD	0 ^{a} (0/6) ± 0	0 ^{a} (0/1) ± 0

AB: antibiotic. AMR: antimicrobial resistance. GEN: gentamicin. CHL: chloramphenicol. CXI: cefoxitin. TRA: trimethoprim–sulfamethoxazole. TIG: tigecycline. CLI: clindamycin. DAP: daptomycin. ERY: erythromycin. NIT: nitrofurantoin. LIN: linezolid. AMP: ampicillin. PEN: penicillin. LEV: levofloxacin. CIP: ciprofloxacin. MOX: marbofloxacin. TET: tetracycline. RIF: rifampicin. QUD: quinupristin/dalfopristin. ^{a,b}: different superscripts in each column indicate statistically significant differences (p-value $\leq$ 0.05) for the resistance found against the different antibiotics studied. ±: standard error.

Table 6. AMR in commensal *Staphylococcus pseudintermedius* isolated from healthy cats.

AB Group	AB	% AMR/AB in Commensal *S. pseudintermedius*
Aminoglycosides	GEN	0 ^{a} (0/3) ± 0
Amphenicols	CHL	33.3 ^{a,b,c} (1/3) ± 27.2
Folate inhibitor pathway	TRS	0 ^{a} (0/3) ± 0
Glycopeptides	VAN	0 ^{a} (0/3) ± 0
Glycylcyclines	TIG	0 ^{a} (0/3) ± 0
Lincosamides	CLI	66.7 ^{b,c} (2/3)
Lipopeptides	DAP	33.3 ^{a,b,c} (1/3) ± 27.2
Macrolides	ERY	66.7 ^{b,c} (2/3) ± 27.2
Nitrofurans	NIT	0 ^{a} (0/3) ± 0
Oxazolidinones	LIN	0 ^{a} (0/3) ± 0
Penicillins	AMP	33.3 ^{a,b} (1/3) ± 27.2
	OXA+	33.3 ^{a,b} (1/3) ± 27.2
	PEN	100 ^{c} (3/3) ± 0
Quinolones	LEV	0 ^{a} (0/3) ± 0
	CIP	33.3 ^{a,b} (1/3) ± 27.2
	MOX	0 ^{a} (0/3) ± 0
Tetracyclines	TET	0 ^{a} (0/3) ± 0
Ansamycins	RIF	0 ^{a} (0/3) ± 0
Streptogramins	QUD	0 ^{a} (0/3) ± 0

AB: antibiotic. AMR: antimicrobial resistance. GEN: gentamicin. CHL: chloramphenicol. TRA: trimethoprim–sulfamethoxazole. TIG: tigecycline. CLI: clindamycin. DAP: daptomycin. ERY: erythromycin. NIT: nitrofurantoin. LIN: linezolid. AMP: ampicillin. OXA+: oxacillin + 2% NaCl. PEN: penicillin. LEV: levofloxacin. CIP: ciprofloxacin. MOX: marbofloxacin. TET: tetracycline. RIF: rifampicin. QUD: quinupristin/dalfopristin. $^{a-c}$: in each column, different superscripts indicate statistically significant differences (p-value $\leq$ 0.05) for the resistance found against the different antibiotics studied. ±: standard error.

Overall, the AMR trends did not follow any pattern, as 126 different AMR patterns were observed in the 187 *Staphylococcus* spp. strain isolates in this study. Of all the AMR patterns, 62.7% (79/126) belonged to *Staphylococcus* spp. isolated from dogs, and 37.3% (47/126) belonged to *Staphylococcus* spp. isolated from cats. The most common AMR pattern was observed in the penicillin group alone in dogs (4%, 5/126) and in cats (3.2%,

4/126), followed by the macrolides group alone in cats (3.2%, 4/126). All AMR patterns are attached in the Supplementary Materials, Table S2.

4. Discussion

The emergence of AMR and MDR strains in companion animals represents a new challenge for global public health, which must be addressed through a One Health strategy [36,37]. This is not only crucial due to therapeutic failures in veterinary medicine but also in human medicine. Studies have shown that these strains can circulate in the environment and be transmitted from animals to humans and vice versa [38]. Therefore, it is essential to assess the presence of this resistance in both commensal and pathogenic bacteria.

In this study, a new genus of the *Staphylococceae* family has been studied, as, due to new phylogenomic studies of this family, some *Staphylococcus* have been relocated to other genera [39]. The study that proposed this taxonomic reassignment was published relatively recently, and therefore MALDI-TOF and other biochemical analyses continue to identify these bacteria as *Staphylococcus*, as in the case of *S. sciuri* (former CoNS) that now belongs to the genus *Mammaliicoccus sciuri* [39,40]. Nevertheless, the implications that *M. sciuri* has on public health remains the same, as it is considered one of the most ancient species in natural history capable of carrying virulence and AMR genes similar to those identified in other *Staphylococcal* species. However, the scientific community has increasingly focused on *M. sciuri*, mainly because this species is believed to be the most likely evolutionary reservoir of the *mecA* gene, which has subsequently spread to *S. aureus* and other *Staphylococcus* species [41].

The overall observed prevalence of *Staphylococcus* spp. and *Mammaliicoccus sciuri* in our study (69%) aligns with that reported in previous studies [42,43]. In line with this, it has also been seen that both healthy and diseased companion animals harbor both CoPS and CoNS, although certain species showed a stronger association with each animal species [43]. In this study, CoPS, such as *S. pseudintermedius*, were more frequently isolated from dogs, while CoNS, including *S. felis*, were more commonly isolated from cats, as reported previously [42–45]. This finding is significant, given the widespread observation of methicillin resistance not only in CoPS but also in CoNS, as both groups are now acknowledged as important pathogens [43,46,47]. Nevertheless, the pathogenic potential of CoNS has not been as well studied as the virulence factors involved in CoPS, although it has been recognised that they cause some important diseases, such as endocarditis or urinary tract infections in at-risk populations, and that they are important reservoirs of AMR genes [48,49]. Thus, it is essential to monitor the commensal microbiota of animals, and not only the most "relevant" bacterial species, to detect the risks of human exposure to animal species [50].

Regarding the AMR obtained, similar rates of AMR and MDR were found in both commensal and infection-causing *Staphylococcus* isolates from dogs and cats. However, upon comparison, high levels of AMR were observed in dogs, consistent with findings published in other studies [51]. One hypothesis that could explain these results is that dogs have more contact with other animals and humans, given their daily walks and shared public spaces. In contrast, cats typically live indoors, as observed in our study, where only 16% of cats went outdoors.

Antimicrobial agents effective in treating infections caused by these organisms are limited, particularly for *Staphylococcus* strains that exhibit MDR, including MRS [52]. In this study, the highest prevalence of MRS was observed in dog *S. pseudintermedius* (37% and 33.3% in commensal and infection-causing isolates, respectively). Similar results were reported in a study conducted in Tennessee (USA), where 30.8% of the isolates were MRSP [52].

Overall, the antibiotics with the highest percentages of AMR were those in the penicillins group (almost 50% in dogs and cats), chloramphenicol (≈50% in dogs and 25% in cats), erythromycin (≈47% for both species), clindamycin (≈40% for both), and tetracycline (≈40% for both). Similar results have been observed in other studies conducted in the

Iberian Peninsula [51] and Canada [53]. However, lower AMR profiles were observed in another study conducted in the USA [52]. In some Scandinavian countries, the main resistance observed was to penicillins in different proportions, with 65% in Denmark [54], ≈70% in Finland [55], 14% in Norway [56], and 19.5% in Sweden [57]. These data have been published in their latest national reports on antimicrobial resistance, but the difference in these results may be due to the *Staphylococcus* species studied in each programme, as only *S. pseudintermedius* were studied in Denmark, *S. aureus* and *S. pseudintermedius* in Finland, and only *S. felis* in Norway and Sweden (as only MRSP and MRSA strains of these two species were evaluated in these countries). For the other antibiotics studied, their results varied between countries, but in all countries, the AMR rates were lower than in this study. This variation may be attributed to the geographical area or choice of antibiotics for treating infections influenced by regional legislation [58]. In addition, it is important to highlight that the observed AMR to the penicillins group was significantly higher in dogs than in cats. These findings could be linked to the administration of penicillins in our study population, as it emerged as the most commonly prescribed antibiotic group for both dogs and cats, with dogs receiving it twice as frequently (56.6%) as cats (25.2%).

In terms of the AMR observed in some of the most important public health species, due to their pathogenic capacity and ability to harbor resistance genes, both *S. pseudintermedius* and *S. aureus* showed similar patterns in the commensal and infection-causing isolates from dogs, with the highest AMR observed against penicillins (≈80%), chloramphenicol (≈57%), erythromycin (≈56%), tetracycline (≈50%), and clindamycin (≈48%), in accordance with Lord et al. (2022) [59]. Although there are not many isolates of *S. aureus* in our study, the results align with those observed by other authors in different geographical areas, such as Nepal [60], Italy [61], India [62], Bangladesh [63], or the USA [64], highlighting the concerning emergence of AMR in these strains, posing a threat to public health. The high levels of AMR to chloramphenicol are particularly concerning, not only for *S. pseudintermedius* (almost 70%) but also for *S. aureus* (more than 80%), given its usefulness for the treatment of MRS infections [59]. Something similar happens with erythromycin and clindamycin, two antibiotics of choice in the treatment of MRSA and MRSP [65,66]. Therefore, the D-test was performed in this study. The observed results were slightly higher in dogs than in cats. In both cases, this inducible phenotype was observed to a greater extent in infection-causing strains, which may lead to treatment failure due to the development of constitutive resistance [60]. Thus, all strains with a positive D-test should be reported as being resistant to clindamycin [67].

In the new WHO medically important antimicrobial list, quinolones belonged to the highest priority critically important antimicrobials (HPCIAs), antibiotics that should only be used in veterinary medicine when all others have failed [28]. Overall, the AMR of quinolones was around 30%, varying from one *Staphylococcus* species to another, as seen in different studies [68–70]. In particular, levofloxacin and moxifloxacin had the highest AMR in the quinolones group in both dog and cat *S. aureus* and *S. pseudintermedius*, regardless of whether they were commensal or infection-causing strains. Even though these quinolones are not authorised for veterinary use in the EU but only for human use [71], these antimicrobials are experiencing an increase in both human [72,73] and animal [61,74] strains worldwide.

Finally, regarding the antibiotics that are only authorised for human use and not intended for animals, commonly known as last resort antibiotics, five of them were tested in this study: vancomycin, tigecycline, daptomycin, lincomycin, and nitrofurantoin. The results on tigecycline resistance are particularly alarming, as high rates have been observed in dog isolates, especially in commensal *S. pseudintermedius* (46.3%). Similar results have been observed in human medicine, ranging from 5.6% [75,76] to almost 30% [77], and up to 88% in CoNS of other animal species, such as turkeys [78]. This high acquired AMR for this antibiotic represents a major public health concern, as it is one of the newest last resort antibiotics used to treat MDR infections caused by *Staphylococcus* strains [79,80]. On the other hand, all cat isolates were susceptible to tigecycline, as reported in other studies in isolates from cats and other animal species [81–83], as well as humans [83,84].

Another of these antibiotics with high AMR found in this study was lincomycin, with the highest resistance observed in commensal *S. aureus* (83.3%) and *S. pseudintermedius* (25.9%), as well as in infection-causing *S. aureus* (33.3%) isolated from dogs. Most of the cases reporting lincomycin resistance in *Staphylococcus* strains are from intensive care units [85] and hospitals [84,86], so the expected results would have been that, as in other studies carried out in different countries such as Portugal [72,78], Brazil [68], Italy [73], and China [79], no resistance to this antibiotic would be observed.

Regarding vancomycin, it is often used for severe infections caused by MRSA and MR-CoNS strains, among other complicated infections [87]. In the present study, of all the *Staphylococcus* spp. isolated, only eight showed AMR to vancomycin, four from dogs and four from cats. The findings in other studies regarding this antibiotic suggest that its AMR is rare in companion animals [70,77,88]. However, higher VAN-resistant *S. aureus* strains have been previously reported in bovine mastitis [89] and human medicine in hospitals [84]. In addition, a low range of AMR has also been seen against daptomycin (4%), rifampicin (8%), and quinupristin/dalfopristin (10%). These antibiotics are reserved to treat fastidious and MDR infections, mainly caused by Gram-positive bacteria [90,91]. In fact, DAP is mostly reserved to treat VAN-resistant infections [92]. In the case of RIF and QUD, these are antibiotics that previously also belonged to the same category as DAP but have now been relocated to the CIA and HIA categories, respectively. Some studies show higher percentages of AMR to DAP and RIF from human isolates [84]. Different AMR rates have been observed in companion animals [93,94], as reported by Burke et al. (2023) in a 10-year study, where only one *S. schleiferi* and one *S. pseudintermedius* isolated from dogs were resistant to RIF, while all cat strains were susceptible [69]. The same has been reported for DAP by Bellato et al. (2022) but with a higher AMR to RIF (12.5%) [82].

In contrast to the previous results mentioned, it is relevant to highlight that the AMR to NIT (around 5% for all the strains) in this study was among the lowest observed. In previous WHO and EMA classifications, this antibiotic was placed in the least important category [10]. However, in the latest WHO report, this antibiotic has been moved to the list of those not authorised for animal use [28]. Therefore, despite the limited knowledge on the mechanism of action of this antibiotic, it represents a potential tool in the fight against antimicrobial resistance, and further studies on this molecule are needed [95]

5. Conclusions

In conclusion, the results obtained in this study highlight the importance of following the new WHO categorisation when prescribing antibiotics for companion animals, as the highest resistance observed in this study is against the first treatments of choice for infection-causing *Staphylococcus* (amphenicols, macrolides, lincosamides, and tetracyclines). Nevertheless, no significant statistical differences were observed among epidemiological clusters. This is particularly concerning since AMR and MDR seem to be extensively disseminated, even in cases where animals have not undergone prior antibiotic treatments, including HPCIAs, and are not authorised antibiotics for animal use. These findings underscore the need to control companion animals as potential reservoirs and transmitters of resistance to both humans and the environment, following a One Health strategy. Moreover, further in-depth epidemiological studies of the transmission of AMR between companion animals and humans are needed to establish adequate control tools.

Supplementary Materials: The following supporting information can be downloaded at: https://www.mdpi.com/article/10.3390/vetsci11020054/s1, Part A: Questionnaire. AMR study in companion animals; Table S1: AMR of all the strains isolated; Table S2: AMR patterns.

Author Contributions: Conceptualization, C.M., L.M.-D. and S.V.; methodology, C.M., L.M.-D. and S.V.; software, A.M.-F.; validation, L.M.-D., C.G.-C., V.A.-M. and C.M.; formal analysis, A.M.-F. and L.M.-D.; investigation, A.M.-F. and L.M.-D.; resources, C.G.-C., C.M. and S.V.; data curation, A.M.-F. and V.A.-M.; writing—original draft preparation, A.M.-F. and L.M.-D.; writing—review and editing, A.M.-F., C.M., L.M.-D., S.V. and C.G.-C.; visualization, A.M.-F.; supervision, C.M., L.M.-D. and S.V.; project administration, C.M., L.M.-D. and S.V.; funding acquisition, C.M. and C.G.-C. All authors have read and agreed to the published version of the manuscript.

Funding: This research was funded by the UNIVERSIDAD CARDENAL HERRERA-CEU, grant numbers INDI22/34 and INDI23/39.

Institutional Review Board Statement: The animal study protocol was approved by the Animal Ethics Committees of the UNIVERSIDAD CARDENAL HERRERA-CEU (CEEA 22/04; 2 March 2022).

Informed Consent Statement: Informed consent was obtained from the animal owners.

Data Availability Statement: The data presented in this study are available on request from the corresponding author.

Acknowledgments: The authors wish to thank the companion animals' owners and the veterinarians, who voluntarily agreed to participate in the study, for allowing us to carry out this work in the following veterinary clinics and hospitals: AniCura Constitución Hospital Veterinario, AniCura Benipeixcar Hospital Veterinario, AniCura Valencia Sur Hospital Veterinario, AniCura Sant Antoni Clínica Veterinaria, Clínica Veterinaria La Graveta, Clínica Veterinaria Nueve Vidas, Clínica Veterinaria Vicent Colechà, and Clínica Veterinaria Silla. The authors also wish to express their gratitude to the Microbiology Service of the Consorcio Hospital General de Valencia for their invaluable collaboration and technical support. Finally, the authors wish to thank the "Improvement of Production System-related Food Safety and End Products" research group (Veterinary Faculty, University Cardenal Herrera-CEU) for their technical and financial support.

Conflicts of Interest: The authors declare no conflicts of interest.

References

1. Graham, D.W.; Bergeron, G.; Bourassa, M.W.; Dickson, J.; Gomes, F.; Howe, A.; Kahn, L.H.; Morley, P.S.; Scott, H.M.; Simjee, S.; et al. Complexities in Understanding Antimicrobial Resistance across Domesticated Animal, Human, and Environmental Systems. *Ann. N. Y. Acad. Sci.* **2019**, *1441*, 17–30. [CrossRef]
2. FEDIAF. The European Pet and Food Industry. Annual Report 2023. Available online: https://europeanpetfood.org/about/annual-report/ (accessed on 7 July 2023).
3. FEDIAF. The European Pet and Food Industry. Annual Report 2010. Available online: http://www.stray-afp.org/nl/wp-content/uploads/sites/2/2012/08/facts_and_figures_2010.pdf (accessed on 25 September 2023).
4. Overgaauw, P.A.M.; Vinke, C.M.; van Hagen, M.A.E.; Lipman, L.J.A. A One Health Perspective on the Human-Companion Animal Relationship with Emphasis on Zoonotic Aspects. *Int. J. Environ. Res. Public Health* **2020**, *17*, 3789. [CrossRef]
5. Marco-Fuertes, A.; Marin, C.; Lorenzo-Rebenaque, L.; Vega, S.; Montoro-Dasi, L. Antimicrobial Resistance in Companion Animals: A New Challenge for the One Health Approach in the European Union. *Vet. Sci.* **2022**, *9*, 208. [CrossRef]
6. World Health Organization. *Global Antimicrobial Resistance and Use Surveillance System (GLASS). Report 2022*; WHO: Geneva, Switzerland, 2022; Volume 2.
7. Murray, C.J.; Ikuta, K.S.; Sharara, F.; Swetschinski, L.; Robles Aguilar, G.; Gray, A.; Han, C.; Bisignano, C.; Rao, P.; Wool, E.; et al. Global Burden of Bacterial Antimicrobial Resistance in 2019: A Systematic Analysis. *Lancet* **2022**, *399*, 629–655. [CrossRef]
8. Ye, M.; Su, J.Q.; An, X.L.; Zhu, Y.G. Silencing the Silent Pandemic: Eliminating Antimicrobial Resistance by Using Bacteriophages. *Sci. China Life Sci.* **2022**, *65*, 1890. [CrossRef] [PubMed]
9. Mendelson, M.; Sharland, M.; Mpundu, M. Antibiotic Resistance: Calling Time on the 'Silent Pandemic'. *JAC Antimicrob. Resist.* **2022**, *4*, dlac016. [CrossRef] [PubMed]
10. EMA Categorisation of Antibiotics Used in Animals Promotes Responsible Use to Protect Public and Animal Health. European Medicines Agency-EMA. Available online: https://www.ema.europa.eu/en/news/categorisation-antibiotics-used-animals-promotes-responsible-use-protect-public-animal-health (accessed on 6 December 2022).
11. Mader, R.; Demay, C.; Jouvin-Marche, E.; Ploy, M.C.; Barraud, O.; Bernard, S.; Lacotte, Y.; Pulcini, C.; Weinbach, J.; Berling, C.; et al. Defining the Scope of the European Antimicrobial Resistance Surveillance Network in Veterinary Medicine (EARS-Vet): A Bottom-up and One Health Approach. *J. Antimicrob. Chemother.* **2022**, *77*, 816–826. [CrossRef]
12. Mader, R.; Damborg, P.; Amat, J.P.; Bengtsson, B.; Bourély, C.; Broens, E.M.; Busani, L.; Crespo-Robledo, P.; Filippitzi, M.E.; Fitzgerald, W.; et al. Building the European Antimicrobial Resistance Surveillance Network in Veterinary Medicine (EARS-Vet). *Eurosurveillance* **2021**, *26*, 2001359. [CrossRef] [PubMed]

13. European Centre for Disease Prevention and Control European Antimicrobial Resistance Surveillance Network (EARS-Net). Available online: https://www.ecdc.europa.eu/en/about-us/networks/disease-networks-and-laboratory-networks/ears-net-data (accessed on 19 July 2023).
14. Hesp, A.; van Schaik, G.; Wiegel, J.; Heuvelink, A.; Mevius, D.; Veldman, K. Antimicrobial Resistance Monitoring in Commensal and Clinical Escherichia Coli from Broiler Chickens: Differences and Similarities. *Prev. Vet. Med.* **2022**, *204*, 105663. [CrossRef]
15. European Food and Safety Authority. The European Union Summary Report on Antimicrobial Resistance in Zoonotic and Indicator Bacteria from Humans, Animals and Food in 2020/2021. *EFSA J.* **2023**, *21*, e07867. [CrossRef]
16. Kowalewicz, C.; Timmermans, M.; Fretin, D.; Wattiau, P.; Boland, C. An In-House 45-Plex Array for the Detection of Antimicrobial Resistance Genes in Gram-Positive Bacteria. *Microbiologyopen* **2023**, *12*, e1341. [CrossRef] [PubMed]
17. Abusleme, F.; Galarce, N.; Quezada-Aguiluz, M.; Iragüen, D.; González-Rocha, G. Characterization and Antimicrobial Susceptibility of Coagulase-Positive *Staphylococcus* Isolated in a Veterinary Teaching Hospital in Chile. *Rev. Argent. Microbiol.* **2022**, *54*, 192–202. [CrossRef] [PubMed]
18. Citron, L.E.; Cain, C.L.; Dietrich, J.; Cole, S.D. Genomic and Clinical Case Characterisation of *Staphylococcus haemolyticus* Isolated from Dogs and Cats in the United States, Including Strains with High-Level Mupirocin Tolerance. *Vet. Dermatol.* **2023**, *34*, 298–309. [CrossRef] [PubMed]
19. Platenik, M.O.; Archer, L.; Kher, L.; Santoro, D. Prevalence of MecA, MecC and Panton-Valentine-Leukocidin Genes in Clinical Isolates of Coagulase Positive Staphylococci from Dermatological Canine Patients. *Microorganisms* **2022**, *10*, 2239. [CrossRef] [PubMed]
20. Wilkinson, D.A.; Rogers, L.E.; Bell, A.; Benschop, J.; Midwinter, A.C. Carriage of *Staphylococcus pseudintermedius* by Clinically Normal Dogs in Canterbury, New Zealand. *N. Z. Vet. J.* **2022**, *71*, 33–36. [CrossRef] [PubMed]
21. Tabatabaei, S.; Najafifar, A.; Askari Badouei, M.; Zahraei Salehi, T.; Ashrafi Tamai, I.; Khaksar, E.; Abbassi, M.S.; Ghazisaeedi, F. Genetic Characterisation of Methicillin-Resistant *Staphylococcus aureus* and *Staphylococcus pseudintermedius* in Pets and Veterinary Personnel in Iran: New Insights into Emerging Methicillin-Resistant *S. pseudintermedius* (MRSP). *J. Glob. Antimicrob. Resist.* **2019**, *16*, 6–10. [CrossRef] [PubMed]
22. Tamakan, H.; Gocmen, H. Genetic Characterization of Methicillin Resistant *Staphylococcus pseudintermedius* in Dogs and Cats in Cyprus: Comparison of MRSP and MRSA Results. *Pak. J. Zool* **2022**, *54*, 1511–1519. [CrossRef]
23. Becker, K.; Heilmann, C.; Peters, G. Coagulase-Negative Staphylococci. *Clin. Microbiol. Rev.* **2014**, *27*, 870–926. [CrossRef]
24. Marco-Fuertes, A.; Jordá, J.; Marin, C.; Lorenzo-Rebenaque, L.; Montoro-Dasi, L.; Vega, S. Multidrug-Resistant Escherichia Coli Strains to Last Resort Human Antibiotics Isolated from Healthy Companion Animals in Valencia Region. *Antibiotics* **2023**, *12*, 1638. [CrossRef]
25. Frosini, S.M.; Bond, R.; King, R.H.; Loeffler, A. The Nose Is Not Enough: Multi-Site Sampling Is Best for MRSP Detection in Dogs and Households. *Vet. Dermatol.* **2022**, *33*, 576–580. [CrossRef]
26. Nocera, F.P.; Iovane, G.; DE MARTINO, L.D.; Holbein, B.E. Antimicrobial Activity of the Iron-Chelator, DIBI, against Multidrug-Resistant Canine Methicillin-Susceptible *Staphylococcus pseudintermedius*: A Preliminary Study of Four Clinical Strains. *Pathogens* **2022**, *11*, 656. [CrossRef]
27. Ardanuy, C.; María, E.C.; Morosini, I.; Torres, C. Detección Fenotípica de mecanismos de Resistencia Grampositivos. Recomendaciones de La Sociedad Española de Enfermedades Infecciosas y Microbiología Clínica (SEIMC). Available online: https://www.seimc.org/contenidos/documentoscientificos/procedimientosmicrobiologia/seimc-procedimientomicrobiologia39.pdf (accessed on 19 July 2023).
28. WHO. *WHO Medically Important List. A Risk Management Tool for Mitigating Antimicrobial Resistance Due to Non-Human Use. 7 th Revision 2023*; WHO: Geneva, Switzerland, 2023.
29. The European Committee on Antimicrobial Susceptibility Testing. Breakpoint Tables for Interpretation of MICs and Zone Diameters. Version 14.0, 2024. Available online: https://www.eucast.org/fileadmin/src/media/PDFs/EUCAST_files/Breakpoint_tables/v_14.0_Breakpoint_Tables.pdf (accessed on 5 January 2024).
30. Clinical and Laboratory Standards Institute. *Performance Standards for Antimicrobial Susceptibility Testing*, 33rd ed.; CLSI Supplement M100; Clinical and Laboratory Standards Institute: Wayne, PA, USA, 2023.
31. Clinical and Laboratory Standards Institute. *Performance Standards for Antimicrobial Disk and Dilution Susceptibility Tests for Bacteria Isolated from Animals*, 6th ed.; CLSI Supplement VET01S; Clinical and Laboratory Standards Institute: Wayne, PA, USA, 2023.
32. Magiorakos, A.P.; Srinivasan, A.; Carey, R.B.; Carmeli, Y.; Falagas, M.E.; Giske, C.G.; Harbarth, S.; Hindler, J.F.; Kahlmeter, G.; Olsson-Liljequist, B.; et al. Multidrug-Resistant, Extensively Drug-Resistant and Pandrug-Resistant Bacteria: An International Expert Proposal for Interim Standard Definitions for Acquired Resistance. *Clin. Microbiol. Infect.* **2012**, *18*, 268–281. [CrossRef] [PubMed]
33. Searle, S.R.; Speed, F.M.; Milliken, G.A. Population Marginal Means in the Linear Model: An Alternative to Least Squares Means. *Am. Stat.* **1980**, *34*, 216–221. [CrossRef]
34. Fox, J.; Weisberg, S. *An R Companion to Applied Regression*, 3rd ed.; Sage: Thousand Oaks, CA, USA, 2019.
35. Graves, S.; Piepho, H.; Selzer, L.; Dorai-Raj, S. MultcompView: Visualizations of Paired Comparisons. R Package Version 0.1-9. Available online: https://cran.r-project.org/web/packages/multcompView/index.html (accessed on 27 July 2023).
36. Velazquez-Meza, M.E.; Galarde-López, M.; Carrillo-Quiróz, B.; Alpuche-Aranda, C.M. Antimicrobial Resistance: One Health Approach. *Vet. World* **2022**, *15*, 743. [CrossRef] [PubMed]

37. Pomba, C.; Rantala, M.; Greko, C.; Baptiste, K.E.; Catry, B.; van Duijkeren, E.; Mateus, A.; Moreno, M.A.; Pyörälä, S.; Ružauskas, M.; et al. Public Health Risk of Antimicrobial Resistance Transfer from Companion Animals. *J. Antimicrob. Chemother.* **2017**, *72*, 957–968. [CrossRef] [PubMed]
38. Despotovic, M.; de Nies, L.; Busi, S.B.; Wilmes, P. Reservoirs of Antimicrobial Resistance in the Context of One Health. *Curr. Opin. Microbiol.* **2023**, *73*, 102291. [CrossRef] [PubMed]
39. Madhaiyan, M.; Wirth, J.S.; Saravanan, V.S. Phylogenomic Analyses of the *Staphylococcaceae* Family Suggest the Reclassification of Five Species within the Genus *Staphylococcus* as Heterotypic Synonyms, the Promotion of Five Subspecies to Novel Species, the Taxonomic Reassignment of Five *Staphylococcus* Species to *Mammaliicoccus* Gen. Nov., and the Formal Assignment of *Nosocomiicoccus* to the Family *Staphylococcaceae*. *Int. J. Syst. Evol. Microbiol.* **2020**, *70*, 5926–5936. [CrossRef] [PubMed]
40. Taxonomy Browser (Mammaliicoccus Sciuri). Available online: https://www.ncbi.nlm.nih.gov/Taxonomy/Browser/wwwtax.cgi?id=1296 (accessed on 7 January 2024).
41. de Moura, G.S.; de Carvalho, E.; Ramos Sanchez, E.M.; Sellera, F.P.; Marques, M.F.S.; Heinemann, M.B.; De Vliegher, S.; Souza, F.N.; Mota, R.A. Emergence of Livestock-Associated Mammaliicoccus Sciuri ST71 Co-Harbouring MecA and MecC Genes in Brazil. *Vet. Microbiol.* **2023**, *283*, 109792. [CrossRef] [PubMed]
42. Gandolfi-Decristophoris, P.; Regula, G.; Petrini, O.; Zinsstag, J.; Schelling, E. Prevalence and Risk Factors for Carriage of Multi-Drug Resistant *Staphylococci* in Healthy Cats and Dogs. *J. Vet. Sci.* **2013**, *14*, 449–456. [CrossRef]
43. Miszczak, M.; Korzeniowska-Kowal, A.; Wzorek, A.; Gamian, A.; Rypuła, K.; Bierowiec, K. Colonization of Methicillin-Resistant *Staphylococcus* Species in Healthy and Sick Pets: Prevalence and Risk Factors. *BMC Vet. Res.* **2023**, *19*, 1–12. [CrossRef]
44. Bierowiec, K.; Korzeniowska-Kowal, A.; Wzorek, A.; Rypuła, K.; Gamian, A. Prevalence of *Staphylococcus* Species Colonization in Healthy and Sick Cats. *Biomed. Res. Int.* **2019**, *2019*, 4360525. [CrossRef]
45. Sukur, H.; Esendal, O.M. Presence and Antimicrobial Resistance of Coagulase-Negative *Staphylococci* Isolated from Animals in a Veterinary Teaching Hospital in Cyprus. *Vet. Med.* **2020**, *65*, 191–198. [CrossRef]
46. Cengiz, S.; Okur, S.; Oz, C.; Turgut, F.; Gumurcinler, B.; Sevuk, N.S.; Kekec, A.I.; Cepoglu, H.; Sevimli, U.; Adiguzel, M.C. Prevalence and Clonal Diversity of Methicillin-Resistant *Staphylococcus aureus* and Methicillin-Resistant *Staphylococcus pseudintermedius* Isolated from Dogs and Cats with Eye Discharge. *Acta. Microbiol. Immunol. Hung.* **2023**, *70*, 134–141. [CrossRef]
47. França, A.; Gaio, V.; Lopes, N.; Melo, L.D.R. Virulence Factors in Coagulase-Negative Staphylococci. *Pathogens* **2021**, *10*, 170. [CrossRef] [PubMed]
48. Foster, T.J. Coagulase Negative Staphylococci. In *Molecular Medical Microbiology*, 3rd ed.; Academic Press: Cambridge, MA, USA, 2024; pp. 681–704.
49. Abdullahi, I.N.; Lozano, C.; Höfle, Ú.; Cardona-Cabrera, T.; Zarazaga, M.; Torres, C. Antimicrobial Resistome of Coagulase-Negative Staphylococci from Nasotracheal Cavities of Nestlings of *Ciconia ciconia* in Southern Spain: Detection of MecC-SCCmec Type-XI-Carrying *S. Lentus*. *Comp. Immunol. Microbiol. Infect. Dis.* **2023**, *99*, 102012. [CrossRef] [PubMed]
50. Sales, I.; Vieira-da-Motta, O.; Tavares, A.; Ruiz-Miranda, C.R.; de Lencastre, H.; Miragaia, M. Impact of Human Created Environments in the Pathogenic Potential and Antimicrobial Resistance of Staphylococci from Wild Neotropical Primates in Brazil. *Comp. Immunol. Microbiol. Infect. Dis.* **2024**, *104*, 102094. [CrossRef]
51. Li, Y.; Fernández, R.; Durán, I.; Molina-López, R.A.; Darwich, L. Antimicrobial Resistance in Bacteria Isolated From Cats and Dogs From the Iberian Peninsula. *Front. Microbiol.* **2021**, *11*, 621597. [CrossRef]
52. Lord, J.; Millis, N.; Jones, R.D.; Johnson, B.; Kania, S.A.; Odoi, A. An Epidemiological Study of the Predictors of Multidrug Resistance and Methicillin Resistance among *Staphylococcus* spp. Isolated from Canine Specimens Submitted to a Diagnostic Laboratory in Tennessee, USA. *PeerJ* **2023**, *11*, e15012. [CrossRef]
53. Awosile, B.B.; Mcclure, J.T.; Saab, M.E.; Heider, L.C. Antimicrobial Resistance in Bacteria Isolated from Cats and Dogs from the Atlantic Provinces, Canada from 1994–2013. *Can. Vet. J.* **2018**, *59*, 885. [PubMed]
54. DANMAP. *Use of Antimicrobial Agents and Occurrence of Antimicrobial Resistance in Bacteria from Food Animals, Food and Humans in Denmark 2022*; DANMAP: Copenhagen, Denmark, 2023.
55. FINRES-Vet. *Finnish Veterinary Antimicrobial Resistance Monitoring and Consumption of Antimicrobial Agents 2022*; FINRES-Vet: Helsinki, Finland, 2023.
56. NORM/NORM-Vet. *Usage of Antimicrobial Agents and Occurrence of Antimicrobial Resistance in Norway*; NORM/NORM-Vet: Tromsø/Oslo, Norway, 2023.
57. Swedres-Svarm. *Sales of Antibiotics and Occurrence of Resistance in Sweden 2022*; Swedres-Svarm: Solna/Uppsala, Sweden, 2023.
58. Tong, Y.C.; Zhang, Y.N.; Li, P.C.; Cao, Y.L.; Ding, D.Z.; Yang, Y.; Lin, Q.Y.; Gao, Y.N.; Sun, S.Q.; Fan, Y.P.; et al. Detection of Antibiotic-Resistant Canine Origin *Escherichia coli* and the Synergistic Effect of Magnolol in Reducing the Resistance of Multidrug-Resistant *Escherichia coli*. *Front. Vet. Sci.* **2023**, *10*, 1104812. [CrossRef]
59. Lord, J.; Millis, N.; Jones, R.D.; Johnson, B.; Kania, S.A.; Odoi, A. Patterns of Antimicrobial, Multidrug and Methicillin Resistance among *Staphylococcus* spp. Isolated from Canine Specimens Submitted to a Diagnostic Laboratory in Tennessee, USA: A Descriptive Study. *BMC Vet. Res.* **2022**, *18*, 91. [CrossRef]
60. Thapa, D.; Pyakurel, S.; Thapa, S.; Lamsal, S.; Chaudhari, M.; Adhikari, N.; Shrestha, D. *Staphylococcus aureus* with Inducible Clindamycin Resistance and Methicillin Resistance in a Tertiary Hospital in Nepal. *Trop. Med. Health* **2021**, *49*, 99. [CrossRef]
61. Nocera, F.P.; Pizzano, F.; Masullo, A.; Cortese, L.; De Martino, L. Antimicrobial Resistant *Staphylococcus* Species Colonization in Dogs, Their Owners, and Veterinary Staff of the Veterinary Teaching Hospital of Naples, Italy. *Pathogens* **2023**, *12*, 1016. [CrossRef]

62. Sudhakara Reddy, B.; Sivajothi, S. Methicillin-Resistant *Staphylococcus aureus* (MRSA) Isolated from Dogs with Recurrent Pyoderma. *Artic. J. Dairy Vet. Anim. Res.* **2016**, *3*, 62–65. [CrossRef]
63. Das, S.; Kabir, A.; Chouhan, C.S.; Shahid, M.A.H.; Habib, T.; Kobir, M.A.; Hossain, M.Z.; Rahman, M.; Nazir, K.H.M.N.H. Antimicrobial Resistance Patterns of *Staphylococcus aureus* Isolated from Apparently Healthy Pet Cats of Bangladesh. *J. Adv. Vet. Anim. Res.* **2023**, *10*, 545–553. [CrossRef]
64. Morris, D.O.; Rookt, K.A.; Shofer, F.S.; Rankin, S.C. Screening of *Staphylococcus aureus*, *Staphylococcus intermedius*, and *Staphylococcus schleiferi* Isolates Obtained from Small Companion Animals for Antimicrobial Resistance: A Retrospective Review of 749 Isolates (2003–2004). *Vet. Dermatol.* **2006**, *17*, 332–337. [CrossRef]
65. Werth, B.J. Macrolides-Infectious Diseases-MSD. Manual Professional Edition. Available online: https://www.msdmanuals.com/professional/infectious-diseases/bacteria-and-antibacterial-drugs/macrolides (accessed on 9 November 2023).
66. Werth, B.J. Clindamycin-Infectious Diseases-MSD. Manual Professional Edition. Available online: https://www.msdmanuals.com/professional/infectious-diseases/bacteria-and-antibacterial-drugs/clindamycin (accessed on 9 November 2023).
67. Shrestha, B. D Test: A Simple Test with Big Implication for *Staphylococcus aureus* Macrolide-Lincosamide-Streptogramin B Resistance Pattern. *Rev. Artic. Nepal. Med. Coll J.* **2014**, *16*, 88–94. [CrossRef]
68. Almuhayawi, M.S.; Alruhaili, M.H.; Gattan, H.S.; Alharbi, M.T.; Nagshabandi, M.; Al Jaouni, S.; Selim, S.; Alanazi, A.; Alruwaili, Y.; Faried, O.A.; et al. *Staphylococcus aureus* Induced Wound Infections Which Antimicrobial Resistance, Methicillin-and Vancomycin-Resistant: Assessment of Emergence and Cross Sectional Study. *Infect. Drug Resist.* **2023**, *16*, 5335–5346. [CrossRef] [PubMed]
69. Burke, M.; Santoro, D. Prevalence of Multidrug-Resistant Coagulase-Positive Staphylococci in Canine and Feline Dermatological Patients over a 10-Year Period: A Retrospective Study. *Microbiol. N. Y.* **2023**, *169*, 001300. [CrossRef] [PubMed]
70. Srednik, M.E.; Perea, C.A.; Giacoboni, G.I.; Hicks, J.A.; Foxx, C.L.; Harris, B.; Schlater, L.K. Genomic Features of Antimicrobial Resistance in *Staphylococcus pseudintermedius* Isolated from Dogs with Pyoderma in Argentina and the United States: A Comparative Study. *Int. J. Mol. Sci.* **2023**, *24*, 11361. [CrossRef] [PubMed]
71. Sitovs, A.; Sartini, I.; Giorgi, M. Levofloxacin in Veterinary Medicine: A Literature Review. *Res. Vet. Sci.* **2021**, *137*, 111–126. [CrossRef]
72. Kanafani, Z.A.; Sleiman, A.; Frem, J.A.; Doumat, G.; Gharamti, A.; El Hafi, B.; Doumith, M.; AlGhoribi, M.F.; Kanj, S.S.; Araj, G.F.; et al. Molecular Characterization and Differential Effects of Levofloxacin and Ciprofloxacin on the Potential for Developing Quinolone Resistance among Clinical *Pseudomonas aeruginosa* Isolates. *Front. Microbiol.* **2023**, *14*, 1209224. [CrossRef]
73. González, J.; Hernandez, L.; Tabera, A.; Bustamante, A.V.; Sanso, A.M. Methicillin-Resistant *Staphylococcus aureus* and Coagulase-Negative *Staphylococcus* from School Dining Rooms in Argentina. *Foodborne Pathog. Dis.* **2023**, *21*, 44–51. [CrossRef] [PubMed]
74. Kang, M.H.; Chae, M.J.; Yoon, J.W.; Lee, S.Y.; Yoo, J.H.; Park, H.M. Resistance to Fluoroquinolones and Methicillin in Ophthalmic Isolates of *Staphylococcus pseudintermedius* from Companion Animals. *Can. Vet. J.* **2014**, *55*, 678.
75. Hamzah, A.M.C.; Yeo, C.C.; Puah, S.M.; Chua, K.H.; Rahman, N.I.A.; Abdullah, F.H.; Othman, N.; Chew, C.H. Tigecycline and Inducible Clindamycin Resistance in Clinical Isolates of Methicillin-Resistant *Staphylococcus aureus* from Terengganu, Malaysia. *J. Med. Microbiol.* **2019**, *68*, 1299–1305. [CrossRef]
76. Yousefi, M.; Fallah, F.; Arshadi, M.; Pourmand, M.R.; Hashemi, A.; Pourmand, G. Identification of Tigecycline- and Vancomycin-Resistant *Staphylococcus aureus* Strains among Patients with Urinary Tract Infection in Iran. *New Microbes New Infect.* **2017**, *19*, 8–12. [CrossRef] [PubMed]
77. Dabul, A.N.G.; Camargo, I.L.B.C. Molecular Characterization of Methicillin-Resistant *Staphylococcus aureus* Resistant to Tigecycline and Daptomycin Isolated in a Hospital in Brazil. *Epidemiol. Infect.* **2014**, *142*, 479–483. [CrossRef]
78. Moawad, A.A.; Hotzel, H.; Awad, O.; Roesler, U.; Hafez, H.M.; Tomaso, H.; Neubauer, H.; El-Adawy, H. Evolution of Antibiotic Resistance of Coagulase-Negative Staphylococci Isolated from Healthy Turkeys in Egypt: First Report of Linezolid Resistance. *Microorganisms* **2019**, *7*, 476. [CrossRef]
79. Papich, M.G. Selection of Antibiotics for Meticillin-Resistant *Staphylococcus pseudintermedius*: Time to Revisit Some Old Drugs? *Vet. Dermatol.* **2012**, *23*, 352-e64. [CrossRef] [PubMed]
80. Heidary, M.; Sholeh, M.; Koupaei, M.; Asadi, A.; Khah, S.M.; Kheirabadi, F.; Saeidi, P.; Darbandi, A.; Taheri, B.; Ghanavati, R. Prevalence of Tigecycline Resistance in Methicillin-Resistant *Staphylococcus aureus*: A Systematic Review and Meta-Analysis. *Diagn. Microbiol. Infect. Dis.* **2024**, *108*, 116088. [CrossRef]
81. Costa, S.S.; Ribeiro, R.; Serrano, M.; Oliveira, K.; Ferreira, C.; Leal, M.; Pomba, C.; Couto, I. *Staphylococcus aureus* Causing Skin and Soft Tissue Infections in Companion Animals: Antimicrobial Resistance Profiles and Clonal Lineages. *Antibiotics* **2022**, *11*, 599. [CrossRef]
82. Bellato, A.; Robino, P.; Stella, M.C.; Scarrone, L.; Scalas, D.; Nebbia, P. Resistance to Critical Important Antibacterials in *Staphylococcus pseudintermedius* Strains of Veterinary Origin. *Antibiotics* **2022**, *11*, 1758. [CrossRef]
83. Glajzner, P.; Szewczyk, E.M.; Szemraj, M. Pathogenic Potential and Antimicrobial Resistance of *Staphylococcus pseudintermedius* Isolated from Human and Animals. *Folia Microbiol.* **2023**, *68*, 231–243. [CrossRef]
84. Ayodo, C.; Mugoh, R.; Ita, T.; Ouma, C.; Jepleting, M.; Oduor, B.; Guyah, B.; Omulo, S. Nasal Carriage of Methicillin-Resistant *Staphylococcus sciuri* Group by Residents of an Urban Informal Settlement in Kenya. *East Afr. Health Res. J.* **2023**, *7*, 76. [CrossRef]
85. Balandin, B.; Lobo, B.; Orden, B.; Román, F.; García, E.; Martínez, R.; Valdivia, M.; Ortega, A.; Fernández, I.; Galdos, P. Emergence of Linezolid-Resistant Coagulase-Negative Staphylococci in an Intensive Care Unit. *Infect. Dis.* **2016**, *48*, 343–349. [CrossRef]

86. Kosowska-Shick, K.; Julian, K.G.; McGhee, P.L.; Appelbaum, P.C.; Whitener, C.J. Molecular and Epidemiologic Characteristics of Linezolid-Resistant Coagulase-Negative Staphylococci at a Tertiary Care Hospital. *Diagn. Microbiol. Infect. Dis.* **2010**, *68*, 34–39. [CrossRef]
87. Werth, B.J. Vancomycin-Infectious Diseases-MSD Manual Professional Edition. Available online: https://www.msdmanuals.com/professional/infectious-diseases/bacteria-and-antibacterial-drugs/vancomycin (accessed on 9 November 2023).
88. Ma, S.; Chen, S.; Lyu, Y.; Huang, W.; Liu, Y.; Dang, X.; An, Q.; Song, Y.; Jiao, Y.; Gong, X.; et al. China Antimicrobial Resistance Surveillance Network for Pets (CARPet), 2018 to 2021. *One Health Adv.* **2023**, *1*, 7. [CrossRef]
89. Muzammil, I.; Ijaz, M.; Saleem, M.H.; Ali, M.M. Molecular Characterization of Vancomycin-Resistant *Staphylococcus aureus* Isolated from Bovine Milk. *Zoonoses Public Health* **2023**, *70*, 420–433. [CrossRef] [PubMed]
90. Werth, B.J. Quinupristin and Dalfopristin-Infectious Diseases-MSD Manual Professional Edition. Available online: https://www.msdmanuals.com/professional/infectious-diseases/bacteria-and-antibacterial-drugs/quinupristin-and-dalfopristin (accessed on 9 November 2023).
91. Werth, B.J. Rifamycins-Infectious Diseases-MSD Manual Professional Edition. Available online: https://www.msdmanuals.com/professional/infectious-diseases/bacteria-and-antibacterial-drugs/rifamycins (accessed on 9 November 2023).
92. Werth, B.J. Daptomycin-Infectious Diseases-MSD Manual Professional Edition. Available online: https://www.msdmanuals.com/professional/infectious-diseases/bacteria-and-antibacterial-drugs/daptomycin (accessed on 9 November 2023).
93. Zhou, Y.; Ji, X.; Liang, B.; Jiang, B.; Li, Y.; Yuan, T.; Zhu, L.; Liu, J.; Guo, X.; Sun, Y. Antimicrobial Resistance and Prevalence of Extended Spectrum β-Lactamase-Producing Escherichia Coli from Dogs and Cats in Northeastern China from 2012 to 2021. *Antibiotics* **2022**, *11*, 1506. [CrossRef] [PubMed]
94. Shariati, A.; Dadashi, M.; Chegini, Z.; Van Belkum, A.; Mirzaii, M.; Khoramrooz, S.S.; Darban-Sarokhalil, D. The Global Prevalence of Daptomycin, Tigecycline, Quinupristin/Dalfopristin, and Linezolid-Resistant *Staphylococcus aureus* and Coagulase–Negative Staphylococci Strains: A Systematic Review and Meta-Analysis. *Antimicrob. Resist. Infect. Control* **2020**, *9*, 56. [CrossRef] [PubMed]
95. Vercelli, C.; Amadori, M.; Gambino, G.; Re, G. Does Nitrofurantoin Improve the Portfolio of Vets against Resistant Bacteria in Companion Animals? *Antibiotics* **2023**, *12*, 911. [CrossRef] [PubMed]

veterinary sciences

MDPI

Article

Antimicrobial Peptide Reduces Cytotoxicity and Inflammation in Canine Epidermal Keratinocyte Progenitor Cells Induced by *Pseudomonas aeruginosa* Infection

Jae-Eun Hyun [1,*] and Cheol-Yong Hwang [2]

[1] Department of Veterinary Internal Medicine, College of Veterinary Medicine, Konkuk University, Seoul 05029, Republic of Korea

[2] Laboratory of Veterinary Dermatology, Research Institute for Veterinary Science, College of Veterinary Medicine, Seoul National University, Seoul 08826, Republic of Korea

* Correspondence: jaeeunhyun@konkuk.ac.kr

Simple Summary: *Pseudomonas aeruginosa* is a representative Gram-negative bacterial species that causes chronic deep infections in the skin and ears of dogs. Increasing *P. aeruginosa* antibiotic resistance in human and veterinary medicine requires the identification of new antibacterial substances. In this study, we demonstrated the antibiotic and antibiofilm activities of synthetic canine antimicrobial peptides (AMPs) against *P. aeruginosa*. In addition, it was confirmed that AMPs significantly reduced the cell toxicity induced by *P. aeruginosa* and reduced the *P. aeruginosa* lipopolysaccharide (LPS)-induced inflammation in canine keratinocytes. These findings suggest the potential of AMPs as a new antibacterial agent for the *P. aeruginosa* infection of canine skin.

Abstract: The direct effects and antimicrobial activity of synthetic antimicrobial peptides (AMPs) obtained from dogs, including cBD, cBD103, and cCath, against *P. aeruginosa* wild-type strain PAO1 and canine keratinocytes were analyzed. Antibacterial effects on planktonic bacteria were assessed by determining the minimum bactericidal concentrations (MBCs) of AMPs and by a time-kill assay. Antibiofilm effects were assessed using the microtiter plate assay. We also evaluated the effects of AMPs on cell cytotoxicity and host immune response induced by stimulating canine epidermal keratinocyte progenitor (CPEK) cells with PAO1 and its LPS. cBD, cBD103, and cCath all exhibited dose-dependent antimicrobial and antibiofilm effects. In particular, 25 μg/mL cBD103 showed rapid bactericidal activity within 60 min and inhibited biofilm formation. In addition, pretreatment with cBD103 (25 μg/mL) and cCath (50 μg/mL) 1 h before stimulation significantly reduced the cytotoxicity of the CPEK cells by PAO1 and LPS-induced IL-6 and TNF-a expressions. cBD had little effect on the response to PAO1 and LPS in the cells. These results indicate the therapeutic potential of AMPs in *P. aeruginosa* skin infections. However, further studies on the mechanism of action of AMPs in keratinocytes and clinical trials are needed.

Keywords: antimicrobial peptides; pseudomonas aeruginosa; lipopolysaccharides; antibacterial activity; antibiofilm effect; dogs; keratinocytes

check for updates

Citation: Hyun, J.-E.; Hwang, C.-Y. Antimicrobial Peptide Reduces Cytotoxicity and Inflammation in Canine Epidermal Keratinocyte Progenitor Cells Induced by *Pseudomonas aeruginosa* Infection. *Vet. Sci.* **2024**, *11*, 235. https://doi.org/10.3390/vetsci11060235

Academic Editor: George Valiakos

Received: 16 April 2024
Revised: 20 May 2024
Accepted: 22 May 2024
Published: 23 May 2024

1. Introduction

Antimicrobial peptides (AMPs) are small endogenous peptides produced by cells in various animal tissues, and numerous types have been reported [1]. In the skin, host defense molecules, such as AMPs, synthesized by resident skin cells, such as keratinocytes, and recruited inflammatory cells participate in the innate cutaneous immune defense [2,3]. Cationic AMPs, in particular, contain a high concentration of positively charged amino acids that exert antibacterial action [4]. Positively charged AMPs initiate antibacterial activity by engaging negatively charged bacterial structures, such as lipopolysaccharides (LPSs), phospholipids, and teichoic acid. Cutaneous barrier defects result in abnormal

AMP secretion, consequently rendering skin vulnerable to infection [3,5]. Because of their complex structures and different modes of action against target cells, AMPs make it extremely difficult for pathogens to gain resistance [1]. The versatility of AMPs highlights their potency as alternatives to antibacterial drugs [6]. Several studies were conducted on the effects of AMPs against problematic antibiotic-resistant bacteria [6–8].

Pseudomonas aeruginosa is one of the most problematic bacterial agents that infect the human respiratory system and cause nosocomial infections, especially in immunosuppressed patients [9]. Several studies investigated the mode of action of *P. aeruginosa* in the human respiratory tract, including the effects of various virulence factors on epidermal cells [10,11]. In *P. aeruginosa* infection, the bacteria not only directly affect the epidermal cells but also induce an inflammatory response through the explosive formation of inflammatory intermediaries through cell signaling pathways. However, the inflammatory response caused by host immunity against *P. aeruginosa* is limited to removing the bacteria to overcome the infection [7]. In addition, the formation of a pseudomonal biofilm facilitates infection through several mechanisms, including protecting the bacteria within the biofilm and the acquisition of antibiotic resistance [12,13]. *P. aeruginosa* is a major bacterial causative agent of deep infections of the skin and ears in dogs [14,15]. An increase in *P. aeruginosa* antibiotic resistance and the emergence of multidrug-resistant *P. aeruginosa* have been reported in dogs [16]. Due to resistance to existing antibiotics, the development of new antibacterial treatments is an emerging necessity.

This study evaluated the antibacterial activity against *P. aeruginosa* of AMPs synthesized from the sequence of beta-defensin and cathelicidin in dogs. We also analyzed the direct antibacterial properties of synthetic peptides against *P. aeruginosa* and their inhibitory effects on pseudomonal biofilm formation. Furthermore, we investigated the effects of AMPs on cellular and inflammatory changes in canine keratinocytes infected with *P. aeruginosa* and its bacterial components.

2. Materials and Methods

2.1. Bacterial Strains and Reagents

The *P. aeruginosa* wild-type strain PAO1 used in this study was kindly provided by Professor Dr. Sang Sun Yoon of the Yonsei University College of Medicine, Seoul, Republic of Korea. For the in vitro bactericidal and antibiofilm assay, *P. aeruginosa* was cultured overnight in Luria Bertani (LB; Becton Dickson, Sparks, MD, USA) broth at 37 °C with shaking until the stationary phase was reached. Bacterial suspensions for infecting cells were prepared as described previously [7]. The concentration of the overnight culture was adjusted to an optical density (OD) of 0.1 (1×10^9 colony forming units (CFUs)/mL) at 600 nm in a Beckman spectrophotometer (Beckman Coulter, Brea, CA, USA), and the supernatant was removed after centrifugation at $450 \times g$ for 10 min. After washing three times with sterile phosphate-buffered saline (PBS), the pellet was resuspended in antibiotic-free cell culture medium. The resuspension was diluted to 10^6–10^8 CFU/mL immediately before the cell infection.

Preparation of LPS isolated from *P. aeruginosa* (L9143, Sigma-Aldrich, St. Louis, MO, USA) was performed as described previously with some modifications [7]. The LPS was dissolved at a concentration of 1 mg/mL and stored at 4 °C until use. Endotoxin-free water (InvivoGen, San Diego, CA, USA) was used for the initial dissolution, and antibiotic-free culture medium was used for further dilution for cell experiments.

2.2. Peptide Synthesis

All peptides were synthesized by solid-phase F-moc chemistry at Lugen Sci Co., Ltd. (Bucheon, Republic of Korea). Each synthetic peptide was subsequently purified to greater than 95% on a reverse-phase high-performance liquid chromatography system. Then, each peptide mass was determined by mass spectroscopy. The sequences of the peptides were derived from canine beta-defensin and cathelicidin as previously described [8] and are presented in Table 1. The peptides were stored as desiccated powders before use. For the in vitro bactericidal experiments, the desiccated powders were suspended in 10 mm of 0.01% acetic acid (Sigma-Aldrich) to a final concentration of 1 mg/mL and then further diluted with 10 mM sodium phosphate buffer (SPB, pH 7.4). However, for the cell experiments, desiccated powders were suspended in endotoxin-free water (InvivoGen) immediately before use [6]. All peptide dilutions were stored at −20 °C in 100 µL aliquots until further use.

Table 1. Peptide sequences used in this study.

Peptide Name	Sequence	Molecular Weight *	NCBI Reference Sequence
cBD	KCWNLRGSCREKCIKNEKLYIFCTSGKLCCLKPK	3994.92	NM_001313788.1 (202–303, 102 bp)
cBD103	GIINTLQRYYCRIRSGRCALLSCLPKEEQIGRCSSTGRKCCRRKK	5206.23	NM_001129980.1 (180–314, 135 bp)
cCath	RLKELITTGGQKIGEKIRRIGQRIKDFFKNLQPREEKS	4512.29	NM_001003359.1 (443–556, 114 bp)

* Molecular weights were determined by mass spectroscopy.

2.3. *Evaluation of Direct Antibacterial and Antibiofilm Activities of Synthetic Peptides against P. aeruginosa PAO1*

2.3.1. Effects of Antimicrobial Peptides on Planktonic *P. aeruginosa* PAO1

To determine the antimicrobial activity of AMPs against *P. aeruginosa*, the minimum bactericidal concentration (MBC) was determined, and the time-kill assay was performed as described previously with some modifications [17]. Briefly, the cultures were grown overnight, centrifuged at 5000 rpm for 5 min, washed once with PBS, and the pellet obtained was suspended in LB medium. The OD of the bacterial suspension was adjusted to 0.1 at 600 nm. Each well of 96-well round-bottom plates (SPL, Seoul, Republic of Korea) was inoculated at 5×10^5 CFU/mL. Peptides were added by serial dilution to the bacterial suspension at concentrations from 6.25 to 100 µg/mL. After incubation at 37 °C for 2 h, 20 µL of each culture medium was subcultured in tryptic soy agar with 5% sheep blood (Hangang, Gunpo, Republic of Korea). The agar plates were incubated aerobically at 37 °C for 24 h, and the number of colonies was counted. The MBC was defined as the lowest concentration at which 99.9% of the test bacteria were killed.

To evaluate the time-killing effect of AMPs, each one was inoculated at its MBC into 5×10^5 CFU/mL of PAO1 and incubated for a specified time to evaluate the change in CFUs as described above. The CFUs were measured every 10 min for 60 min, and every 30 min for 180 min thereafter. The peptide-free SPB solution described earlier was used as a negative control in both assays. All experiments were performed independently in triplicate.

2.3.2. Biofilm Formation Assay

The antibiofilm effects of AMPs were evaluated using the 96-well microtiter plate assay as previously described with some modifications [18]. Briefly, PAO1 was incubated in LB medium overnight. When the culture reached a stationary phase, it was diluted 1:100 in fresh LB medium. The culture (100 µL) was dispensed into four replicate wells in 96-well microtiter plates (SPL), and AMPs were added at serially diluted concentrations (range, 6.25–50 µg/mL). After incubation for 22 h at 37 °C, planktonic bacteria were removed. The biofilm was stained with crystal violet, and the absorbance was measured at 595 nm using a microplate absorbance reader (Bio-Rad, Munich, Germany). All experiments were performed independently in triplicate.

2.4. *Effects of AMPs of Keratinocytes*

2.4.1. Cell Culture

Canine epidermal keratinocyte progenitor (CPEK) cells were purchased from CELL-nTEC Advanced Cells Systems (Bern, Switzerland). The cells were cultured in keratinocyte culture medium (CnT-09, CELLnTEC) according to the manufacturer's instructions. CPEK cells were plated into 12- or 24-well tissue culture plates (SPL) at a density of approximately 1×10^5 cells/cm^2 and incubated at 37 °C in a 5% CO_2 humidified atmosphere until the cells reached 80–90% confluency. Cells between the fifth and seventh passages were used.

2.4.2. Cytotoxicity of AMPs to CPEK Cells

The cytotoxicity of AMPs to CPEK cells was measured using the EZ-Cytox cell viability kit (Daeil Laboratories, Seoul, Republic of Korea) based on the water-soluble tetrazolium salt (WST) assay according to the manufacturer's instructions. CPEK cells were seeded in 96-well plates at a density of 1×10^5 cells/cm^2 and incubated for 24 h at 37 °C under 5% CO_2. The cells were treated with various concentrations of AMPs and incubated for 24 h at 37 °C under 5% CO_2. Subsequently, 10 μL of EZ-Cytox reagent was added to each well. After further incubation for 4 h at 37 °C, the absorbance was measured at 450 nm using a microplate reader (Bio-Rad). The culture medium and 2% (v/v) Triton X-100 (Sigma-Aldrich, St. Louis, MO, USA) were used as negative and positive controls, respectively.

2.4.3. Effects of AMPs on the Cytotoxicity of *P. aeruginosa*

CPEK cells were seeded in 96-well culture plates at a density of 1.3×10^4 cells/well and incubated for 24 h. To determine the best multiplicity of infection (MOI) and infection timepoint of PAO1 in CPEK cells, CPEK cells were incubated with bacterial suspensions at MOIs of 0.1, 1, 10, and 100. After incubation (2, 4, or 6 h), the cytotoxicity of *P. aeruginosa* to CPEK cells was evaluated by measuring the release of lactate dehydrogenase (LDH) using the EZ-LDH cell cytotoxicity assay kit (Daeil Laboratories) according to the manufacturer's instructions. The cell-free supernatants were collected and centrifuged at 1000 rpm for 5 min. Subsequently, a 10 μL aliquot of each supernatant was reacted with 100 μL of the reaction mixture for 30 min in the dark. The cell culture medium was used as a negative control (0% toxicity), and 2% v/v Triton X-100 was used as a positive control (100% toxicity). The absorbance of the reaction was measured at 450 nm using a microplate reader (Bio-Rad). The effect of AMPs on the cytotoxicity of PAO1 was evaluated. The CPEK cells were seeded in 96-well culture plates at a density of 1.3×10^4 cells/well and incubated for 24 h. One hour before the cells were infected with *P. aeruginosa*, cells were pretreated with cBD (50 μg/mL), cBD103 (25 μg/mL), or cCath (50 μg/mL). Then, the cells were co-cultured with PAO1 at a MOI of 1 for 4 h. After 4 h incubation, the cell-free supernatants were collected, and the LDH assay was performed as above. All assays were performed in three independent experiments.

2.4.4. Determination of Cytokine Expression

CPEK cells were seeded in 96-well plates at a density of 1.0×10^6 cells/well and incubated at 37 °C in 5% CO_2. After 24 h incubation, the cell culture medium was replaced with fresh medium, and AMPs were added at various concentrations. After 1 h, *P. aeruginosa* LPS (1 μg/mL) was added to the cells and incubated for a designated time (6 and 24 h). The supernatants were removed and stored at −20 °C until used in an enzyme-linked immunosorbent assay (ELISA). Cytokines secreted by CPEK cells in supernatants were quantified using ELISA kits (R&D Systems, Minneapolis, MN, USA) according to the manufacturer's instruction.

2.5. *Statistical Analysis*

Data were statistically analyzed using the software program IBM SPSS Statistics version 23 for Windows (IBM Corp., Armonk, NY, USA). GraphPad Prism version 8 (GraphPad Software, Inc., La Jolla, CA, USA) was used to perform one- or two-way analysis of variance. Post hoc analysis was performed using Tukey's and Dunnett's multiple comparisons tests. Data are expressed as mean ± standard deviation. A p-value < 0.05 was considered statistically significant.

3. Results

3.1. AMPs Exerted Bactericidal Activity on Planktonic P. aeruginosa

The treatment of *P. aeruginosa* cultures with AMPs for 24 h showed that all AMPs exerted dose-dependent inhibition of the bacterial growth (Figure 1A). Differences were found in the concentrations of *P. aeruginosa* inhibited by the AMPs. The bacterial growth was completely inhibited at 25 µg/mL of cBD103, whereas, for cBD and cCath, the bacterial growth was inhibited at 50 µg/mL. The bactericidal kinetics of AMPs were evaluated using the time-kill assay at the concentration of complete inhibition (Figure 1B). cBD103 completely inhibited the bacterial growth at 25 µg/mL within 60 min. In contrast, 50 µg/mL cCath and 50 µg/mL CBD took 90 min and 150 min to inhibit the bacterial growth, respectively.

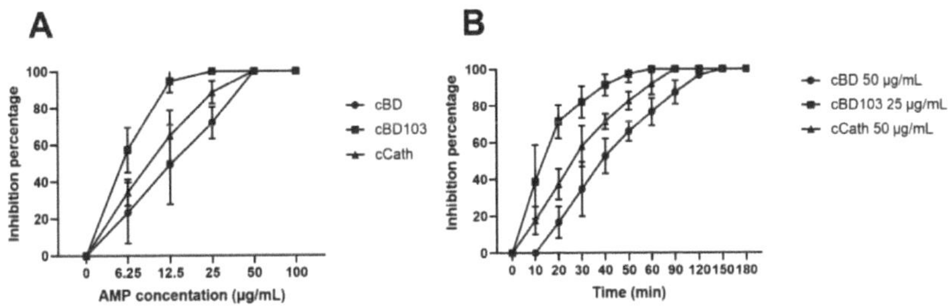

Figure 1. (**A**) Antimicrobial activity of cBD, cBD103, and cCath on PAO1. (**B**) Time-kill assay against Pseudomonas aeruginosa.

3.2. AMPs Suppressed Pseudomonal Biofilm Formation

All AMPs dose-dependently reduced the biofilm formation by PAO1 (Figure 2). However, no significant reduction in biofilm formation was observed at any cBD concentration used in this experiment (Figure 2A). In contrast, cBD103 and cCath significantly reduced the biofilm viability at concentrations of 25 µg/mL and 50 µg/mL, respectively (Figure 2B,C).

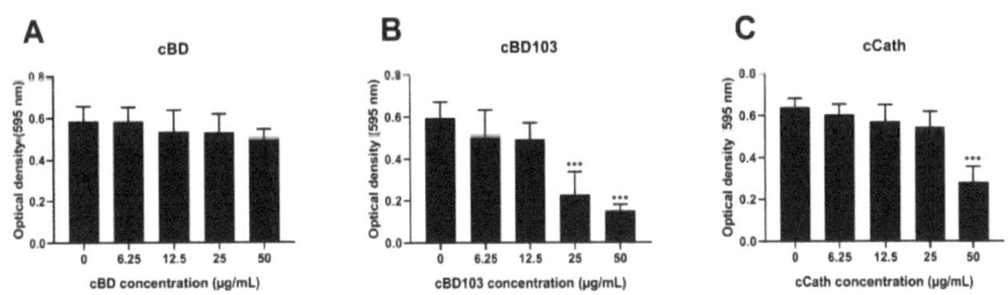

Figure 2. Inhibitory activity against pseudomonal biofilm formation of (**A**) cBD, (**B**) cBD103, and (**C**) cCath. Biofilm formation was evaluated by crystal violet staining. The absorbance was measured at 595 nm using a microplate absorbance reader. All experiments were performed independently in triplicate and analyzed using one-way analysis of variance with Tukey's multiple comparisons. Results are expressed as mean ± standard deviation. *** $p < 0.001$.

3.3. AMPs Alleviated the Cytotoxicity of Canine Keratinocytes Induced by PAO1

After the CPEK cells were treated with AMPs at various concentrations, the cell viability was verified by a WST assay. The AMPs themselves exhibited little to no cytotoxic effects on the CPEK cells (Figure 3). Interestingly, however, after the treatment with cCath at 50 µg/mL, the CPEK cell viability decreased to 80% (Figure 3C). PAO1 was inoculated into CPEK cells at an MOI of 1 for 4 h based on the results that determined the cell infectivity of PAO1 in the CPEK cells (Figure 4A). The cytotoxicity of keratinocytes after bacterial stimulation was further confirmed using the LDH assay. cBD103 at a 25 μg/mL concentration significantly reduced the cell toxicity induced by PAO1, whereas 50 μg/mL cCath also decreased the cytotoxicity but not significantly (Figure 4B). In contrast, cBD had little effect on the cytotoxicity.

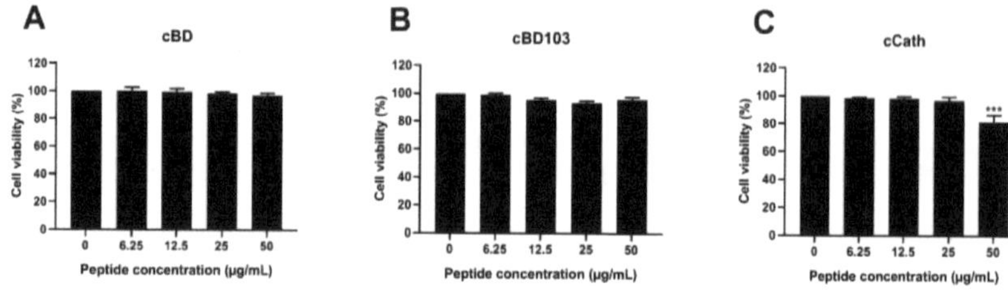

Figure 3. Effects of (**A**) cBD, (**B**) cBD103, and (**C**) cCath on the viability of CPEK cells. Cell viability and proliferation were evaluated using an EZ-Cytox cell viability kit. Values are expressed as the mean ± standard deviation and were analyzed using one-way analysis of variance and Tukey's multiple comparisons test in three independent experiments. *** $p < 0.001$.

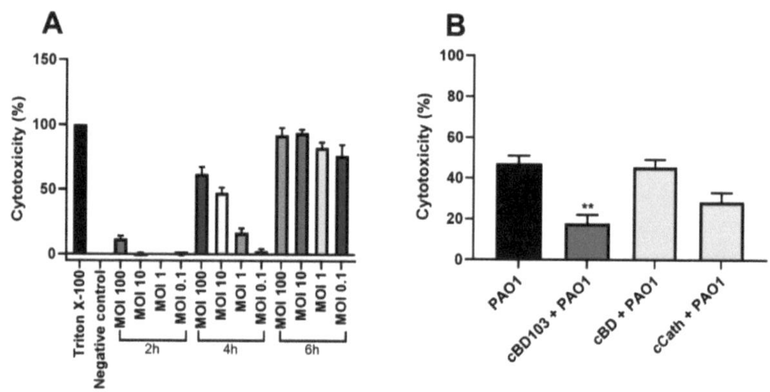

Figure 4. *Pseudomonas aeruginosa*-induced cytotoxicity of CPEK cells. (**A**) Infectivity of *P. aeruginosa* in CPEK cells was established by inoculating PAO1 at multiples of infection of 0.1, 1, 10, or 100 for a designated time (2, 4, or 6 h). (**B**) The effect of antimicrobial peptides on *P. aeruginosa*-induced cytotoxicity on CPEK cells. Cell cytotoxicity was assessed using an EZ-LDH cell cytotoxicity assay kit. Results are expressed as mean ± standard deviation and were analyzed using one-way analysis of variance and Tukey's multiple comparisons test in three independent experiments. ** $p < 0.01$.

3.4. AMPs Mitigated *P. aeruginosa* LPS-Induced Inflammation in Canine Keratinocytes

The expression of proinflammatory cytokines was analyzed by ELISA after 6 or 24 h of stimulation with 1 μg/mL *P. aeruginosa* LPS. LPS significantly increased IL-6 and TNF-α expression in keratinocytes, with no difference in the degree of cytokine expression between the two time points (Figure 5). After the cells were pretreated with AMPs 1 h before LPS stimulation, the changes in the expression of proinflammatory cytokines were analyzed. The addition of cBD103 and cCath significantly reduced both IL-6 and TNF-α expressions. However, cBD had no significant effect on the cytokine expression.

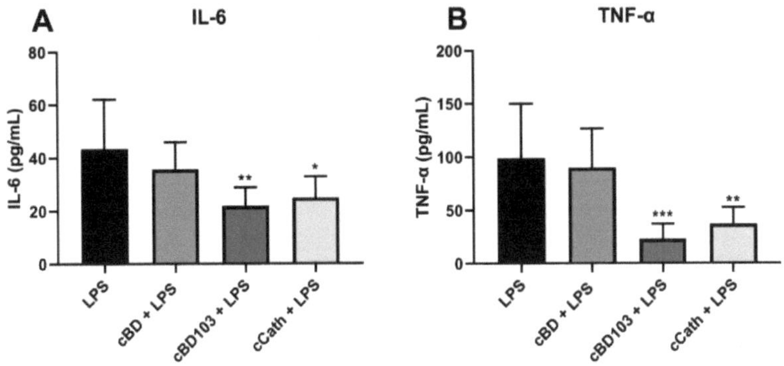

Figure 5. Cytokine release following stimulation with lipopolysaccharide (LPS) derived from *Pseudomonas aeruginosa*. The releases of (**A**) IL-6 and (**B**) TNF-α were quantified in the cell supernatants using an enzyme-linked immunosorbent assay. Antimicrobial peptides were pretreated 1 h before the LPS stimulation. Data are expressed as the mean ± standard deviation. Data were analyzed in three independent experiments conducted in triplicate using one-way analysis of variance with Tukey's multiple comparisons. * $p < 0.05$; ** $p < 0.01$; *** $p < 0.001$.

4. Discussion

Several studies investigated the potential application of AMPs as therapeutic agents and their effects on the immune response of the hosts [1,7]. Although the excellent antimicrobial effects of AMPs for pathogens, such as bacteria, viruses, and fungi, were demonstrated [3,19–21], studies on AMPs in veterinary medicine are limited. In the present study, we investigated the effects of AMPs on *P. aeruginosa* and on canine keratinocytes infected with *P. aeruginosa*. cBD, cBD103, and cCath, which were the synthetic cationic peptides used in this experiment, were derived from canine beta-defensin sequences and canine cathelicidin, which are produced and secreted in the skin of dogs [8,22]. All three peptides showed bactericidal effects against *P. aeruginosa*, as previously reported [8] (Figure 1A). Several studies showed that the antimicrobial effects of synthetic peptides vary with the type of peptide and strains and phenotype of bacteria used in the experiments [17,23,24]. The time-kill assay in this study showed that the onset times of the bactericidal effects of cBD and cCath were slightly slower than those reported previously [8]. However, cBD showed excellent antibacterial activity against *P. aeruginosa* within 1 h (Figure 1B). It has been reported that the antibacterial activity of AMPs is mainly caused by selective disruption of the cell membrane of the pathogens and by pore formation [17,25]. However, further research is required to address the mechanism of action of canine AMPs on pathogens. Furthermore, it has been reported that culture media can influence the chemical stability and minimum inhibitory concentration of compounds [26]. Therefore, additional research is necessary to verify the stability and antibacterial efficacy of canine AMPs in various culture media.

A pseudomonal biofilm not only acts as a barrier against antimicrobial agents but also promotes the acquisition of antibiotic resistance genetically [27]. The characteristic biofilm formation of *P. aeruginosa* is particularly problematic because it causes a treatment-refractory alert detection in dogs [28,29]. Similar to their antibacterial effect on planktonic *P. aeruginosa*, cBD103 and cCath inhibited biofilm formation, with cBD103 especially demonstrating excellent inhibitory activity at lower doses (12.25 μg/mL) (Figure 2). In contrast, cBD exhibited poor biofilm suppression, even at higher doses (Figure 2A). Further studies are needed to elucidate the mechanism of suppressing pseudomonal biofilm formation by AMPs.

Various synthetic AMPs have earned significant research attention for the past decade or so. Low cytotoxicity and high permeability to tissues and suitability for a wide range of microbial diseases are the key characteristics to the therapeutic efficacy of AMPs [30]. This study also demonstrated the low cytotoxicity of AMPs to keratinocytes; however, a slight decrease in cell viability was observed at higher concentrations of cCath (Figure 3C). A study described the synergistic cytotoxic effect of human cathelicidin LL-37 when combined with a *P. aeruginosa* strain 103 [6]. The synergistic cytotoxic effect of cCath and PAO1 was not demonstrated in this study. Although cCath at 50 μg/mL decreased the viability of the CPEK cells, it reduced the cytotoxicity of the CPEK cells inoculated with PAO1. However, further investigation into the cytotoxic potential of high concentrations of cCath on canine keratinocytes over an extended period is necessary before clinical application. The cytotoxicity of the CPEK cells was significantly decreased with the pretreatment of keratinocytes with cBD103 1 h before *P. aeruginosa* exposure (Figure 4). cBD itself did not exhibit cytotoxicity to the CPEK cells or affect the cytotoxicity of keratinocytes induced by PAO1.

We demonstrated that LPS isolated from *P. aeruginosa* increased the expressions of IL-6 and TNF-α, which are major proinflammatory cytokines that activate and coordinate the skin immune response against bacteria, in canine keratinocytes. However, the cytokine levels significantly decreased with AMP pretreatment at 1 h before the LPS was added. cBD103 and cCath reduced the inflammatory response in the CPEK cells, with similarities between the 6 and 24 h time points. Since the intracellular uptake of LPS occurs within 1 h [31], AMPs may have influenced the signaling pathways that induce the inflammatory reactions of keratinocytes rather than directly exerting activity against LPS, as was demonstrated in mouse macrophage cells and human bronchial epithelial cells [6]. As demonstrated in the antibacterial effects or cytotoxicity assay, cBD had little effect on the cytokine secretion.

The *P. aeruginosa* wild-type PAO1 strain was used in this study. Most of the pathogens isolated from patients with pulmonary cystic fibrosis were wild-type *P. aeruginosa*; however, in chronic infection, the bacteria showed diversification, such as conversion to a mucoid phenotype [17,32], which can induce a more vigorous inflammatory reaction in cells [7]. Although bacteria are very unlikely to acquire resistance to AMPs [1], a previous study reported that mucoid *P. aeruginosa* shielded nonmucoid variants and enhanced the resistance to AMPs [33]. Although AMPs exert superior antimicrobial effects against various pathogens, including multidrug-resistant bacteria, the effects of AMPs can vary slightly depending on the bacterial phenotype [17,23]. In addition, there are several phenotypic differences between the PAO1 strain and clinical isolates from humans and dogs [34]. Further studies are still needed to fully understand the efficacy of AMPs against clinical isolates from canine skin, which hold various virulence factors.

5. Conclusions

We demonstrated the direct effects of synthetic AMPs derived from canine AMP sequences on the *P. aeruginosa* wild-type strain PAO1 and elucidated the effect of PAO1 infection on cells. cBD, cBD103, and cCath all demonstrated antibiotic and antibiofilm activities comparable with those reported for human AMPs, with cBD103 exhibiting particularly rapid action at low doses. cBD103 was also noncytotoxic in itself but significantly reduced the cell toxicity of the CPEK cells inoculated with *P. aeruginosa*. cCath exerted some cytotoxicity at relatively higher concentrations (50 μg/mL) but also reduced the cytotoxicity of the CPEK cells infected with *P. aeruginosa*. CBD showed weak or no cytotoxicity effects on the keratinocytes induced by *P. aeruginosa*, as well as low antibiofilm suppression. These results demonstrate the therapeutic potential of AMPs with little to no cytotoxic effects on canine keratinocytes in the treatment of cutaneous *P. aeruginosa* infection in dogs. However, further studies of the chemical stability of AMPs and the optimal conditions for maximal effects, as well clinical trials in animal models, are required to confirm the current findings.

Author Contributions: Conceptualization, J.-E.H. and C.-Y.H.; methodology, J.-E.H.; formal analysis, J.-E.H.; investigation, J.-E.H. and C.-Y.H.; data curation, J.-E.H.; writing—original draft preparation, J.-E.H.; review and editing of the writing process, C.-Y.H.; supervision, J.-E.H.; project administration and funding acquisition, C.-Y.H. All authors have read and agreed to the published version of the manuscript.

Funding: This research received no external funding.

Institutional Review Board Statement: Not applicable.

Informed Consent Statement: Not applicable.

Data Availability Statement: The data presented in this study are available upon request from the corresponding author.

Conflicts of Interest: The authors declare no conflicts of interest.

References

1. Jenssen, H.; Hamill, P.; Hancock, R.E. Peptide antimicrobial agents. *Clin. Microbiol. Rev.* **2008**, *19*, 491–511. [CrossRef] [PubMed]
2. Braff, M.H.; Di Nardo, A.; Gallo, R.L. Keratinocytes store the antimicrobial peptide cathelicidin in lamellar bodies. *J. Investig. Dermatol.* **2005**, *124*, 394–400. [CrossRef] [PubMed]
3. Schauber, J.; Gallo, R.L. Antimicrobial peptides and the skin immune defense system. *J. Allergy Clin. Immunol.* **2008**, *122*, 261–266. [CrossRef] [PubMed]
4. Scott, M.G.; Yan, H.; Hancock, R.E. Biological properties of structurally related α-helical cationic antimicrobial peptides. *Infect. Immun.* **1999**, *67*, 2005–2009. [CrossRef] [PubMed]
5. van Damme, C.M.; Willemse, T.; van Dijk, A.; Haagsman, H.P.; Veldhuizen, E.J. Altered cutaneous expression of beta-defensins in dogs with atopic dermatitis. *Mol. Immunol.* **2009**, *46*, 2449–2455. [CrossRef]
6. Wuerth, K.C.; Falsafi, R.; Hancock, R.E.W. Synthetic host defense peptide IDR-1002 reduces inflammation in *Pseudomonas aeruginosa* lung infection. *PLoS ONE* **2017**, *12*, e0187565. [CrossRef] [PubMed]
7. Dosunmu, E.F.; Emeh, R.O.; Dixit, S.; Bakeer, M.K.; Coats, M.T.; Owen, D.R.; Pillai, S.R.; Singh, S.R.; Dennis, V.A. The antimicrobial peptide TP359 attenuates inflammation in human lung cells infected with *Pseudomonas aeruginosa* via TLR5 and MAPK pathways. *PLoS ONE* **2017**, *12*, e0176640. [CrossRef]
8. Santoro, D.; Maddox, C.W. Canine antimicrobial peptides are effective against resistant bacteria and yeasts. *Vet. Dermatol.* **2014**, *25*, 35-e12. [CrossRef] [PubMed]
9. Kerr, K.G.; Snelling, A.M. *Pseudomonas aeruginosa*: A formidable and ever-present adversary. *J. Hosp. Infect.* **2009**, *73*, 338–344. [CrossRef]
10. Golovkine, G.; Reboud, E.; Huber, P. *Pseudomonas aeruginosa* takes a multi-target approach to achieve junction breach. *Front. Cell. Infect. Microbiol.* **2017**, *7*, 532.
11. Lau, G.W.; Hassett, D.J.; Britigan, B.E. Modulation of lung epithelial functions by *Pseudomonas aeruginosa*. *Trends. Microbiol.* **2005**, *13*, 389–397. [CrossRef] [PubMed]
12. Høiby, N.; Ciofu, O.; Bjarnsholt, T. *Pseudomonas aeruginosa* biofilms in cystic fibrosis. *Future Microbiol.* **2010**, *5*, 1663–1674. [CrossRef] [PubMed]
13. Sauer, K.; Camper, A.K.; Ehrlich, G.D.; Costerton, J.W.; Davies, D.G. *Pseudomonas aeruginosa* displays multiple phenotypes during development as a biofilm. *J. Bacteriol.* **2002**, *184*, 1140–1154. [CrossRef] [PubMed]
14. Hillier, A.; Alcorn, J.R.; Cole, L.K.; Kowalski, J.J. Pyoderma caused by *Pseudomonas aeruginosa* infection in dogs: 20 cases. *Vet. Dermatol.* **2006**, *17*, 432–439. [CrossRef] [PubMed]
15. Nuttall, T.; Cole, L.K. Evidence-based veterinary dermatology: A systematic review of interventions for treatment of Pseudomonas otitis in dogs. *Vet. Dermatol.* **2007**, *18*, 69–77. [CrossRef] [PubMed]
16. Hyun, J.E.; Chung, T.H.; Hwang, C.Y. Identification of VIM-2 metallo-β-lactamase-producing *Pseudomonas aeruginosa* isolated from dogs with pyoderma and otitis in Korea. *Vet. Dermatol.* **2018**, *29*, 186-e68. [CrossRef] [PubMed]
17. Cabassi, C.S.; Sala, A.; Santospirito, D.; Alborali, G.L.; Carretto, E.; Ghibaudo, G.; Taddei, S. Activity of AMP2041 against human and animal multidrug resistant *Pseudomonas aeruginosa* clinical isolates. *Ann. Clin. Microbiol. Antimicrob.* **2017**, *16*, 17. [CrossRef] [PubMed]
18. Merritt, J.H.; Kadouri, D.E.; O'Toole, G.A. Growing and analyzing static biofilms. *Curr. Protoc. Microbiol.* **2005**, *1*, 1B.1.1–1B.1.18.
19. Fazakerley, J.; Crossley, J.; McEwan, N.; Carter, S.; Nuttall, T. In vitro antimicrobial efficacy of β-defensin 3 against Staphylococcus pseudintermedius isolates from healthy and atopic canine skin. *Vet. Dermatol.* **2010**, *21*, 463–468. [CrossRef] [PubMed]
20. Fritz, P.; Beck-Jendroschek, V.; Brasch, J. Inhibition of dermatophytes by the antimicrobial peptides human β-defensin-2, ribonuclease 7 and psoriasin. *Med. Mycol.* **2012**, *50*, 579–584. [CrossRef]
21. Baroni, A.; Buommino, E.; De Gregorio, V.; Ruocco, E.; Ruocco, V.; Wolf, R. Structure and function of the epidermis related to barrier properties. *Clin. Dermatol.* **2012**, *30*, 257–262. [CrossRef] [PubMed]

22. Santoro, D.; Bunick, D.; Graves, T.K.; Campbell, K.L. Expression and distribution of antimicrobial peptides in the skin of healthy beagles. *Vet. Dermatol.* **2011**, *22*, 61–67. [CrossRef] [PubMed]
23. Jindal, H.M.; Le, C.F.; Mohd Yusof, M.Y.; Velayuthan, R.D.; Lee, V.S.; Zain, S.M.; Isa, D.M.; Sekaran, S.D. Antimicrobial activity of novel synthetic peptides derived from indolicidin and ranalexin against Streptococcus pneumoniae. *PLoS ONE* **2015**, *10*, e0128532. [CrossRef] [PubMed]
24. Le, C.F.; Yusof, M.Y.; Hassan, H.; Sekaran, S.D. In vitro properties of designed antimicrobial peptides that exhibit potent antipneumococcal activity and produces synergism in combination with penicillin. *Sci. Rep.* **2015**, *5*, 9761. [CrossRef] [PubMed]
25. Sharma, S.; Sahoo, N.; Bhunia, A. Antimicrobial peptides and their pore/ion channel properties in neutralization of pathogenic microbes. *Curr. Top. Med. Chem.* **2016**, *16*, 46–53. [CrossRef] [PubMed]
26. Rad, H.I.; Arzanlou, M.; Omid, M.R.; Ravaji, S.; Doghaheh, H.P. Effect of culture media on chemical stability and antibacterial activity of allicin. *J. Funct. Foods* **2017**, *28*, 321–325.
27. Mah, T.F.; Pitts, B.; Pellock, B.; Walker, G.C.; Stewart, P.S.; O'Toole, G.A. A genetic basis for *Pseudomonas aeruginosa* biofilm antibiotic resistance. *Nature* **2003**, *426*, 306–310. [CrossRef]
28. Roland, P.S.; Stroman, D.W. Microbiology of acute otitis externa. *Laryngoscope* **2002**, *112*, 1166–1177. [CrossRef]
29. Post, J.C.; Hiller, N.L.; Nistico, L.; Stoodley, P.; Ehrlich, G.D. The role of biofilms in otolaryngologic infections: Update 2007. *Curr. Opin. Otolaryngol. Head Neck Surg.* **2007**, *15*, 347–351. [CrossRef]
30. Nuti, R.; Goud, N.S.; Saraswati, A.P.; Alvala, R.; Alvala, M. Antimicrobial Peptides: A Promising Therapeutic Strategy in Tackling Antimicrobial Resistance. *Curr. Med. Chem.* **2017**, *24*, 4303–4314. [CrossRef]
31. Dunzendorfer, S.; Lee, H.-K.; Soldau, K.; Tobias, P.S. TLR4 is the signaling but not the lipopolysaccharide uptake receptor. *J. Immunol.* **2004**, *173*, 1166–1170. [CrossRef] [PubMed]
32. Sousa, A.M.; Pereira, M.O. *Pseudomonas aeruginosa* diversification during infection development in cystic fibrosis lungs—A review. *Pathogens* **2014**, *3*, 680–703. [CrossRef]
33. Malhotra, S.; Limoli, D.H.; English, A.E.; Parsek, M.R.; Wozniak, D.J. Mixed communities of mucoid and nonmucoid *Pseudomonas aeruginosa* exhibit enhanced resistance to host antimicrobials. *mBio* **2018**, *9*, e00275-18. [CrossRef] [PubMed]
34. Milivojevic, D.; Šumonja, N.; Medić, S.; Pavic, A.; Moric, I.; Vasiljevic, B.; Senerovic, L.; Nikodinovic-Runic, J. Biofilm-forming ability and infection potential of *Pseudomonas aeruginosa* strains isolated from animals and humans. *Pathog. Dis.* **2018**, *76*, fty041. [CrossRef] [PubMed]

veterinary sciences

MDPI

Article

In Vitro Antimycotic Activity and Structural Damage against Canine *Malassezia pachydermatis* Strains Caused by Mexican Stingless Bee Propolis

Diana Berenice Fuentes Esquivel [1], Betsabé Rodríguez Pérez [1], Nelly Tovar Betancourt [1], Carlos Gerardo García Tovar [2], José Guillermo Penieres Carrillo [3], Florentina Hernández Galindo [4], Javier Pérez Flores [5] and Tonatiuh Alejandro Cruz Sánchez [1,*]

1 Laboratorio de Servicio de Análisis de Propóleos (LASAP®), Facultad de Estudios Superiores Cuautitlán, Campo 4, National Autonomous University of México, Km 2.5 Carretera Cuautitlán-Teoloyucan, San Sebastián Xhala, Cuautitlán Izcalli 54714, State of México, Mexico; db_ciani@hotmail.com (D.B.F.E.); betsarguez79@gmail.com (B.R.P.); nelly.tovar@cuautitlan.unam.mx (N.T.B.)
2 Laboratorio de Morfología Veterinaria y Biología Celular, Facultad de Estudios Superiores Cuautitlán, Campo 4, National Autonomous University of México, Km 2.5 Carretera Cuautitlán-Teoloyucan, San Sebastián Xhala, Cuautitlán Izcalli 54714, State of México, Mexico; cgarciatov@yahoo.com.mx
3 Laboratorio de Química Orgánica, Facultad de Estudios Superiores Cuautitlán, Campo 1, National Autonomous University of México, Av. 1o de Mayo s/n, Santa María las Torres, Cuautitlán Izcalli 54714, State of México, Mexico; penieres@unam.mx
4 Colegio de la Frontera Sur (ECOSUR) San Cristóbal de las Casas, Panamerican Highway Periférico Sur s/n, Barrio María Auxiliadora, San Cristóbal de las Casas 29290, Chiapas, Mexico; florentina.hega@gmail.com
5 Laboratorio de Espectrometría de Masas, Instituto de Química, National Autonomous University of México, Circuito exterior s/n Circuito de la Investigación Científica, Ciudad Universitaria, México City 04510, Mexico; japeflo10@hotmail.com
* Correspondence: tonatiuhcruz@comunidad.unam.mx; Tel.: +52-5556231999 (ext. 39416)

check for updates

Citation: Fuentes Esquivel, D.B.; Pérez, B.R.; Betancourt, N.T.; García Tovar, C.G.; Penieres Carrillo, J.G.; Galindo, F.H.; Flores, J.P.; Sánchez, T.A.C. In Vitro Antimycotic Activity and Structural Damage against Canine *Malassezia pachydermatis* Strains Caused by Mexican Stingless Bee Propolis. *Vet. Sci.* **2024**, *11*, 106. https://doi.org/10.3390/vetsci11030106

Academic Editors: Karl Pedersen and George Valiakos

Received: 18 December 2023
Revised: 8 February 2024
Accepted: 21 February 2024
Published: 28 February 2024

Simple Summary: Surprisingly, there is little research on the antimicrobial activity of stingless bee propolis on disease-causing microorganisms in animals. The present work demonstrates the effect of propolis from two native Mexican bees, *Scaptotrigona mexicana* and *Tetragonisca angustula*, on the yeast *Malassezia pachydermatis*, the main agent of canine otitis externa, and its damage to the cellular structure. The chemical analysis showed that the most abundant components are some sesquiterpenes. The antifungal activity of the propolis was evaluated using both strains isolated from clinical cases and a reference strain. Both types of propolis inhibited all Malassezia pachidermatis strains. Cell damage was assessed by fluorescence microscopy with calcofluor white, which specifically stains the fungal cell wall, and propidium iodide, which has the ability to stain the interior of the cell, only if the cell wall or membrane has been damaged. The propidium iodide staining of the yeasts treated with both types of propolis revealed the penetration of this marker, which suggests the destruction of the cell wall and plasma membrane of the fungi. These results suggest that these types of propolis could be used as alternative treatments for canine external otitis. This seems to be the first scientific report that has demonstrated structural damage in *Malassezia pachydermatis* by Mexican stingless bee propolis.

Abstract: This work describes the antimycotic activity of propolis from the stingless bees *Scaptotrigona mexicana* and *Tetragonisca angustula*, collected from two Mexican regions (Veracruz and Chiapas, respectively), against three clinical isolates and the reference strain ATCC 14522 of *Malassezia pachydermatis*, the causative agent of canine otitis. The chemical components of the ethanolic extracts of propolis were determined by gas chromatography coupled with mass spectrometry (GC-MS), and sesquiterpenes were the predominant compounds. The antimycotic activity was evaluated by plate microdilution. The induced changes in the yeasts were evaluated by fluorescence microscopy and staining with calcofluor white and propidium iodide. The minimum inhibitory concentration (MIC) was 7.11 mg/mL, and the minimum fungicidal concentration was 21.33 mg/mL for both extracts. The EPPs of *Scaptotrigona mexicana* and *Tetragonisca angustula* caused substantial damage to yeast morphology, where the propidium iodide staining of the yeasts treated with both EEPs revealed the penetration of this marker, which indicates the destruction of the cell wall and plasma membrane of the fungi. This result suggests that these types of propolis could be used as alternative treatments for canine external otitis. To the best of our knowledge, this seems to be the first scientific report that has demonstrated structural damage in *Malassezia pachydermatis* by Mexican stingless bee propolis.

Keywords: Mexican stingless bee; propolis; antimicotic activity; structural damage; *Malassezia pachydermatis*

1. Introduction

Propolis is a natural resinous substance produced by bees. Propolis is derived from substances collected by bees from vegetation and shows antifungal, antibacterial, antiviral, and antiparasitic activities. Propolis shows variation in its biological activity depending on its geographical origin [1–3]. In Latin America, there is a great variety of ecosystems with diverse vegetation from which native bees extract propolis, which in turn results in exceptional medicinal richness. For this reason, chemotaxonomic studies on native bees in Mexico are scarce, despite the fact that there are 46 species [4].

In general, identifying the origin of the material with which bees produce propolis should be carried out, and the behaviour of each bee species in a region should be observed [5–7]. Since ancient times, products elaborated by the stingless bees *Scaptotrigona mexicana* and *Tetragonisca angustula* have been used in Central America; however, there is little scientific evidence demonstrating the medicinal efficacy of these products. In comparison, the propolis produced by *Apis mellifera* has been extensively studied, and its fungicidal effects have been reported [8–10].

Moreover, the application of propolis can prove beneficial in veterinary medicine, for instance, with dogs. Propolis from *Apis mellifera* has been used as a prophylactic agent against gastrointestinal and respiratory diseases and mycoses, as well as a wound-healing agent, and its therapeutic use has spread to many areas [11,12], such as the treatment of canine external otitis. This condition can be defined as the inflammation of the external auditory canal and represents between 5% and 20% of consultations. The main causative agent of canine otitis externa is the yeast *Malassezia pachydermatis*, which is part of the normal microbiota of the external auditory canal in dogs [13–16].

Propolis can be an alternative to conventional antifungals for the treatment of canine otitis in patients with a high incidence of relapse because of its antifungal, antiinflammatory, and wound-healing properties. However, only the propolis obtained from *Apis mellifera* has been evaluated so far [17]. The antifungal activities of the propolis of other bee species and that of *Melipona beecheii* have been reported in Mexico; however, only their antifungal activities against *Candida albicans* were evaluated. It is important to mention that this activity is attributed to compounds such as sesquiterpenes and flavonones [2,3].

In addition to the inhibitory effect of a compound, possible damage to the microbial cell structure should be evaluated. As we demonstrated in a previous study by scanning electron microscopy, *Apis mellifera* propolis produced the formation of pores and promoted the destruction of the cell structure of *Malassezia pachydermatis* [18]. In this work, we will use fluorescence microscopy using calcofluor white and propidium iodide stains to detect cell damage, as other authors have done with fungi [19].

Research on the antimicrobial activity of stingless bee propolis on disease causing microorganisms in animals is scarce; therefore, studies on the antifungal potential of propolis in this type of bee may be of veterinary interest. Therefore, the aim of this study is to provide data regarding the antimycotic properties of propolis from native bees.

We expect that this study will provide scientific evidence that supports the use of propolis as an alternative treatment for canine otitis.

2. Materials and Methods

2.1. Ethanolic Extract of Propolis (EEP)

Propolis samples were obtained from *Scaptotrigona mexicana* and *Tetragonisca angustula* stingless bees. The *S. mexicana* sample was from Yecuatla, Veracruz, Mexico, located at 19°51 N and 96°46 W, at an altitude of 432 m.a.s.l. The *T. angustula* sample was from Chalchihuitan, Chiapas, located at 16°57 N and 92°37 W, at an altitude of 1461 m.a.s.l. The collected material was evaluated for its physical properties according to Mexican regulations regarding colour, odour, taste and consistency [20]. Propolis from *S. mexicana* (30 g) and *T. angustula* (12 g) was weighted, and any present impurities were eliminated. Thereafter, 100 mL of 70% ethanol was added to each sample, and the obtained mixture was subjected to ultrasonic extraction (Branson, CPX1800H, Danbury, CT, USA). Each sample was then vacuum-filtered, and the obtained filtrates were concentrated using a rotary evaporator (Science MED, SM100-PRO, Helsinki, Finland) and dried with a vacuum pump. Then, both dried extracts were placed in light-resistant containers and kept at 4 °C until use [21].

2.2. Gas Chromatography–Mass Spectrometry (GC-MS)

A chromatographic analysis of ethanolic extracts was performed using a gas chromatograph (6850) coupled to a mass spectrometer (7890 model, JEOL MC-GC-Mate II, Tokyo, Japan). A HP-5MS (30 m × 0.32 mm) capillary column and a film thickness of 0.25 μm were used. Helium gas was used as the carrier gas. The elected injection method was split mode with an injection volume of 1 μL. The separation conditions were as follows: 70 °C at the beginning for two minutes, followed by two ramp increments. The first one was an increase of 20 °C per minute until a temperature of 230 °C was reached; the second one was an increase of 8 °C per minute until a temperature of 290 °C was reached, keeping this temperature for a period of 5 min. The total analysis time was 21.25 min. The detected mass range was 35 m/z to 750 m/z, and each sample was subjected to electron impact ionisation at 70 eV, with the ionisation source reaching a temperature of 230 °C. Compound identification was carried out by comparison with the library database from the equipment [22].

2.3. Evaluation of Antimycotic Activity

2.3.1. Inoculum Preparation

Four *Malassezia pachydermatis* strains were used: ATCC 14522 and three clinical isolates from three German Shepherd dogs, two females and a male of three years of age, which presented symptoms of otitis externa with an accumulation of abundant foul-smelling ceruminous secretion. The patients were sampled on the premise that they had not received any treatment for otitis externa. The sample was obtained with a sterile swab and placed in a tube with Sabouraud Dextrose broth as a transport medium. A first seeding was carried out on modified Dixon Agar and incubated at 33 °C for 72 h, after which a reseeding was carried out on Sabouraud Dextrose Agar (SDA) and incubated at 33 °C for 72 h.

All strains were identified by biochemical testing [23]. Microorganisms were provided by the Laboratorio de Servicio de Análisis de Propóleos (LASAP®) of Facultad de Estudios Superiores Cuautitlán, Universidad Nacional Autónoma de México. To activate the *M. pachydermatis* strains, each type of yeast was seeded in modified Dixon agar (mDA). Each type of yeast was seeded in a different Petri dish and incubated for 72 h at 33 °C. Then, samples were reseeded in other mDA-containing plates and incubated for 48 h at 33 °C to rule out strain contamination [23,24]. A roast of the colonies sown with yeast was taken on Sabouraud Dextrose Agar (SDA) supplemented with 2% glucose (Bioxon, Monterrey, Mexico). It was incubated at 33 °C for 48 h. The inoculum density was adjusted according to the 0.5 tube of the MacFarland Nephelometer (1.5×10^6 cells/mL), comparing turbidity in a spectrophotometer at 625 nm with an absorbance between 0.08 and 0.10. From this inoculum (1.5×10^6 cells/mL), 100 μL was taken and placed in a tube with 9.9 mL of Dextrose Sabouraud broth to obtain a concentration of 1.5×10^3 CFU/mL [10].

2.3.2. Determination of Minimum Inhibitory Concentration and Minimum Fungicidal Concentration

The procedure of the microtechnique of dilution in broth was carried out according to the M27-A3 microdilution protocol for *Candida* spp., with adaptations for *Malassezia pachydermatis* [25,26]. Microdilution in broth was performed by determining the minimum inhibitory concentration (MIC) and the minimum fungicidal concentration (MFC). To this end, serial double dilutions of each EEP were performed to evaluate concentrations from 0.0001 to 21.33 mg/mL. Then, 50 μL of the inoculum of 10^3 CFU/mL was added to each well. The positive control was broth with microorganisms, and as a negative control, only broth was used, and the plate was subsequently incubated at 33 °C for 48 h. To detect the respiratory activity of *Malassezia pachydermtis*, a 0.08% solution of 2,3,5-Triphenyltetrazolium chloride for microbiology (TTC) (MERK, Darmstadt, Germany) was used, which generates a red pigment (formazan) in the presence of microorganisms. This procedure was performed as follows: 50 μL of TCC was added to each well, inoculated, mixed using a plate stirrer, and incubated at 33 °C for 30 min. After this time, the formation of an insoluble red precipitate was observed, representing the MIC. The CMF was determined in the well where no color developed, indicating that there was no yeast growth. To confirm the results, it was determined whether the effect was fungicidal by taking a sample of the crop with a loop and seeding it in an SDA plate that was kept in incubation at 33 °C for 48 h. The growth on the plate was considered to be indicative of a fungistatic effect, while its absence corresponds to a fungicidal effect [10].

2.4. Structural Damage

To evaluate the structural changes induced by the EEP on *Malazassia pachidermatis*, fluorescence microscopy was used, and the reference strain and one clinical strain were employed. A concentration of 21.3 mg/mL of each EEP of *Scaptotrigona mexicana* and *Tetragonisca angustula* was added to each strain. The concentration used was the minimum fungicide concentration obtained in the antifungal evaluation. Incubation was carried out at 33 °C for 48 h. When the incubation ended, the yeast was stained with calcofluor white (M2R 1 g/L, (Sigma Aldrich, St. Louis, MO, USA) and propidium iodide (2.4 mmol/L, (Sigma Aldrich, St. Louis, MO, USA)), and as a negative control, a culture without EEP was used. Calcofluor white staining stains the yeast wall blue and allows its integrity to be evaluated; propidium iodide binds to DNA, stains it red, and only penetrates the cells if there is damage to the cell wall, so cells damaged with EEP do not stain with calcofluor white due to damage to the cell wall, and they are stained with propidium iodide. Preparations were viewed on a microscopy Zeiss Axioscop 40, coupled to an Evolution VF Cooled Color camera from Media Cibernetics (Silver Spring, MD, USA). All experiments were performed in triplicate [19,27].

3. Results

3.1. Gas Chromatography–Mass Spectrometry (GC-MS)

The chemical composition of *S. mexicana* propolis extract is shown in Table 1. The database identified five compounds, two sesquiterpenes with antimicrobial activity, a heterocyclic compound called pyridazine with antioxidant properties, a macrocycle, and a compound of the furan class which has no information of any biological activity so far. The database detected other peaks but failed to identify them (Figure 1).

In the case of the *T. angustula* sample, only five compounds with biological activity were identified (Table 2). Terpenes with antifungal activity can be appreciated. It is also a compound with antibacterial activity (1,3-Benzenediol, 5-hexyl) as well as one with nematicide capacity (Hexadecanoic acid ethyl ester) The main peaks identified are shown in Figure 2.

Table 1. Constituents of Mexican *Scaptotrigona mexicana* propolis characterised by CG-MS.

Retention Time (min)	Compound Proposed by the Database	Chemical Classification	Biological Activity	Reference
30.60	(1.alpha.,4.alpha.,4a.alpha.,10a.alpha)-1,4,4a,5,6,9,10,10a-octahydro-11,11-dimethyl-1,4-methanocycloocta[d]pyridazine	Pyridazine (heterocyclic compound)	Antioxidant	[28]
31.58	Farnesol Isomer a	Sesquiterpene	Antimicrobial	[29]
32.48	Ethanone,1-(1,3a,4,5,6,7-hexahydro-4-hydroxy-3,8-dimethyl-5-azulenyl)-	Sesquiterpene ketone	Antimicrobial	[30]
33.68	3,4,5,6,7,8,9,10,11,12,13,14-Dodecahydro-18,18a-benzoxacyclohexadecin-16(18aH)-one dihydroxy methyl-2*H*-1--2-	Macrocycle	Activity not reported	[31]
33.73	Furan-2,5-dicarbaldehyde	Heterocyclic compound with aldehyde groups	Antioxidant, antimicrobial	[32,33]

Main compounds identified with an accurate identification (>90%) as related to the equipment database.

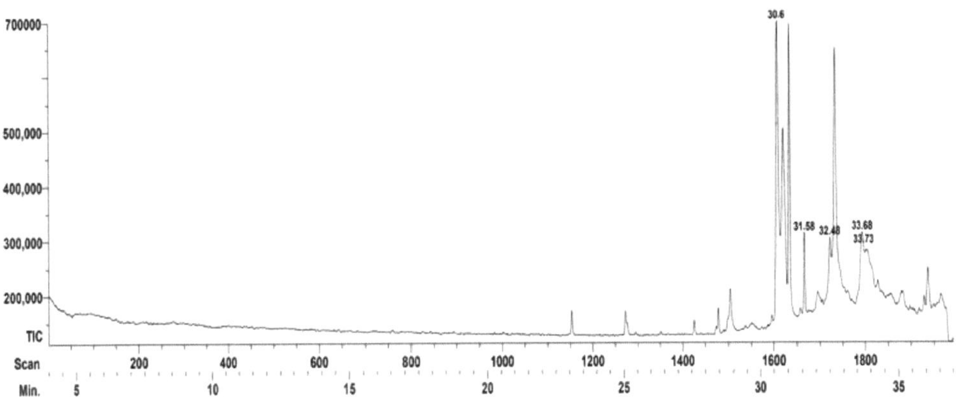

Figure 1. Gas chromatogram corresponding to the *Scaptotrigona mexicana* propolis. The retention time of each compound identified by the database is indicated (minutes): 30.6: 1,4-Methanocycloocta[d]piridazine, 1,4, 4a,5,6,9,10,10a-octahydro-11,11-dimethyl-,(1-alpha.,4-alpha,4a-alfa,10a-alfa)-; 31.58: Farnesol isomer a; 32.48: Ethanone, 1-(1,3a,4,5,6,7-hexahydro-4-hydroxy-3,8-dimethyl-5-azulenyl)-; 33.68: 2*H*-1-Benzoxacyclohexadecin-16(18a*H*)-one,3,4,5,6,7,8,9,10,11,12,13,14-dodecahydro-18,18a-dihidroxy-2-methyl; and 33.73: Furan-2,5-dicarbaldehyde. The characteristics of the compounds can be found in Table 1. Spikes that are not numbered were not identified by the team's database.

Table 2. Main constituents of propolis of *Tetragonisca angustula* characterised by CG-MS.

Retention Time (min)	Compound Proposed by the Database	Chemical Classification	Biological Activity	Reference
23.17	Hexadecanoic acid ethyl ester	Fatty acid	Antioxidant, hypocholesterolemic, nematicide pesticide	[34]
25.20	9-Octadecenoic acid, ethyl ester	Fatty acid	Antiinflammatory	[35]
28.98	1-(1,1-dimethylethoxy)-4-methylbenzene	Impurity	Not founded information	
30.68	Solavetivone	Sesquiterpenoid and a cyclic ketone	Antifungal, antiinflammatory	[36,37]
30.90	Benzene, 1-(1,1-dimethylpropoxy)-4-methyl	Impurity	No information found	
31.65	2-Methyl-3-(3-methyl-but-2-enyl)-2-(4-methyl-pent-3-enyl)-oxethane	Heterocycle Compound derivative	No information found	
32.75	(1S,6R,9S)-5,5,9,10-Tetramethyltricycle[7.3.0.0(1,6)]dodec-10(11)-ene	Sesquiterpene	Antibacterial antifungal	[38,39]
33.75	1,3-Benzenediol, 5-hexyl	Resorcinol derivative	Antibacterial, anthelmintic, local anaesthetic	[40]

Main compounds identified with an accurate identification (>90%) as related to the equipment database.

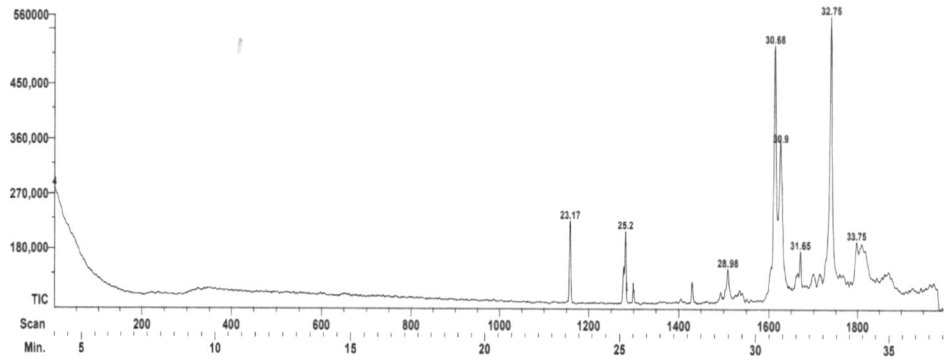

Figure 2. Gas chromatogram corresponding to the *Tetragonisca angustula* propolis. The retention time of each compound identified by the database is indicated (minutes): 23.17: Hexadecanoic acid ethyl ester; 25.20: 9-Octadecenoic acid, ethyl ester; 28.98: 1-(1,1-dimethylethoxy)-4-methylbenzene; 30.68: Solavetivone; 30.90: Benzene, 1-(1,1-dimethylpropoxy)-4-methyl; 31.65: 2-Methyl-3-(3-methyl-but-2-enyl)-2-(4-methyl-pent-3-enyl)-oxethane; and 32.75: (1S,6R,9S)-5,5,9,10-etramethyltricy-cle[7.3.0.0(1,6)]dodec-10(11)-ene. The characteristics of the compounds can be found in Table 2. Spikes that are not numbered were not identified by the team database.

3.2. Evaluation of the Antimycotic Activity

All the tested *Malassezia pachydermatis* strains were susceptible to the propolis extracts. The minimum inhibitory concentration (MIC) was 7.11 mg/mL, and the minimum fungicidal concentration (MFC) was 21.33 mg/mL (Table 3).

Table 3. Minimum inhibitory concentration (MIC) and minimum fungicidal concentration (MFC) values of stingless bee propolis extracts from two regions of the Mexican Republic on the reference strain *M. pachydermatis* ATCC 14522 and strains isolated from clinical samples.

Mexican Stingless Bee Species	Origin	*M. pachydermatis* ATCC 14522		Isolation Clinical *		Number of Isolates Clinical Inhibited *
		MIC (mg/mL)	MFC (mg/mL)	Media MIC (mg/mL)	Media MFC (mg/mL)	
Scaptotrigona mexicana	Yecuatla, Veracruz	7.11	21.33	7.11	21.33	3
Tetragonisca angustula	Chalchihuitan Chiapas	7.11	21.33	7.11	21.33	3

* $n = 3$.

3.3. Structural Damage

The structural damage to the yeasts was determined by using different stains: calcofluor white and propidium iodide. The structural damage of yeasts was determined by using different stains: calcofluor white and propidium iodide. Yeasts stained blue (calcofluor white) indicate integrity of the cell wall (control), those stained red (propidium iodide) indicate damage to the cell wall, since the damage wall allows the entry of propidium iodide (red) (EEP exposed). Propidium iodide penetration indicates damage to the yeast and with calcofluor-white stain, only morphology deformation was observed. In the untreated control cultures, the yeasts were clearly stained blue with calcofluor-white, but not with propidium iodide, which indicates the integrity of the yeast cell.

Figure 3 illustrates the effects of both propolis extracts on the reference strain of *M. pachydermatis* (ATCC 14522). The yeast samples treated with the EEP of *S. mexicana* did not show staining with calcofluor white but showed staining with propidium iodide in the form of red colouration, with the most severe damage being that produced by the EEP of *S. mexicana*.

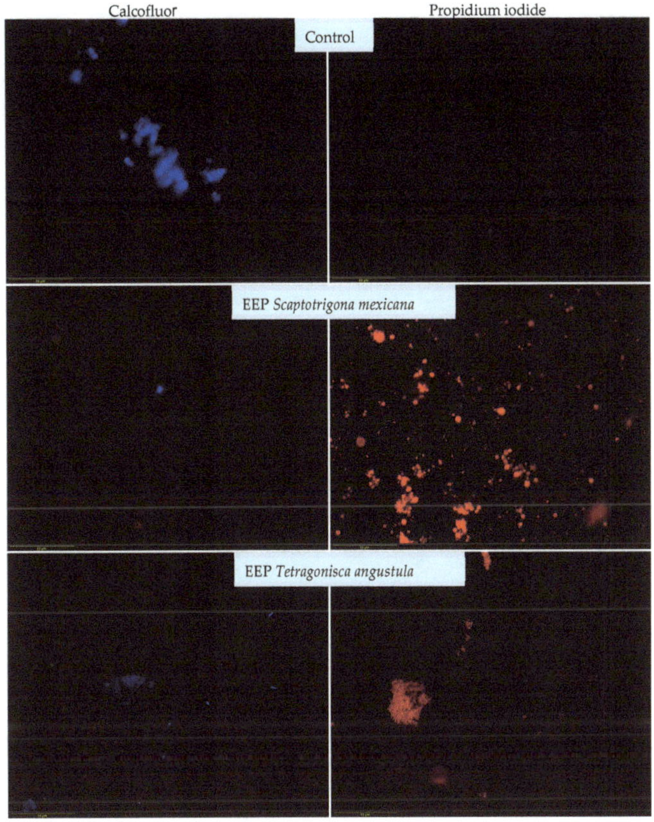

Figure 3. Effects of EEP of *Scaptotrigona mexicana* and *Tetragonisca angustula* on reference strain *Malassezia pachydermatis* ATCC 14522 were obtained by fluorescence microscopy and dyeing with calcofluor white and propidium iodide. Cultures were exposed to EEP at a concentration of 21.3 mg/mL for 48 h at 28 °C. Yeast samples stained in blue (calcofluor white) indicate the integrity of the cell wall (control), while those stained in red (propidium iodide) indicate damage to the cell wall. They do not stain in blue due to damage to the cell wall, which allows the entry of propidium iodide (red) (EEP exposed). Propidium iodide penetration was observed, indicating damage to the yeast. With the calcofluor white stain, only morphology deformation was observed (40× magnification).

Figure 4 shows the effects of both EEPs on the clinical strain and also shows the staining of the untreated yeasts with calcofluor white but not with propidium iodide, which indicates the integrity of the plasma membrane and cell wall. The yeast samples treated with the EEPs of *S. mexicana* and *T. angustula* exhibited a similar effect. In both cases, the yeast samples were not stained with calcofluor white but were stained with propidium iodide, which indicates that the EEP damaged the structure of the fungi. Considering both the reference strain and the clinical one, we can conclude that the EEPs of *S. mexicana* and *T. angustula* affect the structural integrity of the yeasts, and this is more evident in the clinical strain.

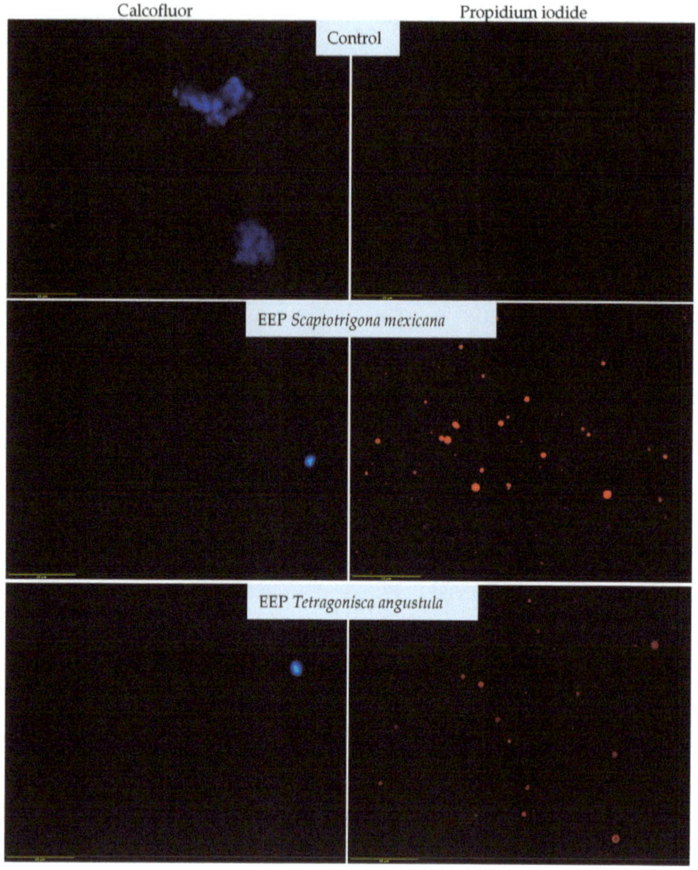

Figure 4. Effects of the EEPs of *Scaptotrigona mexicana* and *Tetragonisca angustula* on the clinical strain of *Malassezia pachydermatis*, as determined by fluorescence microscopy and dyeing with calcofluor-white and propidium iodide. Cultures were exposed to 21.3 mg/mL of EEP for 48 h at 28 °C. Alteration of morphology is observed with the calcofluor white stain. As in Figure 3, yeast samples stained in blue (calcofluor white) indicate the integrity of the cell wall (control), while those stained in red (propidium iodide) indicate damage to the cell wall. They do not stain in blue due to damage to the cell wall, which allows the entry of propidium iodide (red) (EEP exposed). The red colouration of propidium iodide in both extracts indicates greater damage to the yeast cells (40× magnification).

4. Discussion

Many of the identified metabolites of stingless bee propolis were reported to exhibit a myriad of different chemical compositions, which is consistent with other studies that have biological activities, including antimicrobial, antiinflammatory, cytotoxic, antioxidant, hepatoprotective, and antiulcer effects [41–45]. The propolis extracts analysed show a diversity of compounds [46,47]. The floral diversity, time of collection, and bee species are all determinant factors for the final composition of each propolis, where sesquiterpene compounds predominated; these compounds are known for their antimicrobial activity [1,3].

Regarding the propolis of *Scaptotrigona mexicana*, bibliographic research of furan-2,5-dicarbaldehyde (a heterocyclic compound with two aldehyde groups) was unsuccessful using that exact denomination; however, we found mention of a similar compound, 2-acetyl-5-methylfuran, which was reported to exhibit antimicrobial activity against *Escherichia coli*, *Candida albicans*, and *Staphylococcus aureus* [32,33,39].

On the other hand, as the propolis of *Tetragonisca angustula*, the antibacterial and antifungal activity of solavetivone has been reported [35,36]. There is no specific information on the activity of l-(1S,6R,9S)-5,5,9,10-tetramethyltricyclo[7.3.0.0(1,6)]dodec-10(11)-ene; however, antifungal and antibacterial activities have been reported for a similar compound, 3,3,7,7-tetramethyl-5-(2-methyl-1-propenyl)-tricyclo[4.1.0.0(2,4)]heptane [38].

A notable difference was observed between the chemical composition of the analysed propolis and that of Mexican *Apis mellifera.* The antimicrobial activity of *Apis mellifera* propolis is related to the presence of flavonoids such as pinocembrin, tectochrysin (flavone), and the flavonoid precursor cardamomin (chalcone), as well as 2-methoxy-4-vinylphenol and a few terpenoids [10]. On the other hand, in this work, sesquiterpenes were the most important compounds, which is in agreement with a study by Bankova [2], which showed that terpenoids predominate in the propolis of native bees from various parts of the world. This seems to be a marked difference between the propolis of honeybees and native bees.

The antimycotic activity of propolis extracts, mainly from *Apis mellifera*, was demonstrated against *Candida albicans*. Moreover, the fungicidal and fungistatic properties of green and red propolis extracts from Brazil against other fungi genera were reported [43]. In addition, the inhibition and morphologic alterations of *Cryptococcus neoformans* when exposed to propolis were described [46].

The antimycotic activity of propolis from stingless bees, mainly against *Candida albicans*, has been reported for the following species: *Lestrimellata* spp., *Melipona favora orlinge, Melipona marginata, Melipona quadrifasciata, Melipona scutellaris, Nannotrigona testaceicornis, Plebeia droryana, Plebeia remota, Scaptotrigona bipunctata, Tetragona clavipes, Tetragonisca angustula*, and *Tetragonisca fiebrigi* (against *Candida glabrata*) [2,47]. The propolis from the Malaysian stingless bee *Trigona thoracica* was demonstrated to act against *Cryptococcus neoformans* [47]. Furthermore, an Indonesian propolis from *Tetragonula* sp. was evaluated as a possible therapeutic agent for the treatment of vaginal candidiasis [48]. However, no studies on the use of the propolis from this stingless bee for antifungal applications in animals were found.

Some reports have evaluated the activity of the propolis from *Apis mellifera* against *Malassezia pachydermatis*, but there are no such reports focusing on native bees. In a recent study, a correlation was established between the antimycotic activity of the ethanolic extract of Brazilian green and red propolis against *M. pachydermatis*, with an MIC between 4 and 8 mg/mL and an MFC of 8–16 mg/mL [49]. The study reported that, as the total content of phenols and flavonoids increased, propolis exhibited an enhanced biological effect, suggesting that the mechanism of action of EEP is based on the rupture of the cell wall. This idea is reinforced by the observation that some azole-resistant *M. pachydermatis* strains were inhibited by the EEP. The efficacy of an Argentinian propolis against *M. pachydermatis* was evaluated by different in vitro techniques; the results demonstrated that the yeast was vulnerable to all tested propolis concentrations, with an MIC of 0.30 mg/mL; however, the researchers were unable to determine the MFC [17]. The efficacy of a 2.5% EEP solution against 48 clinical strains of *M. pachydermatis* isolated from dogs diagnosed with otitis externa was also proven, as it was found that all the strains were susceptible to the EEP solution [17]. An EEP from Rio Grande do Soul, Brazil, demonstrated antimycotic activity against clinical isolates from dogs with otitis externa, with an MIC of 2.6 mg/mL and an MFC of 5.3 mg/mL. In this work, an MFC of 21.3 mg/mL was determined, being higher than what was found with *Apis mellifera* propolis, but the MIC of 7.11 mg/mL determined was similar to what was reported for this propolis; however, it is unclear whether high EPP concentrations could induce cytotoxicity. Therefore, more research is needed to identify the active principles of propolis as well as their action mechanisms. Currently, there are two theories aiming to explain the antifungal activity of propolis: the first proposes that propolis elicits cellular wall lysis. and the other proposes that propolis damages the plasma membrane by inhibiting ergosterol synthesis [16].

The photomicrographs obtained in the present report revealed that the EEP was able to penetrate the plasma membrane, which was found using the minimum fungicidal concentration (21.33 mg/mL), causing severe damage and eventually the death of yeast samples through structural and functional damage caused by membrane disruption. Calcofluor white exhibits a high affinity for fungal wall components [50,51]. The alterations observed with this stain were mainly deformed morphologies. In some cases, we hypothesise that the complete destruction of the cell wall prevented the observation of the yeasts, which would be a possible effect of the sesquiterpenes present in the EEP, as previously described [52]. It would be advisable to perform computational chemistry studies to establish the extent of the damage to the yeast's cell wall caused by these compounds.

On the other hand, propidium iodide binds to nucleic acids and increases red colouration when there is damage to the cell membranes. This red colouration indicates severe cell damage and death, which was observed in the reference strain and the clinical strains treated with both EEPs, demonstrating the effectiveness of this type of propolis. This effect with propidium iodide has been observed in other fungi, such as *Fusarium*, by evaluating naturally occurring compounds' efficacy against fungal growth [19]. It is likely that using higher concentrations of EPP would have resulted in more damage being detected, so it would be advisable to use higher concentrations in future research.

The above-mentioned stains have been used to detect cellular damage by *Apis mellifera* propolis in yeasts of medical importance, such as *Candida albicans* [53–55] as well as the bacterium *Staphylococcus aureus* [56]. There is scant research on the antimicrobial activity of stingless bee propolis against infectious agents in animals, which represents an opportunity to improve animal health.

Our team demonstrated the antiviral activity of the propolis of the stingless bee *Plebeia frontalis* against the canine distemper virus. This encourages further investigation into its effects on other viruses that affect animals [57].

Therefore, this work demonstrates that the propolis of the Mexican stingless bees *Scaptotrigona mexicana* and *Tetragonisca angustula* has antimycotic effects and causes structural damage to *Malassezia pachydermatis*. This result supports the use of these types of propolis for therapeutic purposes. The findings of this study and their implications should be discussed in the broadest possible context. Future research directions may also be highlighted.

5. Conclusions

In conclusion, the present study has shown the antifungal properties and extent of structural damage caused by two propolis ethanolic extracts from two stingless bee species (*Scaptotrigona mexicana* and *Tetragonisca angustula*) found in the Mexican municipalities of Yecuatla, Veracruz, and Chalchihuitan, Chiapas, against different *Malazessia pachydermatis* strains, one as a reference (ATCC 14522) and three clinical isolates. Sesquiterpenes, along with other compounds, are possibly the reason for the antimycotic activity of both extracts. It is important to mention that, to our knowledge, the present work is the first to demonstrate the structural damage caused by and antifungal effects of Mexican stingless bee propolis against *Malazessia pachydermatis*.

Nonetheless, further research must be undertaken in order to provide a more solid scientific basis for the future employment of propolis as an alternative treatment for canine external otitis.

Author Contributions: Evaluation of antimicotic activity, D.B.F.E.; analysis of propolis samples, B.R.P.; collection of propolis in the meliponaries, F.H.G.; collection of clinical and reference strains and conservation of microorganisms, N.T.B.; gas chromatography–mass spectrometry, J.P.F.; performance of chemical determinations, J.G.P.C.; fluorescence microscopy and revision of manuscript, B.R.P. and C.G.G.T.; funding acquisition and revision of manuscript, T.A.C.S. All authors have read and agreed to the published version of the manuscript.

Funding: This research was funded by National Autonomous University of México Programa de Apoyo a Proyectos de Investigación e Innovación Tecnológica (PAPIIT) IN 223719 project and IN215723 project. Facultad de Estudios Superiores Cuautitán: Research Chairs CI 2237, CI 2222, CI 2267, and the Bee Team from the South Border College (ECOSUR).

Institutional Review Board Statement: Not applicable.

Informed Consent Statement: Not applicable.

Data Availability Statement: Data are contained within the article.

Acknowledgments: The authors would like to thank Lázaro Arroyo and Yliana Delfin Fuentes for their support in propolis sample recollection from stingless bee families, and Amparo Londoño Orozco, Samantha Jardon Xicotencatl, and Francisco Rodolfo González Diaz for their technical support.

Conflicts of Interest: The authors declare no conflicts of interest.

References

1. Bakchiche, B.; Temizer, I.K.; Güder, A.; Gencay, Ö.; Yegin, S.; Bardaweel, S.; Ghareeb, M. Chemical Composition and Biological Activities of Honeybee Products From Algeria. *J. Appl.* **2020**, *7*, 93. [CrossRef]
2. Bankova, V.; Popova, M. PHCOG *REV:* Review Article Propolis of Stingless Bees: A Promising Source of Biologically Active Compounds. *Pharmacogn. Rev.* **2007**, *1*, 88–92.
3. Ramón-Sierra, J.; Peraza-López, E.; Rodríguez-Borges, R.; Yam-Puc, A.; Madera-Santana, T.; Ortiz-Vázquez, E. Partial Characterization of Ethanolic Extract of *Melipona beecheii* Propolis and in vitro Evaluation of Its Antifungal Activity. *Rev. Bras. Farmacogn.* **2019**, *29*, 319–324. [CrossRef]
4. Ayala, R. Revisión de Las Abejas Sin Aguijón de México (Hymenoptera: Apidae: Meliponini). *Folia Entomol. Mex.* **1999**, *10*, 106–123.
5. Arnold, N.; Zepeda, R.; Vázquez, M.; Aldasoro, M. *Las Abejas sin Aguijón y su Cultivo en Oaxaca, México Con Catálogo de Especies*; Libros ECOSUR-CONABIO: San Cristóbal de las Casas, Mexico, 2018; pp. 10–58.
6. Pardini, R.; Rocha, P.; El-Hani, C.; Pardini, F. Challenges and Opportunities for Bridging the Research—Implementation Gap in Ecological Science and Management in Brazil. In *Conservation Biology: Voices from the Tropics*; John Wiley & Sons, Ltd.: Hoboken, NJ, USA, 2013; pp. 75–85. [CrossRef]
7. Hurtado Burillo, M.; Jara, L.; May-Itzá, W.; Quezada-Euán, J.J.; Ruiz, C.; Rúa, P.A. Geometric Morphometric and Microsatellite Analyses of *Scaptotrigona mexicana* and *S. pectoralis* (Apidae: Meliponini) Sheds Light on the Biodiversity of Mesoamerican Stingless Bees. *J. Insect Conserv.* **2016**, *20*, 753–763. [CrossRef]
8. Aldama, J.; Rodríguez Pérez, B.; Gallardo, S.; Sánchez, T. Structural damage in *Cryptococcus neoformans* caused by a Mexican propolis. *Nova Sci.* **2020**, *12*, 1–17. [CrossRef]
9. Quintero-Mora, M.L.; Londoño, O.A.; Hernández, H.F.; Manzano, G.P.; Lópe, M.R.; Soto, Z.C.I.; Carrillo, M.L.; Penieres, C.G.; García, C.G.; Cruz, T.A. Effect of Mexican propolis extracts from *Apis mellifera* on *Candida albicans* in vitro growth. *Rev. Iberoam. Micol.* **2008**, *25*, 22–26. [CrossRef]
10. Rodríguez, P.B.; Canales, M.M.; Penieres, C.J.; Cruz, S.T. Chemical composition, antioxidant properties and antimicrobial activity of Mexican propolis. *Acta Univ.* **2019**, *30*, 1–29. [CrossRef]
11. Flores, I.S.; Moreno, M.; Londoño, A.; Cruz, T.A. Use of mexican propolis for the tropical treatment of dermatomycosis in horses. *Open J. Vet. Med.* **2016**, *6*, 62770. [CrossRef]
12. Tovar, N.; García, L.; Cruz, T.A. Propolis in dogs: Clinical experiences and perspectives (a brief review). *Open J. Vet. Med.* **2016**, *5*, 11–17. [CrossRef]
13. Girão, M.D.; Prado, M.R.; Brilhante, R.S.N.; Cordeiro, R.A.; Monteiro, A.J.; Sidrim, J.J.C.; Rocha, M.F.G. *Malassezia pachydermatis* Isolated from Normal and Diseased External Ear Canals in Dogs: A Comparative Analysis. *Vet. J.* **2006**, *172*, 544–548. [CrossRef]
14. Lozina, L.; Boehringer, S.; Daquino, M.; Acosta, O. Eficacia del Propóleo sobre *Malassezia pachydermatis*. Correlación de distintas Técnicas in vitro. *Acta Farm. Bonaerense* **2006**, *25*, 560.
15. Cabañes, F.J. Diagnosis of *Malassezia* dermatitis and otitis in dogs and cats, is it just a matter of counting? *Rev. Iberoam. Micol.* **2021**, *38*, 3–4. [CrossRef]
16. Deegan, K.R.; Fonseca, M.S.; Oliveira, D.C.P.; Santos, L.M.; Fernandez, C.C.; Hanna, S.A.; Machado, B.A.S.; Umsza-Guez, M.A.; Meyer, R.; Portela, R.W. Susceptibility of *Malassezia pachydermatis* Clinical Isolates to Allopathic Antifungals and Brazilian Red, Green, and Brown Propolis Extracts. *Front. Vet. Sci.* **2019**, *6*, 460. [CrossRef] [PubMed]
17. Lozina, L.A.; Peichoto, M.E.; Boehringer, S.I.; Koscinczuk, P.; Granero, G.E.; Acosta, O.C. Efficacy of Argentine Propolis Formulation for Topical Treatment of Canine Otitis Externa. *Arq. Bras. Med. Vet. Zootec.* **2010**, *62*, 1359–1366. [CrossRef]
18. Tovar Betancourt, N. Evaluación Antimicótica in vitro del Propóleo Mexicano Sobre Malassezia pachydermatis. Master's thesis, Universidad Nacional Autónoma de México, Facultad de Estudios Superiores Cuautitlán, Izcalli, Mexico, 2016. Available online: https://ru.dgb.unam.mx/bitstream/20.500.14330/TES01000753315/3/0753315.pdf (accessed on 5 October 2023).
19. Medina, R.Y.M.; Rodríguez, C.M.; Rodríguez, M.M.A.; Hernández, H.A.B.; Delgado, B.N.L.; Chirino, Y.I.; Cruz, S.T.; Garcia, T.C.G.; Canales, M.M.M. Effect of the Essential Oils of *Bursera morelensis* and *Lippia graveolens* and Five Pure Compounds on the Mycelium, Spore Production, and Germination of Species of *Fusarium*. *J. Fungi* **2022**, *8*, 617. [CrossRef] [PubMed]
20. Diario Oficial de la Federación. Norma Oficial Mexicana NOM-003-AG/GAN-2017: Propóleos, Producción y Especificaciones para su Procesamiento. 2017. Available online: https://www.dof.gob.mx/nota_detalle.php?codigo=5500103&fecha=06/10/2017#gsc.tab=0 (accessed on 15 April 2023).

21. Dos Santos, L.; Hochheim, S.; Boeder, A.; Kroger, A.; Tomazzoli, M.; Neto, R.; Maraschin, M.; Guedes, A.; Cordova, C. Chemical Characterization, Antioxidant, Cytotoxic and Antibacterial Activity of Propolis Extracts and Isolated Compounds from the Brazilian Stingless Bees *Melipona quadrifasciata* and *Tetragonisca angustula*. *J. Apic. Res.* **2017**, *56*, 543–558. [CrossRef]
22. Rivera, Y.C.R.; Terrazas, L.I.; Jiménez, E.M.; Campos, J.E.; Flores, O.C.M.; Hernández, L.B.; Cruz, S.T.; Garrido, F.G.I.; Rodríguez, M.M.A.; Canales, M.M.M. Anti-*Candida* Activity of *Bursera morelensis* Ramírez Essential Oil and Two Compounds, α-Pinene and γ-Terpinene—An in vitro Study. *Molecules* **2017**, *22*, 2095. [CrossRef] [PubMed]
23. Kaneko, T.; Makimura, K.; Abe, M.; Shiota, R.; Nakamura, Y.; Kano, R.; Hasegawa, A.; Sugita, T.; Shibuya, S.; Watanabe, S.; et al. Revised Culture-Based System for Identification of *Malassezia* Species. *J. Clin. Microbiol.* **2007**, *45*, 3737–3742. [CrossRef]
24. Aspíroz, C.; Gilaberte, Y.; Rezusta, A.; Boekhout, T.; Rubio, M.A.C. Gentamycin Inhibits the Growth of *Malassezia pachydermatis* in Culture. *Rev. Iberoam. Micol.* **2010**, *27*, 20–21. [CrossRef] [PubMed]
25. Balouiri, M.; Sadiki, M.; Ibnsouda, S.K. Methods for in vitro Evaluating Antimicrobial Activity: A Review. *J. Pharm. Anal.* **2016**, *6*, 71–79. [CrossRef]
26. Sánchez, A.K.; Fernández, M.R.F.; Moreno, M.M.; Villegas, A.L.; Meneses, G.F.; Arenas, G.R. Sensitivity and specificity of mycological direct examination with calcofluor white for the diagnosis of onychomycosi. *Med. Cut. Ibero-Lat. Am.* **2013**, *6*, 261–266. [CrossRef]
27. Phillips, A.J.; Sudbery, I.; Ramsdale, M. Apoptosis Induced by Environmental Stresses and Amphotericin B in *Candida albicans*. *Proc. Natl. Acad. Sci. USA* **2003**, *100*, 14327–14332. [CrossRef]
28. Urbizu, G.A.L.; Castillo, R.O.; Martínez, A.G.C.; Torres, C.J.A. Natural Variability of Essential Oil and Antioxidants in the Medicinal Plant *Turnera diffusa*. *Asian Pac. J. Trop. Med.* **2017**, *10*, 121–125. [CrossRef] [PubMed]
29. Delmondes, G.A.; Santiago, L.I.C.; Días, D.Q.; Cunha, G.L.D.; Araujo, I.M.; Barbosa, R.; Coutinho, H.D.M.; Felipe, C.F.B.; Barbosa, F.J.M.; Lima, N.T.R.; et al. Pharmacological Applications of Farnesol ($C_{15}H_{26}O$): A patent review. *Expert. Opin. Ther Pat.* **2020**, *30*, 227–234. [CrossRef]
30. Fatnassi, S.; Zarrouk, H.; Chatti, S. Chemical Composition and Antimicrobial Activity of Volatile Fraction of the Peel of *Maclura pomifera* Fruit Growing in Tunisia. *J. Soc. Chim. Tunisia* **2011**, *13*, 1–6.
31. Chen, C.; Change, H.C.; Kirk, K. Betulachrysoquinone Hemiketal: A *p*-Benzoquinone Hemiketal Macrocyclic Compound Produced by *Phanerochaete chrysosporium*. *Phytochemistry* **1977**, *16*, 1983–1985. [CrossRef]
32. Murira, K.G.; Njagi, E.N.M.; Machocho, A.K.; Nyawira, W.L.; Mungiria, J.N.M. Chemical Composition and in vitro Antioxidant Activities of *Ocimum americanum*. *Adv. Anal. Chem.* **2015**, *5*, 42–49.
33. Otieno, A.J. Antimicrobial Activity and Phytochemical Profiles of Warburgia ugandensis Sprague (Canellaceae) Extracts from Different Populations across the Kenyan Rift Valley. Master's thesis, Kenyatta University, Nairobi, Kenia, 2016. Available online: https://ir-library.ku.ac.ke/handle/123456789/17924 (accessed on 6 December 2023).
34. Siswadi, S.; Saragih, G.S. Phytochemical Analysis of Bioactive Compounds in Ethanolic Extract of *Sterculia quadrifida* R. Br. *AIP Conf. Proc.* **2021**, *2353*, 030098. [CrossRef]
35. Xie, C.; Wang, S.; Cao, M.; Xiong, W.; Wu, L. (E)-9-Octadecenoic Acid Ethyl Ester Derived from *Lotus seedpod* Ameliorates Inflammatory Responses by Regulating MAPKs and NF-κB Signalling Pathways in LPS-Induced RAW264.7 Macrophages. *Evid. Based Complement. Altern. Med.* **2022**, *2022*, 6731360. [CrossRef] [PubMed]
36. Takahashi, S.; Yeo, Y.-S.; Zhao, Y.; O'Maille, P.E.; Greenhagen, B.T.; Noel, J.P.; Coates, R.M.; Chappell, J. Functional Characterization of Premnaspirodiene Oxygenase, a Cytochrome P450 Catalyzing Regio- and Stereo-Specific Hydroxylations of Diverse Sesquiterpene Substrates. *Int. J. Biol. Chem.* **2007**, *282*, 31744–31754. [CrossRef] [PubMed]
37. Peng, D.Q.; Yu, Z.X.; Wang, C.H.; Gong, B.; Liu, Y.Y.; Wei, J.H. Chemical Constituents and Anti-Inflammatory Effect of Incense Smoke from Agarwood Determined by GC-MS. *Int. J. Anal. Chem.* **2020**, *2020*, 4575030. [CrossRef]
38. Bruzual Villarroel, H.Y.; Henríquez Guzmán, W.; Crescente, O.; Lanza, J.G. Aceite esencial de *Wedelia calycina* (*Asteraceae*): Composición química, actividad antibacteriana y antifúngica. *Saber* **2015**, *27*, 87–93.
39. Sarvesan, R.; Eganathan, P.; Saranya, J.; Sujanapa, P. Chemical Composition and Antibacterial Activity of Leaf Essential Oil of *Eugenia cotinifolia* ssp. Codyensis (Munro Ex Wight) Ashton. *Int. J. Pharm. Sci.* **2015**, *6*, 3981–3985.
40. Ambriz-Pérez, D.; Leyva-López, N.; Gutiérrez-Grijalva, E.P. Phenolic Compounds: Natural Alternative in Inflammation Treatment. A Review. *Cogent Food Agric.* **2016**, *2*, 1131412. [CrossRef]
41. Badiazaman, A.A.M.; Zin, N.B.M.; Annisava, A.R.; Nafi, N.E.M.; Mohd, K.S. Phytochemical Screening and Antioxidant Properties of Stingless Bee *Geniotrigona thoracica* Propolis. *Malays. J. Fundam. Appl. Sci.* **2019**, *15*, 330–335. [CrossRef]
42. Surek, M.; Fachi, M.M.; de Fátima Cobre, A.; de Oliveira, F.F.; Pontarolo, R.; Crisma, A.R.; de Souza, W.M.; Felipe, K.B. Chemical composition, cytotoxicity and antibacterial activity of propolis from Africanized Honeybees and three different *Meliponini* species. *J. Ethnopharmacol.* **2021**, *269*, 113662. [CrossRef] [PubMed]
43. Torres, A.R.; Sandjo, L.P.; Friedemann, M.T.; Tomazzoli, M.M.; Maraschin, M.; Mello, C.F.; Santos, A.R.S. Chemical characterization, antioxidant and antimicrobial activity of propolis obtained from *Melipona quadrifasciata* and *Tetragonisca angustula* Stingless Bees. *Braz. J. Med. Biol. Res.* **2018**, *51*, e7118. [CrossRef] [PubMed]
44. Yam, P.A.; Santana, H.A.; Yah, P.N.; Ramón, S.J.M.; Cáceres, M.R.; Borges, A.R.L.; Ortiz-Vázquez, E. Pentacyclic triterpenes and other constituents in propolis extract from *Melipona beecheii* collected in Yucatán, México. *Rev. Bras. Farmacogn.* **2019**, *29*, 358–363. [CrossRef]
45. Grajales, C.J.; Elias, C.J.; Lozano, G.E.; Moreno, C.F.; Albores, F.V.; Lópe, G.A. Actividad antimicrobiana de propóleos de abejas sin aguijón en combinación con ajo, *Allium sativum* (*Amaryllidaceae*). *Rev. Biol. Trop.* **2020**, *69*, 23–35. [CrossRef]

46. Hee, Y.C. In vitro evaluation of the antifungal activity of propolis extract on *Cryptococcus neoformans* and *Candida albicans*. *Mycobiology* **2002**, *30*, 93–95. [CrossRef]
47. Shehu, A.; Ismail, S.; Rohin, M.; Harun, A.; Aziz, A.; Haque, M. Antifungal properties of Malaysian Tualang Honey and stingless bee propolis against *Candida albicans* and *Cryptococcus neoformans*. *J. App. Pharm. Sci* **2016**, *6*, 44–50. [CrossRef]
48. Farida, S.; Sahlan, M.; Rohmatin, E.; Adawiyah, R. The beneficial effect of Indonesian Propolis wax from *Tetragonula* sp. as a therapy in limited vaginal candidiasis patients. *Saudi J. Biol. Sci.* **2020**, *27*, 142–146. [CrossRef] [PubMed]
49. Cardoso, R.L.; Maboni, F.; Machado, G.; Alves, S.H.; de Vargas, A.C. Antimicrobial activity of propolis extract against *Staphylococcus* coagulase positive and *Malassezia pachydermatis* of canine otitis. *Vet. Microbiol.* **2010**, *142*, 432–434. [CrossRef] [PubMed]
50. Hageage, G.J.; Harrington, B.J. Use of calcofluor white in clinical mycology. *Lab. Med.* **1984**, *15*, 109–112. [CrossRef]
51. Ram, A.F.J.; Klis, F.M. Identification of fungal cell wall mutants using susceptibility assays based on calcofluor white and Congo red. *Nat. Protoc.* **2006**, *1*, 2253–2256. [CrossRef]
52. Medina-Romero, Y.M.; Hernández-Hernández, A.B.; Rodríguez-Monroy, M.A.; Canales-Martínez, M.M. Essential oils of *Bursera morelensis* and *Lippia graveolens* for the development of a new biopesticides in postharvest control. *Sci. Rep.* **2021**, *11*, 20135. [CrossRef]
53. De Castro, P.A.; Bom, V.L.P.; Brown, N.A.; Almeida, R.S.C.D.; Ramalho, L.N.Z.; Savoldi, M.; Goldman, M.H.S.; Berretta, A.A.; Goldman, G.H. Identification of the cell targets important for propolis-induced cell death in *Candida albicans*. *Fungal Gent. Biol.* **2013**, *60*, 74–86. [CrossRef] [PubMed]
54. Adomavičiūtė, E.; Pupkevičiūtė, S.; Juskaite, V.; Zilius, M.; Stanys, S.; Pavilonis, A.; Briedis, V. Formation and investigation of electrospun PLA Materials with propolis extracts and silver nanoparticles for biomedical applications. *J. Nanomater.* **2017**, *2017*, 8612819. [CrossRef]
55. Alfarrayeh, I.; Pollák, E.; Czéh, Á.; Vida, A.; Das, S.; Papp, G. Antifungal and anti-biofilm effects of caffeic acid phenethyl ester on different *Candida* species. *Antibiotics* **2021**, *10*, 1359. [CrossRef]
56. Grecka, K.; Xiong, Z.R.; Chen, H.; Pełka, K.; Worobo, R.W.; Szweda, P. Effect of ethanol extracts of propolis (EEPs) against *Staphylococcal* biofilm—microscopic studies. *Pathogens* **2020**, *9*, 646. [CrossRef] [PubMed]
57. Jiménez, O.V.D.; Pérez, B.R.; Cruz-Sánchez, T.A.; Tovar, C.G.G.; Bordes, J.L.N.; Zárate, C.I.S. Evaluation of the antiviral activity of propolis from native bees (*Plebeia frontalis*) against canine distemper virus. *Open J. Vet. Med.* **2020**, *10*, 207–218. [CrossRef]

veterinary sciences

MDPI

Article

The Comparative Study of the Antioxidant and Antibacterial Effects of Propolis Extracts in Veterinary Medicine

Dovile Svetikiene [1,*], Gintaras Zamokas [1], Monika Jokubaite [2], Mindaugas Marksa [3], Liudas Ivanauskas [3], Lina Babickaite [1] and Kristina Ramanauskiene [4,*]

1 Department of Dr. L. Kriauceliunas Small Animal Clinic, Faculty of Veterinary Medicine, Lithuanian University of Health Sciences, Tilzes Str. 18, LT-47181 Kaunas, Lithuania; gintaras.zamokas@lsmu.lt (G.Z.); lina.babickaite@lsmu.lt (L.B.)

2 Department of Drug Chemistry, Faculty of Pharmacy, Lithuanian University of Health Sciences, Sukileliai Avenue 13, LT-50162 Kaunas, Lithuania; monika.jokubaite@lsmu.lt

3 Department of Analytical and Toxicological Chemistry, Faculty of Pharmacy, Lithuanian University of Health Sciences, Sukileliai Avenue 13, LT-50162 Kaunas, Lithuania; mindaugas.marksa@lsmu.lt (M.M.); liudas.ivanauskas@lsmu.lt (L.I.)

4 Department of Clinical Pharmacy, Faculty of Pharmacy, Lithuanian University of Health Sciences, Sukileliai Avenue 13, LT-50162 Kaunas, Lithuania

* Correspondence: dovile.svetikiene@lsmu.lt (D.S.); kristina.ramanauskiene@lsmu.lt (K.R.)

Simple Summary: Bacterial congenital diseases are among the most common ailments in dogs and cats within veterinary medicine. Microorganisms are becoming more and more resistant to antibiotic drugs. The search for natural substances has been driven primarily by the irrational and reckless use of available antimicrobials in clinical scenarios. Consequently, there is a growing interest in natural substances for treating animal diseases, aiming to find sources of active compounds of natural origin. Propolis is one such substance that scientists have extensively investigated. It has garnered contemporary interest due to its natural complex of active compounds and broad biological activity. This study demonstrates propolis extracts' chemical analysis and biological activity using different solvents. The bee propolis extracts' antimicrobial and antifungal activities were evaluated using clinical and reference bacterial strains.

Abstract: Antimicrobial resistance (AMR) is one of the biggest threats to human and animal health. Efforts to combat AMR include the introduction of antimicrobial drugs as alternative treatment options. To contribute to an effective plan for the treatment of infectious diseases caused by bacteria, the development of new antimicrobial agents is increasingly being explored. Propolis has garnered significant attention from both scientists and industry due to its extensive spectrum of biological activity. The growing interest in polyphenols of natural origin and their plant sources further encourages the investigation of their chemical composition and biological effects. Propolis serves as a rich source of phenolic compounds. Baltic region propolis, classified as poplar-type propolis, was selected for this study, and extracts were prepared using raw propolis materials from various Baltic countries. The production of liquid extracts utilized a combination of 70 percent ethanol, a mixture of water and poloxamer P407, and DES (deep eutectic solvent). The research aims to produce liquid propolis extracts using different solvents and to assess their chemical composition, antioxidant, and antimicrobial activity against different veterinary pathogens. Antioxidant activity was evaluated using DPPH (2,2-diphenyl-1-picrylhydrazyl), revealing antioxidant activity in all extracts, with results correlating with the total phenolic compound content. It was found that *p*-coumaric acid predominated in the studied propolis extracts (in ethanol extracts 1155.90–1506.65 mg/g, in DES extracts 321.13–954.76 mg/g, and in polymeric extracts 5.34–30.80 mg/g), with smaller amounts of ferulic acid and vanillin detected. Clinical and reference bacterial strains were collected from the Lithuanian University of Health Sciences, the Academy of Veterinary Medicine, and the Institute of Microbiology and Virology. To effectively treat bacterial infections, the antimicrobial activity of propolis extracts was tested against six pathogenic bacterial species and one pathogenic fungus (*S. aureus*, *S. agalactiae*, *B. cereus*, *E. faecalis*, *E. coli*, *P. aeruginosa*, and *C. albicans*). Antimicrobial activity

Citation: Svetikiene, D.; Zamokas, G.; Jokubaite, M.; Marksa, M.; Ivanauskas, L.; Babickaite, L.; Ramanauskiene, K. The Comparative Study of the Antioxidant and Antibacterial Effects of Propolis Extracts in Veterinary Medicine. *Vet. Sci.* **2024**, *11*, 375. https://doi.org/10.3390/vetsci11080375

Academic Editors: Patrick Butaye, George Valiakos and Giovanni Cilia

Received: 19 June 2024
Revised: 20 July 2024
Accepted: 12 August 2024
Published: 15 August 2024

studies demonstrated that DES propolis extracts exhibited stronger antimicrobial activity compared to ethanolic propolis extracts. The minimum inhibitory concentration (MIC) values of DES propolis extracts against the tested strains ranged between 50 and 1000 μg/mL. Considering the study results, it can be concluded that propolis from the Baltic region is abundant in phenolic compounds exhibiting antioxidant and antibacterial activities.

Keywords: animal; antimicrobial; propolis; extract

1. Introduction

Natural products are important as medicines in many treatment areas [1]. With the discovery of antimicrobial agents almost a century ago, several classes of antimicrobial agents of natural origin, such as β-lactams, tetracyclines, and aminoglycosides, have been introduced as therapeutic agents. At that time, the use of antimicrobials had a significant impact on the treatment of infectious diseases in human and veterinary medicine. Domestic animals are considered potential carriers of antimicrobial resistance (AMR) to humans due to the excessive use of antimicrobials and their close contact with humans. The World Health Organization (WHO) considers antimicrobial resistance to be a major threat to human and animal health. These developments are strongly influenced by the misuse or overuse of antimicrobials, leading to the spread of bacterial antimicrobial resistance (AMR) [2–4]. There is now a worldwide increase in the use of plant-based products. Production animals account for around 70% of animals treated with herbal products, poultry (9.1%), dogs (5.3%), and rabbits (4.3%) [5]. One of the threats of antimicrobial resistance (AMR) is the presence of antibiotic residues, which promotes the development of antibiotic-resistant bacteria in humans; toxic metabolites remain in the meat, and the by-products of synthetic substances become a concerning side effect of medication use. These aspects encourage the search for safe alternatives compared to modern animal health systems [6]. The search for natural and safe alternatives to traditional veterinary medicines has become an important area of research in animal and veterinary medicine. This includes natural products or components extracted from plants, microorganisms, insects, or marine organisms. As natural products from different sources have unique structures and properties and do not cause significant side effects, their effect is hoped to reduce the likelihood of drug resistance [7]. The plant kingdom is currently garnering significant attention. Plants are characterized by compounds with a strong antibacterial effect, which are effective in treating bacterial infections without causing adverse side effects. Natural products have a chemical diversity of bioactive compounds so that they can be promising sources for drug development [8–10]. Bacterial skin, eyes, and ear infections are among the most common bacterial infections in dogs and cats in veterinary medicine. Anti-inflammatory and antibacterial agents are commonly included in the treatment regimens for many inflammatory diseases. With the growing popularity of natural substances among pet owners, there is an increasing demand for natural veterinary preparations. In this study, it was decided to focus on propolis, one of the substances extensively studied by scientists. Propolis is a bee product known for its natural complex of active compounds and its broad spectrum of biological effects.

Bees collect propolis from living plants, which contributes to the diversity of its chemical composition and, therefore, presents challenges in standardizing its quality. Between 80 and 100 chemical compounds have been identified in propolis samples, most of which are classified as phenolic compounds. These phenolic compounds account for the wide range of biological effects of propolis and, due to their potential health benefits, have garnered significant attention in the scientific community [11]. Recently, scientists have been focusing extensively on polyphenolic compounds as antioxidants. According to epidemiological studies and related meta-analyses, long-term consumption of phenolic compounds is associated with protection against cancer, cardiovascular diseases, diabetes, osteoporosis, and neurological disorders [12,13]. Scientific research results confirm propolis's antioxidant activity [14–24]. Currently, researchers are working to determine the relationship that the antioxidant activity and individual chemical composition of propolis have. The results of studies on antioxidant activity conducted by Polish scientists have confirmed that the composition of propolis varies depending on the geographical region of collection and plant sources [25–27]. Furthermore, these studies demonstrated a dependence of antioxidant activity on the chemical composition of the samples [28]. Shigenori Kumazawa et al. found that ethanolic propolis extracts from some countries (e.g., Argentina, China, Hungary) exhibited strong antioxidant activity, correlated with polyphenol and flavonoid content [29]. Propolis is widely used in cosmetic, pharmaceutical, veterinary, and medical products due to its antibacterial properties. The antibacterial effects of propolis are examined in different ways. First, the antimicrobial activity directly impacts the microorganism, and second, it boosts the immune system by activating the body's defense mechanisms [30,31]. The antimicrobial activity of propolis is stronger against Gram-positive bacteria than against Gram-negative bacteria [30,32–34]. Ethanol predominantly produces propolis extracts, with aqueous extracts being less common. Propolis has a lower water solubility than ethanol solvents. Solvents such as PEG, propylene glycol, and cyclodextrins have been used in aqueous formulations to improve the solubility of these compounds. Korean researchers utilized poloxamers as solubilizers and gelling agents in topical antimicrobial formulations of propolis [35]. Poloxamers can improve the solubility of active compounds by reducing surface tension and forming micelles, which promote the dissolution of active substances [36]. Deep eutectic solvents (DES) have been increasingly used to produce extracts. The results of antimicrobial studies by Trusheva and other researchers have confirmed that DES can improve the antimicrobial effect of propolis extracts on pathogenic microorganisms [37]. This may be influenced by the presence of natural acids as DES elements. The literature indicates that DES containing organic acids have a higher antimicrobial activity due to their pharmacological properties [38]. However, research data on propolis extraction using deep eutectic solvents have been limited. Turkish researchers found that natural deep eutectic solvents (NADES) can be a viable alternative to conventional solvents for propolis extraction [39]. Funari's results strongly suggest that some NADES extracts can replace hydroethanol, propylene glycol, and aqueous propolis solutions without sacrificing extraction efficiency [40]. Considering that the propolis of the Baltic region belongs to the poplar-type propolis, this study chose to make extracts using raw materials from Lithuanian, Polish, and Latvian propolis. For the production of liquid extracts, 70 percent ethanol, a mixture of water and poloxamer P407, and a deep eutectic solvent (DES) were used. The purpose of this research is to produce liquid extracts of propolis using various solvents and evaluate their chemical composition and antioxidant and antimicrobial activity against different veterinary pathogens.

2. Materials and Methods

2.1. Propolis Extraction

Ethanolic, polymeric, and DES propolis extracts were produced. Unprocessed propolis samples were commercially purchased from various Baltic region countries (Lithuania, Latvia, Poland). Lithuanian propolis was purchased from R. Serksnienes farm, Raseiniu district, Latvian propolis was purchased from a company ("Bites", Vālodzes, Stopiņu province, Latvia, LV-2130), Polish propolis was purchased from a beekeeping company (PROKIT, Halinów, Poland).

2.2. Ethanolic Propolis Extracts Preparation

Crushed propolis was extracted with 70% ethanol (v/v) by maceration method in a ratio of 1:10. The macerated content was stored in dark glass bottles at 21 ± 1 °C for 12 days, with the contents being stirred several times during this period [24].

2.3. DES Propolis Extracts Preparation

Deep Eutectic solvent (DES) was prepared separately, pouring in the same quantity of materials as specified earlier and mixing at a certain temperature. Choline chloride (Sigma-Aldrich, St. Louis, MO, USA), lactic acid (Sigma-Aldrich, St. Louis, MO, USA) and purified water were mixed in a ratio of 1:1:1. Crushed propolis were extracted with DES solvent by maceration method in a ratio of 1:10. The macerated content was stored in dark glass bottles (21 ± 1 °C) for 12 days, while the content being stirred several times [41].

2.4. Polymeric Propolis Extracts Preparation

Polymeric solvent was prepared with poloxomer 407 (Fagron, St. Paul, MN, USA). The mixture was stored in the refrigerator (4 ± 1 °C) for 24 h until the polymer completely dissolved. Crushed propolis was extracted by the polymeric solvent in a ratio of 1:10 by maceration. The stirring macerate content was stored in dark glass bottles (4 ± 1 °C) for 12 days and stirred several times in the meantime [35].

The extracts were filtered through ashless filter paper (retention 8–12 μm, diameter 90 mm, ash content 0.007%) [42].

2.5. HPLC Analysis

HPLC analysis was performed based on certain modifications [43]. Chromatographic equipment used included a Waters 2695 system with a Waters 996 diode array detector and an ACE 5C18 chromatography column with dimensions of 250 × 4.6 mm. The obtained data were processed using Empower 2 Chromatography Data Software. HPLC eluent: acetonitrile and trifluoroacetate acid; temperature: 25 °C; duration time: 81 min; and sample injection volume: 10 μL. The setting for the analyzed compounds was established by comparing them with benchmark materials, considering their retention time and UV absorption within the range of 250 to 400 nm. Standard compounds used in the analysis: *p*-coumaric acid (R^2 = 0.9999), cinnamic acid (R^2 = 0.9999), caffeic acid (R^2 = 0.9999), vanillin (R^2 = 0.9999), vanillic acid (R^2 = 0.9999), and ferulic acid (R^2 = 0.9999).

2.6. Antioxidant Activity in DPPH

The antioxidant activity of the extracts was evaluated using the DPPH method, with some modifications according to Yim et al. [44]. A 60 μM DPPH solution in 96% ethanol (*v*/*v*) was prepared. We mixed 10 μL of propolis extracts with 3000 μL of DPPH working solution. The samples were incubated for 30 min at room temperature. Absorbance was measured with a spectrophotometer (Agilent Technologies 8453 UV-Vis, Santa Clara, CA, USA) at the wavelength of 517 nm.

2.7. Antimicrobial Activity In Vitro

Assessment of antimicrobial activity in vitro was determined based on previous studies [32,45]. The bacteriological properties of propolis extracts were evaluated in vitro using the agar diffusion method. Mueller–Hinton agar (Biolife Mueller Hinton Agar II, Italy), Columbia blood agar (CBA, EO Labs, Bonnybridge, Scotland, UK), and Sabouraud dextrose agar (SDA, EO Labs, Bonnybridge, Scotland, UK), approved by the Clinical and Laboratory Standards Institute (CLSI), were used as standards. Liquid Mueller–Hinton and Sabouraud dextrose agars were prepared according to the standards and poured into Petri dishes with a diameter of 10 cm, approximately 35 mL each. Ready-made Colombia blood agar was used. Each isolate was prepared according to the 0.5 McFarland standard. Clinical and reference bacterial strains were collected from the Lithuanian University of Health Sciences, the Academy of Veterinary Medicine, and the Institute of Microbiology and Virology. Small animals with acute superficial or deep pyoderma, wound infections, abscesses, otitis, conjunctivitis, or complicated corneal ulcers were treated. Clinical and reference bacterial strains, including *Staphylococcus aureus* (ATCC 25923), *Bacillus cereus* (ATCC 11778), *Enterococcus faecalis* (ATCC 29212), *Escherichia coli* (ATCC 25922), and *Pseudomonas aeruginosa* (ATCC 27853), were spread onto Columbia blood agar. Additionally, the *Streptococcus agalactiae* (ATCC 13813) strain was disseminated on Columbia blood agar. Clinical and referential *Candida albicans* (ATCC 10231) strains were spread onto Sabouraud dextrose agar. Wells with a diameter of 7 mm were made in the agar, into which 0.1 mL of the propolis extract research material was added. Plates with bacteria were incubated for 24 h at 37 ± 0.5 °C, and plates with fungi were incubated for 24 h at 35 ± 0.5 °C in the thermostat. The antibacterial and antifungal properties of the propolis extracts in vitro were assessed after 24 h of incubation. The diameter of sterile zones formed around the wells was calculated in millimeters.

2.8. Minimum Inhibitory Concentration

The minimum inhibitory concentration (MIC) was determined using the broth microdilution method as recommended by the Clinical and Laboratory Standards Institute. The MIC was defined as the lowest concentration of the propolis extract that inhibited the growth of the tested microorganisms. The test was performed on clinical and reference bacterial strains, including *Staphylococcus aureus* (ATCC 25923), *Streptococcus agalactiae* (ATCC 13813), *Bacillus cereus* (ATCC 11778), *Enterococcus faecalis* (ATCC 29212), *Escherichia coli* (ATCC 25922), and *Pseudomonas aeruginosa* (ATCC 27). For the study, the bacterial strains were initially inoculated in sterile saline (0.9%). Turbidity was adjusted according to the McFarland 0.5 turbidity standard (Densi-La-Meter II, Erba Lachema, Brno, Czech Republic). A 10 μL aliquot of the diluted bacterial suspension was added to 11 mL of Mueller–Hinton Broth II (MHB, Oxoid Ltd., Basingstone, Hampshire, UK) liquid medium, resulting in a final inoculum volume of 1.5×10^8 CFU/mL. The prepared inoculum was then inoculated into 96-well microplates (VWR International, LLC., Radnor, PA, USA). Fifty microliters of medium and 50 μL of various concentrations of the propolis extract were added. The concentrations of propolis extracts tested ranged from 0.05 to 100 mg/mL. The plates were incubated at 37 °C for 24 h. Controls included a negative control (medium only), positive control (medium with bacteria), and color control (extract dilutions). The MIC of *p*-coumaric acid, the predominant compound in the extracts as determined by HPLC analysis, was also estimated. The concentrations of *p*-coumaric acid tested ranged from 0.1 mg/mL to 1 mg/mL. Bacterial growth was evaluated with a Multiskan FC microplate photometer (Thermo Scientific, Foster City, CA, USA) after 24 h of incubation by measuring the absorbance at 570 nm (OD570). Based on the control samples, thresholds were established to classify the results as either bacterial growth or no growth.

2.9. Statistical Analysis

The results express averages of three measurements and standard deviations. Statistically significant differences between the compared data were determined using one-way ANOVA. Differences were considered statistically significant when $p < 0.05$. The correlation was evaluated based on the Spearman correlation coefficient. Data statistical analysis and visualization were performed using software tools, including OriginPro® 2021 (OriginLab, Northampton, MA, USA), IBM SPSS Statistics 27 (SPSS Inc., Chicago, IL, USA), and SigmaPlot 13.0 (Systat Software, San Jose, CA, USA).

3. Results

3.1. HPLC Analysis of Active Compounds

The chemical composition of the produced extracts was evaluated using HPLC (Table 1). The research data indicate that *p*-coumaric acid was predominant in the studied ethanolic extracts, accounting for 19.73% of the total amount of all determined active compounds. Compared to other active substances found in the tested extracts, higher amounts of ferulic acid (9%) and vanillin (7.68%) of the total active compounds were found. Small amounts of cinnamic acid (0.5%) and vanillic acid (0.38%) were also detected. In eutectic extracts, *p*-coumaric acid continued to dominate, comprising about 33% of the total active compounds. Vanillin and ferulic acid accounted for 14% and 10.9%, respectively. Vanillic acid and cinnamic acid were the least found, accounting for 0.67% and 0.03%, respectively, of all identified compounds in the eutectic extracts. Further data show that in polymeric extracts of propolis, smaller amounts of *p*-coumaric acid (31.32%), vanillin (14.60%), and ferulic acid (11.11%) were found. Compared to other substances, the least amounts of vanillic and cinnamic acids were found, making up a minimal percentage of all active compounds.

Table 1. Propolis HPLC analysis of extracts.

Solvent	Propolis	Vanillic Acid (mg/g)	Vanillin (mg/g)	Ferulic Acid (mg/g)	Cinnamon Acid (mg/g)	*p*-Coumaric Acid (mg/g)
Ethanol	1 LT	28.63	587.49	693.18	36.05	1506.65
	SD	2.37	33.86	27.83	3.98	91.39
	2 PL	23.16	463.10	473.59	21.12	1155.90
	SD	0.20	2.99	15.02	0.99	21.57
	3 LV	22.23	574.51	593.26	38.83	1415.98
	SD	0.94	19.62	22.82	0.23	73.92
Eutectic	4 LT	19.61	404.78	315.54	16.29	954.76
	SD	0.40	13.49	8.30	0.01	51.95
	5 LV	3.19	141.03	112.79	7.41	321.13
	SD	0.19	0.42	1.62	0.28	10.44
	6 PL	5.28	168.84	89.06	5.18	328.05
	SD	0.27	3.46	5.24	0.09	17.33
P407	7 LT	5.30	-	6.05	0.49	14.94
	SD	0.22	-	0.05	0.00	0.03
	8 PL	7.26	-	1.48	-	5.34
	SD	0.40	-	0.08	-	0.37
	9 LV	0.53	14.36	10.93	0.87	30.80
	SD	0.01	1.33	0.12	0.01	1.79

3.2. Antioxidant Activity In Vitro

Antioxidant activity of prepared experimental extracts was analyzed by DPPH method. Research results are presented in Figure 1.

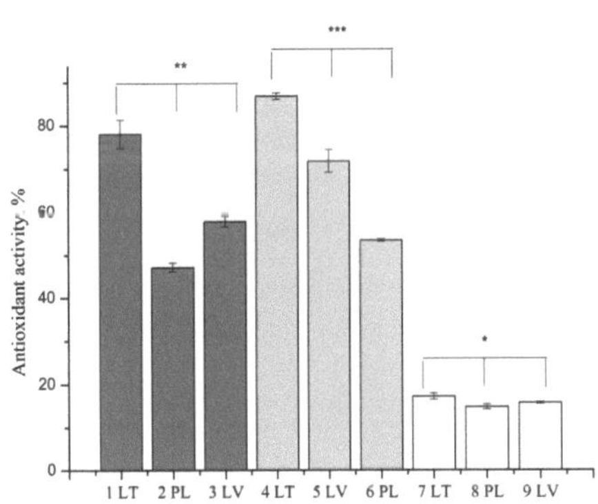

Figure 1. Antioxidant activity of ethanolic, eutectic, and polymeric (P407) propolis extracts. 1 LT, 2 PL, 3 LV—ethanolic extracts, 4 LT, 5 LV, 6 PL—eutectic propolis extracts, 7 LT, 8 PL, 9 LV—polymeric (P407) propolis extracts. Ethanolic propolis extracts were diluted five times before the test. The asterisks indicate statistical significance between polymeric (P407) propolis extracts group compared to other propolis extracts groups, $p < 0.05$.

The data presented in Figure 1 show that all tested extracts exhibit antioxidant activity. The ethanolic extract of Lithuanian propolis showed similar antioxidant activity to the propolis DES extract. Using the DPPH method, polymeric (P407) propolis extracts (7 LT,

8 PL, 9 LV) exhibited statistically significantly weaker antioxidant activity compared to other analyzed propolis extracts ($p < 0.05$). The results of the study indicate a strong correlation between the total sum of active phenolic compounds identified through HPLC analysis and the antioxidant activity results ($p < 0.001$).

3.3. Determination of Antimicrobial Activity

The antimicrobial activity of ethanolic, eutectic, and polymeric extracts of P407 propolis was studied. Research data in Figure 2, show that propolis polymeric extracts did not exhibit inhibitory antibacterial activity against any tested bacterial strains; bacterial growth was observed around the cavities (7 LT, 8 PL, 9 LV).

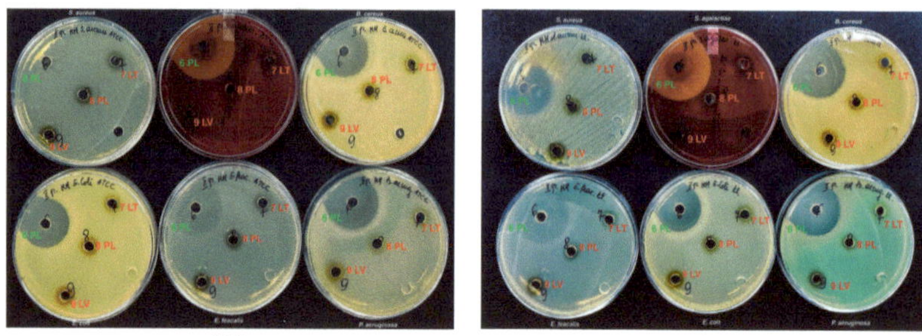

(**A**) (**B**)

Figure 2. Antimicrobial activity of propolis polymeric (P407) extracts. Reference strains (**A**). Wild strains (**B**).

The ethanolic and DES propolis extracts in Figure 3A exhibit antibacterial activity against the reference and clinical bacteria of the *Staphylococcaceae* family. The eutectic propolis extracts were characterized by a stronger inhibitory antibacterial effect compared to the ethanolic propolis extracts. A significant difference ($p < 0.05$) was found between the inhibition zones of these propolis extracts. The tested propolis extracts showed a better antibacterial effect against the reference bacterium of the *Staphylococcaceae* family compared to the clinical strain of *S. aureus*. A statistically significant difference ($p < 0.05$) was found between the inhibition zones of the reference and clinical *S. aureus*. The DES extracts showed a statistically significantly stronger antibacterial inhibitory effect against *S. aureus* compared to the control eutectic solvent ($p < 0.05$) and the positive control. Ethanol propolis extracts exhibited statistically weaker activity compared to the chlorhexidine solution. The ethanolic and DES propolis extracts in Figure 3B demonstrate antibacterial activity against the reference and clinical bacteria of the *Streptococcaceae* family. The DES propolis extracts exhibited a stronger inhibitory antibacterial effect compared to the ethanolic propolis extracts. A significant difference ($p < 0.05$) was observed between the inhibition zones of these propolis extracts. Furthermore, the eutectic propolis extracts were characterized by a statistically significantly stronger antibacterial inhibitory effect against *S. agalactiae* compared to the control eutectic solvent ($p < 0.05$). The DES extracts inhibited the growth of *S. agalactiae* statistically significantly compared to the chlorhexidine control group ($p < 0.05$). The ethanol propolis extracts and chlorhexidine solution exhibited antibacterial activity against the *B. cereus* strain. The ethanol extracts demonstrated a statistically significantly weaker antibacterial effect against bacteria of the *Bacillaceae* family compared to the propolis DES extracts and DES solvents ($p < 0.05$). The inhibition zones of the strains were statistically significantly larger in diameter for the eutectic extracts compared to the ethanol propolis extracts Figure 3C.

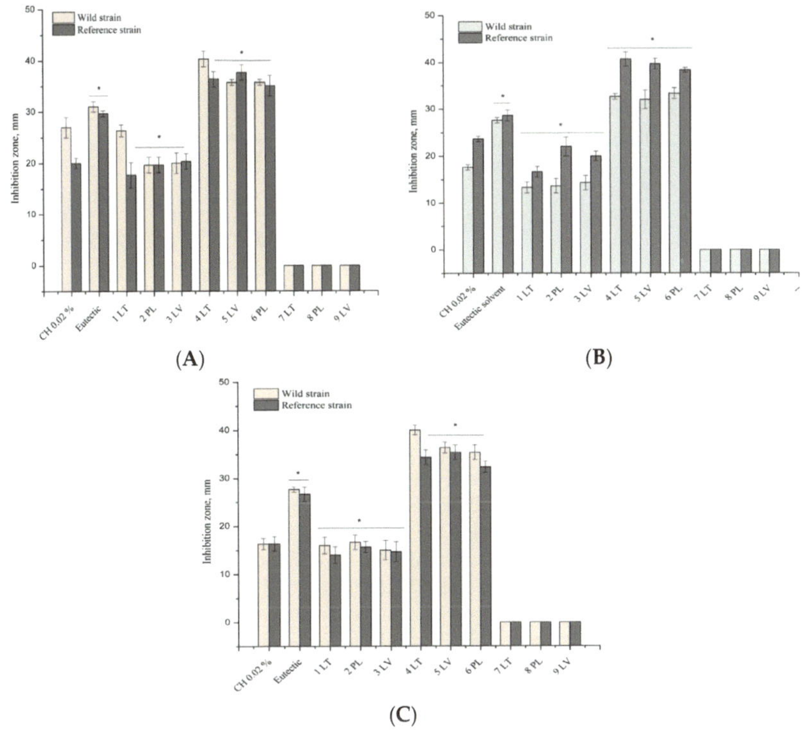

Figure 3. Antibacterial activity of propolis extract on *S. aureus* (**A**), *S. agalactiae* (**B**), and *B. cereus* (**C**) In the figure, an asterisk (*) indicates statistically significant results ($p < 0.05$).

This study's results, depicted in Figure 4, showed that the ethanol propolis extracts did not inhibit bacteria from the *Enterococcaceae*, *Enterobacteriaceae*, and *Pseudomonadaceae* families. Bacterial growth was observed around the cavities (1 LT, 2 PL, 3 LV).

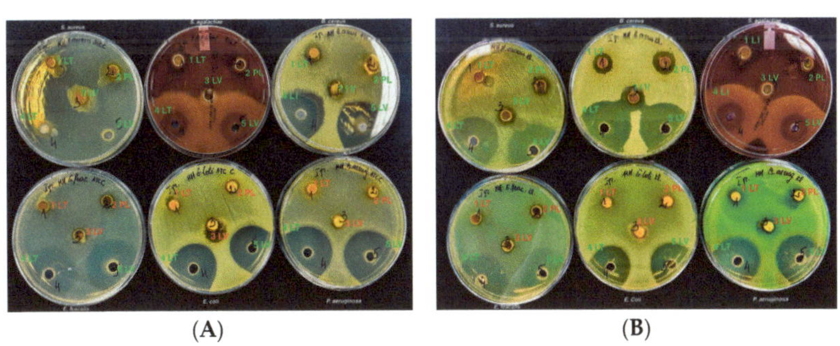

Figure 4. Antimicrobial activity of propolis ethanolic extracts. Reference strains (**A**). Wild strains (**B**). 1—LT ethanolic propolis, 2—PL ethanolic propolis, 3—LV ethanolic propolis, 4—LT eutectic propolis, 5—LV eutectic propolis, 6—PL eutectic propolis, 7—LT polymeric propolis, 8—PL polymeric propolis, 9—LV polymeric propolis.

Figure 5 illustrates that the DES propolis extracts exhibited antibacterial activity against reference and clinical *E. feacalis* bacteria of the *Enterococcaceae* family. Statistically significant strain inhibition zones were larger in diameter in the DES extracts compared to the DES solvent and positive control ($p < 0.05$).

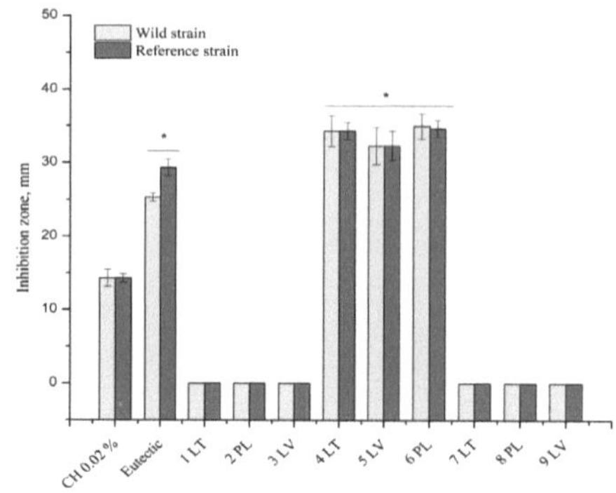

Figure 5. Antibacterial activity of propolis extract on *E. feacalis*. In the figure, an asterisk (*) indicates statistically significant results ($p < 0.05$).

During the research, it was found that the DES propolis extracts have strong antimicrobial activity against the Gram-negative bacteria *E. coli* strain. As depicted in Figure 6A, the eutectic propolis extracts exhibit antibacterial activity against the reference and clinical bacteria of the *Enterobacteriaceae* family. Statistically significantly, the zones of inhibition of the tested strains had a wider diameter in the DES extracts compared to the DES solvent and the positive control ($p < 0.05$). During the research, it was found that the DES propolis extracts have excellent antimicrobial activity against the Gram-negative *P. aeruginosa* bacterial strain (Figure 6B). The growth zone of the reference *P. aeruginosa* strain was statistically significantly inhibited by the DES propolis extracts compared to the control DES solvent. The growth of clinical *Pseudomonadaceae* bacteria was statistically significantly inhibited by the DES propolis extracts compared to the chlorhexidine control group ($p < 0.05$).

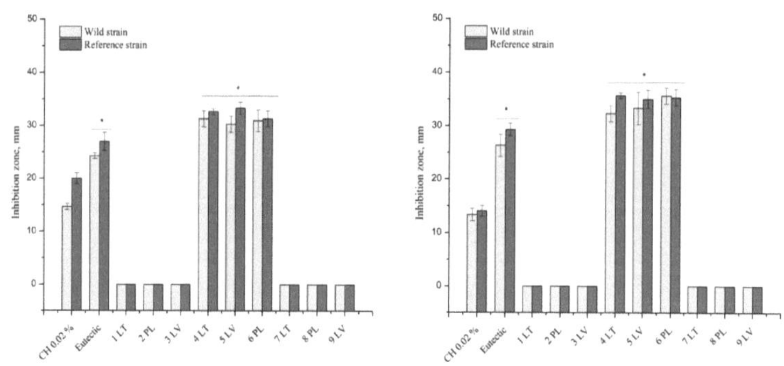

(A) (B)

Figure 6. Antibacterial activity of the propolis extract on *E. coli* (**A**), *P. aeruginosa* (**B**). In the figure, an asterisk (*) indicates results that are statistically significant ($p < 0.05$).

For the bacterial spot studies, Figure 7 was selected as positive controls, consisting of chlorhexidine 0.02% aqueous solution for the control group. From the results of the study, it was observed that the aqueous solution of chlorhexidine exhibited antibacterial effects. It demonstrated antibacterial activity against both Gram-positive and Gram-negative clinical and reference bacterial strains. The selected eutectic solvent served as a negative control and also showed antibacterial activity against the tested clinical and reference bacterial isolates. The components of the DES solvent contributed significantly to the antibacterial activity of the propolis extracts against the tested bacterial isolates.

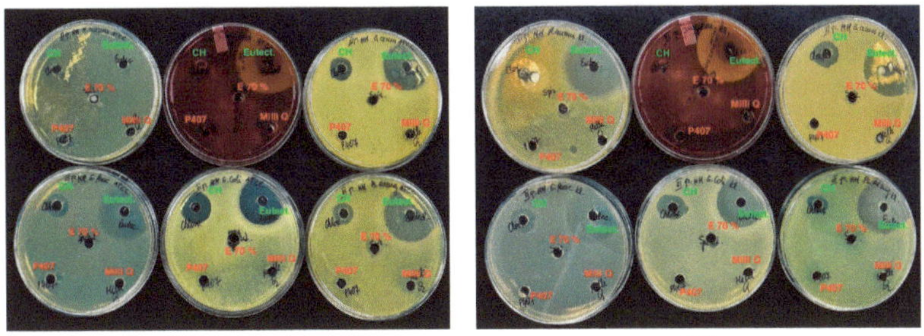

(A) (B)

Figure 7. Antimicrobial activity of control groups. Reference strains (**A**). Wild strains (**B**). CH—chlorhexidine aqueous solution 0.02%, Eutect.—eutectic aqueous solution, P407—poloxomer (P407) aqueous solution, E-70%—ethanol, Milli Q.

The research data depicted in Figure 8A show that propolis polymeric extracts did not exhibit inhibitory antifungal activity against the clinical and reference fungi of the *Saccharomycetaceae* family. Fungal growth was observed around the cavities (7 LT, 8 PL, 9 LV).

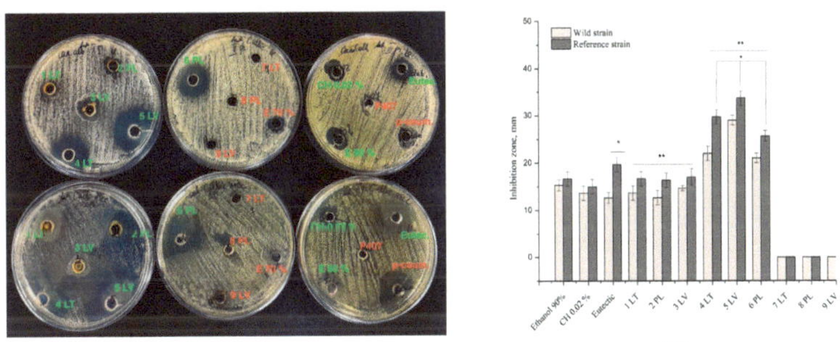

(A) (B)

Figure 8. Antifungal activity of the propolis extracts. Referential and wild *C. albicans*. 1—LT ethanolic propolis, 2—PL ethanolic propolis, 3—LV ethanolic propolis, 4—LT eutectic propolis, 5—LV eutectic propolis, 6—PL eutectic propolis, 7—LT polymeric propolis, 8—PL polymeric propolis, 9—LV polymeric propolis, CH—chlorhexidine aqueous solution 0.02%, Eutec.—eutectic aqueous solution, P407—poloxomer (P407) aqueous solution E—90/70%—ethanol (**A**). Antifungal activity of propolis extract on *C. albicans* (**B**). Asterisks indicate statistical significance between the eutectic control and propolis extract groups, $p < 0.05$.

The DES propolis extracts in Figure 8B demonstrated statistically significant antifungal activity against the reference and clinical fungi of the *Saccharomycetaceae* family ($p < 0.05$). The DES propolis extracts exhibited a stronger inhibitory antifungal effect compared to the ethanolic propolis extracts. A significant difference ($p < 0.05$) was found between the inhibition zones of these propolis extracts. Furthermore, the DES propolis extracts showed a statistically stronger antifungal inhibitory effect against *C. albicans* compared to the control solutions ($p < 0.05$).

3.4. Minimum Inhibitory Concentration

The MIC results of the present study demonstrated that the antimicrobial activity of the tested propolis extracts ranged from 0.05 mg/mL to 5 mg/mL. The minimum inhibitory concentration of *p*-coumaric acid, the most abundant active compound in the tested extracts, ranged from 0.1 mg/mL to 1 mg/mL (Table 2). Comparison of the values in this study with a previous agar diffusion study clearly shows that the propolis extracts exhibit significant antimicrobial properties against the *Staphylococcus aureus*, *Streptococcus agalactiae*, *Bacillus cereus*, *Enterococcus faecalis*, *Escherichia coli*, and *Pseudomonas aeruginosa* clinical and reference strains.

Table 2. Minimum suppressive concentrations of propolis extracts against the bacterial strains tested.

	Eutectic Propolis Extract		Ethanol Propolis Extract		*p*-Coumaric Acid	
	MIC Reference Strains (µg/mL)	MIC Wild Strains (µg/mL)	MIC Reference Strains (µg/mL)	MIC Wild Strains (µg/mL)	MIC Reference Strains (µg/mL)	MIC Wild Strains (µg/mL)
S. aureus	50	100	500	1000	200	500
S. agalactiae	200	200	1000	500	100	200
B. cereus	120	200	1000	500	500	500
E. feacalis	500	500	1000	1000	200	500
E.coli	1000	1000	5000	5000	500	1000
P. aeruginosa	1000	1000	5000	5000	500	1000

4. Discussion

The emergence and spread of AMR are putting global health systems at risk; for example, the 2015 WHO Global Action Plan on AMR has become one of the key roadmaps in the fight against the elements of AMR, which includes the screening of compounds in order to capture new antimicrobials [46]. Advances in many different research areas have fueled the resurgence of natural products. All of this leads to curbing AMR and better understanding [47]. Veterinary medicine focuses more on herbal medicines to curb the spread of AMR. There is a widespread popular belief that medicinal plants are effective and safer than synthetic compounds. The other main reason is economical, as they are cheaper than conventional therapies and useful in treating subclinical or chronic diseases without conventional treatment, as well as disorders that do not require professional diagnosis [48,49]. The selection of natural products must cover the widest and most diverse range of target pathogens/species (e.g., Gram-positive, Gram-negative bacterial species, spore-forming and acid-fast species, extracellular micro-organisms, clinical isolates) to be able to evaluate the potential of the active compounds [50]. The use of natural materials in veterinary medicine has been the subject of a number of studies on the health and treatment of animals. Several veterinary medicine researchers have investigated the antimicrobial activity of natural products against canine Staphylococcus and *Enterobacteriaceae* spp. These studies investigated various natural sources, including herbal extracts, honey products, and bacteriophages [51,52]. Such studies have provided valuable insights into the potential effectiveness of natural compounds as alternative treatments. Organic sulphur derivatives of garlic, such as allicin compounds, have been reported to have antibacterial activity against *Staphylococcus* spp. [53]. Essential oils are another perspective that is being widely explored. Scientists have studied the activity of oregano essential oil (OEO) and found that it has unique antibacterial and antioxidant properties. These properties have brought OEO to the attention of the whole world so that it can replace antimicrobial growth promoters, which have a significant impact on the livestock industry. OEO has been recommended for the treatment of infections caused by *Candida* spp. [54]. Siddique and other researchers have carried out in vitro probiotic studies with lactic acid bacteria as potential probiotic candidates. Probiotics are beneficial bacteria that live in the gut and help improve health; they are also one of the most popular alternatives to antibiotics [55]. Regarding small animal health disorders that can be treated with medicinal plants or their active compounds, a study was carried out in Spain on the number of animals treated with medicinal plants in clinic visits. The results showed that half of the patients treated in the clinic were treated for dermatological diseases, 70.1% were treated for musculoskeletal diseases, and about 51% of the dogs were treated for supraintestinal disorders [56]. Among the natural plant-based treatment options for veterinary diseases is propolis. For example, propolis has been used as an ointment to control mastitis in cows, as a prophylactic in pig herds for diseases of the respiratory tract, as a gastrointestinal tract, as a growth stimulant, and even as a local anaesthetic during surgery [57]. Propolis has also been used to effectively treat eye diseases in humans and animals due to its many healing properties. In cats and dogs suffering from blepharitis, infectious conjunctivitis, corneal edema, lacrimal duct obstruction, dry keratoconjunctivitis, corneal ulcers, and glaucoma, treatment was administered [58]. Propolis is a natural source of phenolic compounds, and choosing the right solvent for its extraction is important [59]. One of the objectives of this study was to produce propolis extracts with different solvents. For the production of extracts, the maceration method was chosen as the most accessible and most widely used extraction method in practice [60]. Our research results confirm data from the scientific literature that the qualitative profile of phenolic compounds and their quantity in extracts depend on the solvent used in the production and the propolis raw material for extraction. DES components can be selected not only to tune the physicochemical properties of the selected solvents but also to enhance the biological activity of the DES extracts [61,62]. The smallest quantities of the studied active compounds were found in extracts produced using water and a mixture of poloxamer (P407). The application of eutectic solvents allowed for the separation of larger quantities of active compounds; however, these solvents demonstrated weaker extraction capabilities compared to 70% (*v*/*v*) ethanol. Additionally, the application of DES propolis raw material for extraction has paved the way for potential non-ethanol liquid extracts as a basis for adaptation [63]. These extracts showed a statistically significant increase in the total quantity of active compounds compared to extracts produced with poloxamer (P407). Propolis extracts were derived from raw propolis collected in various countries across the Baltic region. The profile of active compounds in the investigated propolis extracts exhibited similarity. The scientific literature indicates that the propolis of Lithuania, Poland, and Latvia is categorized as poplar-type propolis [64–67]. The highly dominant compound in propolis extracts, *p*-coumaric acid, is recognized as a hydroxyl compound of cinnamic acid. It is known for its ability to reduce the peroxidation of low-density lipoproteins, exhibit antimicrobial activity, contribute to inhibiting cellular melanogenesis, and positively affect the regulation of the human immune system [68]. Lower amounts of ferulic acid and vanillin were found in the tested extracts compared to *p*-coumaric acid. Ferulic acid in propolis extracts is an important natural antioxidant. There is a wealth of published scientific research data on the antioxidant, antiallergic, hepatoprotective, anticarcinogenic, anti-inflammatory, antimicrobial, antiviral, vasodilatory, and antithrombotic effects of ferulic acid [69]. Vanillin (4-hydroxy-3-methoxybenzaldehyde) is a natural aromatic compound primarily used to impart fragrance and flavour to food products and beverages. In recent years, scientific data have been compiled on vanillin's anticancer, antidiabetic, antioxidant, anti-inflammatory, and antimicrobial effects [70]. Vanillin is one of the components of propolis, contributing not only to a pleasant aroma but also to its biological impact. The smallest quantities of vanillin and cinnamic acids have been detected in the studied extracts. Vanillic acid and cinnamic acid are phenolic acids, which are secondary aromatic products known for their strong antioxidant activity [71].

Our research group found that the investigated propolis extracts exhibited excellent antiradical activity. Overall, our research data indicated that antiradical activity directly correlates with the concentrations of active compounds in the extract. We used a simple and quickly performed method to evaluate antioxidant activity. Our chosen method confirmed the antioxidant activity of the extracts. The DPPH method is suitable for identifying antioxidants that are soluble in organic solvents. The DPPH radical is sensitive to light, oxygen, and pH changes, leading to varying results when different solvents are used [72]. According to the data obtained during this study, the examined extracts were effective antioxidants in vitro.

The results confirm published research data indicating that Baltic propolis extracts exhibit high antioxidant activity attributable to their chemical composition [73,74]. Our research findings demonstrated that the antioxidant activity of propolis extracts directly depended on the quantity of active compounds present. A strong correlation was established between the number of active compounds and antioxidant activity.

After examining the scientific literature, we can conclude that antimicrobial activity is one of propolis's most studied biological effects [17,30,64–66,75–78]. This is particularly relevant given the high resistance of some microorganisms to antibiotics and antifungal drugs, posing a global threat [79–81]. Betancourt and other researchers made a remark that propolis can be successfully used to treat eye diseases in cats and dogs; e.g., propolis drops have been used for the treatment of eye diseases such as conjunctivitis, corneal ulcers, and keratoconjunctivitis, for up to seven days in acute cases and up to ten days in chronic cases [82]. Cardoso and other scientists also used propolis as ear drops when *S. aureus* was isolated from dogs with otitis [83]. The results of these studies confirmed the antimicrobial activity of propolis. They provided additional information on the biological properties of propolis extracts and their potential use in the fight against antimicrobial resistance (AMR) [84,85]. Various studies on the antibacterial efficacy of ethanolic propolis extracts correlate with the active compounds found in the extracts [86].

After conducting an in vitro antimicrobial evaluation study with ethanolic propolis extracts, we observed that the ethanolic propolis extracts did not inhibit *Enterococcaceae*, *Enterobacteriaceae*, and *Pseudomonadaceae* family bacteria. The results of our study showed that the propolis ethanol extracts (Lithuania, Latvia, Poland) are effective against *S. auresus*, *S. agalactiae*, *B. cereus*. The diameters of the sterile zones of the clinical strains were 13.33 ± 1.15–26.33 ± 1.15 mm, and those of the reference were 14 ± 1.15–22 ± 2 mm. The researchers believe that propolis has weaker activity against Gram-negative bacteria due to its multi-layered structure and higher fat content in the cell wall [87,88]. Polish researchers reported that propolis showed activity against various bacterial strains, including Gram-positive bacteria (*S. aureus*, *Staphylococcus* spp., *Bacillus subtilis*, and *Enterococcus faecalis*) and Gram-negative *Enterobacter colica* (*E. coli*, *P. aeruginosa*) [25–27]. However, upon analyzing Nepali propolis, scientists noticed it exhibited similar antibacterial activity against Gram-negative and Gram-positive bacteria [32]. Propolis from the Middle East was also found to be highly active against both Gram-positive (*Staphylococcus aureus*) and Gram-negative (*Escherichia coli*) strains [30]. In general, there is a lack of studies investigating the antimicrobial activity of non-ethanolic extracts. The examined extracts, made from water and poloxamer (P407), did not inhibit the growth of *Staphylococcaceae*, *Streptococcaceae*, *Bacillaceae*, *Enterococcaceae*, *Enterobacteriaceae*, *Pseudomonadaceae*, and *Saccharomycetaceae* family isolates. This could be attributed to the low levels of active compounds isolated. The mixture of poloxamer and water was not an effective solvent for extracting the active components of propolis compared to other solvents we tested and to the aqueous extract of propolis with PEG additive evaluated by other researchers [78]. However, the Lithuanian aqueous propolis extract with a PEG additive exhibited antibacterial effects. Non-ethanolic extracts could serve as an excellent alternative to ethanolic ones for propolis extracts, provided that the solvent chosen for their production effectively separates the active compounds from the raw material. This was confirmed by experimental studies conducted by our research team using a DES solvents to produce propolis extracts. Although DESs are increasingly used for extracting active compounds from plant and animal raw materials, it should be noted that the extraction of propolis active compounds with these solvents is limited. Greek researchers studied the effect of different DESs on the yield of active compounds and antioxidant activity but did not investigate the antimicrobial activity of the extracts [89]. Our results show that the eutectic extracts of propolis DES were effective in inhibiting all Gram-positive and Gram-negative bacterial strains tested. The results of our study showed that propolis DES extracts (Lithuania, Latvia, Poland) effectively inhibited the growth of clinical and reference *S. aureus*, *S. agalactiae*, *B. cereus*, *Enterococcus feacalis*, *E. coli*, and *P. aeruginosa*. The diameters of the zones of inhibition in the clinical strains were 30.33 ± 1.52–40.33 ± 1.52 mm, and those of the reference were 31.33 ± 1.52–40.66 ± 1.52 mm. Radošević et al. investigated the antibacterial activity of DESs solvents with choline chloride and found that this solvent effectively inhibited the growth of *S. aureus* [38]. Bedair et al. also found that a DES with choline chloride has potential antimicrobial activity against *Enterococcus feacalis* strains [90]. According to the researchers, Gram-negative bacteria are less sensitive because of their outer membrane in the cell wall. This may be the reason why eutectic solvents were less inhibitory to *E. coli*, *P. mirabilis*, *S. typhimurium*, and *P. aeruginosa* than *S. aureus* [38].

A review of the scientific literature suggests that the MIC values of propolis can vary due to differences in geographical origin and composition across various countries [91–97]. The antibacterial efficacy of selected propolis extracts (ethanol and a DES) and *p*-coumaric acid at minimum inhibitory concentration (MIC) was determined against *Staphylococcus aureus*, *Streptococcus agalactiae*, *Bacillus cereus*, *Enterococcus faecalis*, *Escherichia coli*, and *Pseudomonas aeruginosa*, as well as clinical and reference bacterial strains. The MIC values of the propolis DES extracts were lower than those of the conventional ethanolic propolis extract. In our study, the minimum inhibitory concentration of the propolis DES extract was lower against Gram-positive bacterial strains (50–500 μg/mL). Higher concentrations of the propolis DES extract were required for Gram-negative bacterial strains, with 1000 μg/mL readings. The differences in antibacterial efficacy against Gram-positive and Gram-negative bacteria reported in the literature [96,98,99] corroborate our findings, with the MIC values for Gram-positive bacteria ranging from 3.125 to 400 μg/mL. Comparison of our data with those of Trusheva et al., who also studied the antimicrobial activity of NADES solvents, showed similar MIC values against *S. aureus* [100]. The MIC values of the ethanolic propolis extract were also found to be lower against Gram-positive bacterial strains (500–1000 μg/mL) compared to Gram-negative bacterial strains (5000 μg/mL). Polish researchers conducted a study on the antimicrobial activity of ethanolic propolis extract (EEP). They reported the highest activity against *S. aureus* in EEP from Turkey, Taiwan, and Oman, with the MIC values of 8, 10, and 81 μg/mL, respectively, and the lowest activity in propolis samples from Chile, Australia, and Germany, with the MIC values of 1445, 1200, and 750 μg/mL, respectively. Propolis extracts in ethanol from Turkey, Oman, and Slovakia were the most active against *E. coli*, with the MICs of 116, 302, and 510 μg/mL, respectively. Propolis samples from Germany, Korea, and Ireland had the lowest activity, with the MICs of 1200–5000 μg/mL [30]. Romanian and Brazilian researchers have also found that propolis ethanol extracts are considered viable synthetic products for treating canine superficial dermatitis and otitis caused by staphylococci [97,101]. However, a study by Grecka et al. revealed significant differences in the activity of ethanolic extracts of Polish propolis against Gram-positive and Gram-negative bacteria. Up to a 4096 μg/mL concentration, the researchers observed no activity against the *E. coli* and *P. aeruginosa* strains studied [92]. In our study, the MIC of *E. coli* and *P. aeruginosa* was 5000 μg/mL. Al-Ani et al. investigated European propolis samples from different geographical origins, finding both antimicrobial properties (the MICs against Gram-positive microorganisms ranged from 0.08 mg/mL to 2.5 mg/mL) and similarity to the minimum inhibitory concentration (0.5–1 mg/mL) of the ethanolic propolis extract studied in our group [102]. The results with *p*-coumaric acid demonstrated antimicrobial activity as a single active compound, with the readings ranging from 100 to 1000 μg/mL. Comparing these results with the MIC of Brazilian propolis prenylated *p*-coumaric acid, similar efficacy was observed against Gram-positive bacterial strains (*S. aureus* 100–300 μg/mL) [103]. A group of US researchers found, similar to our study, that *p*-coumaric acid had inhibitory effects on *E. coli* at 1000 μg/mL, *S. aureus* at 500 μg/mL, and *B. cereus* at 500 μg/mL [104]. The primary reason for the differing results compared to the literature is propolis's geographically variable chemical composition.

To assess the efficacy of propolis extracts, we evaluated their antifungal activity using the agar diffusion method against *C. albicans* [61,78]. The results of our study showed that both ethanolic and DES propolis extracts (Lithuania, Latvia, Poland) had antifungal activity. The diameters of the inhibitory zones of the ethanolic propolis extracts have been measured at 12.66 ± 1.52 to 17 ± 1.73 mm, and those of the DES propolis extracts have been measured at 21 ± 1 to 33.66 ± 1.52 mm. Turkish scientists reported that propolis also inhibits the growth of some clinical strains of bacteria and yeasts [105]. Fuentes Esquivel and other researchers investigating the antifungal activity of Mexican propolis in canine otitis showed that propolis extracts exhibited antifungal effects [106]. A study was also conducted in which dogs with dermatophytosis were treated with propolis-based soap at weekly bathing intervals of three to eight baths. After the study, it was observed that after two weeks of treatment, the dogs recovered from the lesions [107].

This study's results confirmed that the eutectic solvent components possess antimicrobial properties. Portuguese scientists examined extracts prepared based on NADES, which included lactic acid in their composition. During the study, they discovered that lactic acid introduced into the composition of eutectic solvents exhibits antimicrobial activity against Gram-negative and Gram-positive bacterial strains [108]. Furthermore, Greek scholars attempted to compare the antimicrobial efficacy of natural organic acid and found that lactic acid (LA) also exhibits antimicrobial efficacy [109]. The latest research data from scientists showed that the use of DES extracts with choline chloride and lactic acid not only resulted in the highest extraction efficiency but also demonstrated antimicrobial efficacy against *Salmonella*, *S. aureus*, *E. coli*, *P. aeruginosa*, and *B. subtilis* [41]. A eutectic solvent with choline chloride and lactic acid emerges as a potential carrier of active compounds in producing non-ethanol propolis extracts.

5. Conclusions

The findings of this study indicate that the isolation of phenolic compounds is feasible in producing propolis extracts using a eutectic solvent. A eutectic solvent based on choline chloride and lactic acid emerges as a potential carrier of propolis active substances, capable of enhancing the antibacterial effect of propolis active compounds. The antioxidant activity of propolis extracts directly depends on the extracted amounts of active substances. Propolis extracts from the Baltic region exhibit a similar chemical composition and demonstrate consistent biological effects. Our study confirms that *p*-coumaric acid dominates the chemical composition of propolis extracts from the Baltic region. Based on the results of this study regarding the antibacterial and antioxidant activities, further development in this area holds promise for integrating and expanding the practical application of ethanolic and DES propolis extracts in veterinary medicine, particularly for treating antibacterial and anti-inflammatory diseases.

Author Contributions: Conceptualization, D.S., K.R. and M.J.; methodology, K.R., D.S., G.Z., L.I. and L.B.; investigation, K.R., D.S., M.J., L.B. and M.M.; data curation, K.R., D.S. and M.J.; writing—original draft preparation, K.R., D.S. and M.J.; writing—review and editing, K.R., D.S. and M.J.; visualization, K.R., D.S. and M.J.; supervision, K.R.; project administration, K.R. and M.J. All authors have read and agreed to the published version of the manuscript.

Funding: This research received no external funding.

Institutional Review Board Statement: Not applicable.

Informed Consent Statement: Not applicable.

Data Availability Statement: The data presented in this study are openly available in the article.

Conflicts of Interest: The authors declare no conflicts of interest.

References

1. Patridge, E.; Gareiss, P.; Kinch, M.S.; Hoyer, D. An analysis of FDA-approved drugs: Natural products and their derivatives. *Drug Discov. Today* **2016**, *21*, 204–207. [CrossRef] [PubMed]
2. Cui, C.Y.; Li, X.J.; Chen, C.; Wu, X.T.; He, Q.; Jia, Q.L.; Zhang, X.J.; Lin, Z.Y.; Li, C.; Fang, L.X.; et al. Comprehensive analysis of plasmid-mediated *tet*(X4)-positive *Escherichia coli* isolates from clinical settings revealed a high correlation with animals and environments-derived strains. *Sci. Total Environ.* **2022**, *806*, 150687. [CrossRef]
3. He, T.; Wang, R.; Liu, D.J.; Walsh, T.R.; Zhang, R.; Lv, Y.; Ke, Y.B.; Ji, Q.J.; Wei, R.C.; Liu, Z.H.; et al. Emergence of plasmid-mediated high-level tigecycline resistance genes in animals and humans. *Nat. Microbiol.* **2019**, *4*, 1450–1456. [CrossRef]

4. Li, Y.; Sun, X.R.; Xiao, X.; Wang, Z.Q.; Li, R.C. Global distribution and genomic characteristics of *tet*(X)-positive *Escherichia coli* among humans, animals, and the environment. *Sci. Total Environ.* **2023**, *887*, 164148. [CrossRef]
5. Kuralkar, P.; Kuralkar, S.V. Role of herbal products in animal production—An updated review. *J. Ethnopharmacol.* **2021**, *278*, 114246. [CrossRef]
6. Smith, J.; Doe, A.; Brown, B. The Impact of Antibiotic Residues on Antimicrobial Resistance. *Antibiotics* **2022**, *11*, 123–135.
7. Liao, F.; Han, C.; Deng, Q.; Zhou, Z.; Bao, T.; Zhong, M.; Tao, G.; Li, R.; Han, B.; Qiao, Y.; et al. Natural Products as Mite Control Agents in Animals: A Review. *Molecules* **2023**, *28*, 6818. [CrossRef]
8. Niu, G.; Li, W. Next-generation drug discovery to combat antimicrobial resistance. *Trends Biochem. Sci.* **2019**, *44*, 961–972. [CrossRef]
9. Pancu, D.F.; Scurtu, A.; Macasoi, I.G.; Marti, D.; Mioc, M.; Soica, C.; Coricovac, D.; Horhat, D.; Poenaru, M.; Dehelean, C. Antibiotics: Conventional Therapy and Natural Compounds with Antibacterial Activity—A Pharmaco-Toxicological Screening. *Antibiotics* **2021**, *10*, 401. [CrossRef] [PubMed]
10. Rizzo, A.; Piccinno, M.; Lillo, E.; Carbonari, A.; Jirillo, F.; Sciorsci, R.L. Antimicrobial Resistance and Current Alternatives in Veterinary Practice: A Review. *Curr. Pharm. Des.* **2023**, *29*, 312–322. [PubMed]
11. Rahman, M.M.; Rahaman, M.S.; Islam, M.R.; Rahman, F.; Mithi, F.M.; Alqahtani, T.; Almikhlafi, M.A.; Alghamdi, S.Q.; Alruwaili, A.S.; Hossain, M.S.; et al. Role of Phenolic Compounds in Human Disease: Current Knowledge and Future Prospects. *Molecules* **2022**, *27*, 233. [CrossRef]
12. Graf, B.A.; Millbury, P.E.; Blumberg, J.B. Flavonols, flavones, flavanones, and human health: Epidemiological evidence. *J. Med. Food* **2005**, *8*, 281–290. [CrossRef]
13. Arts, I.C.W.; Hollman, P.C.H. Polyphenols and disease risk in epidemiologic studies. *Am. J. Clin. Nutr.* **2005**, *81*, 317–325. [CrossRef]
14. Farooqui, T. Beneficial effects of propolis he human health and neurological diseases. *Front. Biosci.* **2012**, *4*, 779.
15. Wang, K.; Zhang, J.; Ping, S.; Ma, Q.; Chen, X.; Xuan, H.; Shi, J.; Zhang, C.; Hu, F. Anti-inflammatory effects of ethanol extracts of Chinese propolis and buds from poplar (*Populus canadensis*). *J. Ethnopharmacol.* **2014**, *155*, 300–311. [CrossRef]
16. Peng, S.; Wei, P.; Lu, Q.; Liu, R.; Ding, Y.; Zhang, J. Beneficial Effects of poplar Buds on Hyperglycemia, Dyslipidemia, Oxidative Stress, and Inflammation in Streptozotocin-Induced Type-2 Diabetes. *J. Immunol. Res.* **2018**, *2018*, 7245956. [CrossRef] [PubMed]
17. Abd Rashid, N.; Mohammed, S.N.; Syed Abd Halim, S.A.; Ghafar, N.A.; Abdul Jalil, N.A. Therapeutic Potential of Honey and Propolis on Ocular Disease. *Pharmaceuticals* **2022**, *15*, 1419. [CrossRef]
18. El Adham, E.C.; Hassan, A.I.; Dawoud, M.M.A. Evaluating the Role of Propolis and Bee Venom on the Oxidative Stress Induced by Gamma Rays in Rats. *Sci. Rep.* **2022**, *12*, 2656. [CrossRef] [PubMed]
19. Kurek-Górecka, A.; Rzepecka-Stojko, A.; Górecki, M.; Stojko, J.; Sosada, M.; Swierczek-Zieba, G. Structure and antioxidant activity of polyphenols derived from propolis. *Molecules* **2014**, *19*, 78–101. [CrossRef]
20. Pang, Z.; Chen, J.; Wang, T.; Gao, C.; Li, Z.; Guo, L.; Xu, J.; Cheng, Y. Linking Plant Secondary Metabolites and Plant Microbiomes: A Review. *Front. Plant Sci.* **2021**, *12*, 300. [CrossRef]
21. Stanciauskaite, M.; Marksa, M.; Ivanauskas, L.; Perminaite, K.; Ramanauskiene, K. Ophthalmic In Situ Gels with Balsam Poplar Buds Extract: Formulation, Rheological Characterization, and Quality Evaluation. *Pharmaceutics* **2021**, *13*, 953. [CrossRef] [PubMed]
22. Stanciauskaite, M.; Marksa, M.; Babickaite, L.; Majiene, D.; Ramanauskiene, K. Comparison of Ethanolic and Aqueous *Populus balsamifera* L. Bud Extracts by Different Extraction Methods: Chemical Composition, Antioxidant and Antibacterial Activities. *Pharmaceuticals* **2021**, *14*, 1018. [CrossRef] [PubMed]
23. Aboulghazi, A.; Fadil, M.; Touzani, S.; Hibaoui, L.; Hano, C.; Lyoussi, B. Phenolic Screening and Mixture Design Optimization for In Vitro Assessment of Antioxidant and Antimicrobial Activities of Honey, Propolis, and Bee Pollen. *J. Food Biochem.* **2024**, *1*, 8246224. [CrossRef]
24. Stanciauskaite, M.; Marksa, M.; Rimiekene, L.; Ramanauskiene, K. Evaluation of Chemical Composition, Sun Protection Factor and Antioxidant Activity of Lithuanian Propolis and Its Plant Precursors. *Plants* **2022**, *11*, 3558. [CrossRef] [PubMed]
25. Popova, M.; Giannopoulou, E.; Skalicka-Wózniak, K.; Graikou, K.; Widelski, J.; Bankova, V.; Kalofonos, H.; Sivolapenko, G.; Gaweł-Bęben, K.; Antosiewicz, B.; et al. Characterization and biological evaluation of propolis from Poland. *Molecules* **2017**, *22*, 1159. [CrossRef]
26. Kubina, R.; Kabała-Dzik, A.; Dziedzic, A.; Bielec, B.; Wojtyczka, R.D.; Bułdak, R.J.; Wyszyńska, M.; Stawiarska-Pięta, B.; Szaflarska-Stojko, E. The Ethanol Extract of Polish Propolis Exhibits Anti-Proliferative and/or Pro-Apoptotic Effect on HCT 116 Colon Cancer and Me45 Malignant Melanoma Cells In Vitro Conditions. *Adv. Clin. Exp. Med.* **2015**, *24*, 203–212. [CrossRef]
27. Szliszka, E.; Krol, W. Polyphenols isolated from propolis augment TRAIL-induced apoptosis in cancer cells. *Evid.-Based Complement. Altern. Med.* **2013**, *2013*, 731940. [CrossRef] [PubMed]
28. Martínez-Lorente, S.E.; Martí-Guillén, J.M.; Pedreño, M.Á.; Almagro, L.; Sabater-Jara, A.B. Higher Plant-Derived Biostimulants: Mechanisms of Action and Their Role in Mitigating Plant Abiotic Stress. *Antioxidants* **2024**, *13*, 318. [CrossRef]
29. Shigenori, K.; Tomoko, H.; Tsutomu, N. Antioxidant activity of propolis of various geographic origins. *Food Chem.* **2004**, *84*, 329–339.
30. Przybyłek, I.; Karpiński, T.M. Antibacterial Properties of Propolis. *Molecules* **2019**, *24*, 2047. [CrossRef]
31. Smith, J.; Doe, A.; Brown, B. The Dual Mechanisms of Propolis in Antimicrobial Activity. *J. Nat. Prod.* **2023**, *15*, 100–110.

32. Okińczyc, P.; Paluch, E.; Franiczek, R.; Widelski, J.; Wojtanowski, K.K.; Mroczek, T.; Krzyżanowska, B.; Skalicka-Woźniak, K.; Sroka, Z. Antimicrobial activity of *Apis mellifera* L. and *Trigona* sp. propolis from Nepal and its phytochemical analysis. *Biomed. Pharmacother.* **2020**, *129*, 110435. [CrossRef]
33. Oryan, A.; Alemzadeh, E.; Moshiri, A. Potential role of propolis in wound healing: Biological properties and therapeutic activities. *Biomed. Pharmacother.* **2018**, *98*, 469–483. [CrossRef] [PubMed]
34. Torres, A.; Sandjo, L.; Friedemann, M.; Tomazzoli, M.; Maraschin, M.; Mello, C.; Santos, A. Chemical characterization, antioxidant and antimicrobial activity of propolis obtained from *Melipona quadrifasciata quadrifasciata* and *Tetragonisca angustula* stingless bees. *Br. J. Med Biol. Res.* **2018**, *51*, e7118. [CrossRef]
35. An, S.-H.; Ban, E.; Chung, I.-Y.; Cho, Y.-H.; Kim, A. Antimicrobial Activities of Propolis in Poloxamer Based Topical Gels. *Pharmaceuticals* **2021**, *13*, 2021. [CrossRef] [PubMed]
36. Dugar, R.P.; Gajera, B.Y.; Dave, R.H. Fusion Method for Solubility and Dissolution Rate Enhancement of Ibuprofen Using Block Copolymer Poloxamer 407. *AAPS PharmSciTech* **2016**, *17*, 1428–1440. [CrossRef]
37. Trusheva, B.; Petkov, H.; Chimshirova, R.; Popova, M.; Dimitrova, L.; Zaharieva, M.M.; Ilieva, Y.; Vasileva, B.; Tsvetkova, I.; Bankova, V.; et al. Insight into the influence of natural deep eutectic solvents on the extraction of phenolic compounds from poplar type propolis: Composition and in vitro biological activity. *Heliyon* **2024**, *10*, e28621. [CrossRef]
38. Radošević, K.; Čanak, I.; Panić, M.; Markov, K.; Bubalo, M.C.; Frece, J.; Srček, V.G.; Redovniković, I.R. Antimicrobial, cytotoxic and antioxidative evaluation of natural deep eutectic solvents. *Environ. Sci. Pollut. Res.* **2018**, *25*, 14188–14196. [CrossRef]
39. Alpat, U.; Nar, T.; Karasu, S.; Sagdic, O. Optimization of propolis extraction with natural deep eutectic solvents using central composite design. *Phytochem. Lett.* **2023**, *58*, 49–59. [CrossRef]
40. Funari, C.S.; Sutton, A.T.; Carneiro, R.L.; Fraige, K.; Cavalheiro, A.J.; da Silva Bolzani, V.; Hilder, E.F.; Arrua, R.D. Natural deep eutectic solvents and aqueous solutions as an alternative extraction media for propolis. *Food Res. Int.* **2019**, *125*, 108559. [CrossRef] [PubMed]
41. Yu, T.; Yang, L.; Shang, X.; Bian, S. Recovery of Cembratrien -Diols from Waste Tobacco (*Nicotiana tabacum* L.) Flowers by Microwave-Assisted Deep Eutectic Solvent Extraction: Optimization, Separation, and In Vitro Bioactivity. *Molecules* **2024**, *29*, 1563. [CrossRef]
42. Zhang, J.; Cao, X.; Xueping, C.; Wang, K.; Shi, J.; Zhang, C.; Zheng, H.; Huoqing, Z. Comparisons of Ethanol Extracts of Chinese Propolis (Poplar Type) and poplar Gums Based he the Antioxidant Activities and Molecular Mechanism. *Evid. Based Complement. Altern. Med.* **2015**, *2015*, 307594.
43. Ramanauskienė, K.; Savickas, A.; Ivanauskas, L.; Kalveniene, Z.; Kasparaviciene, G.; Banionyte, I.; Amsiejus, A.; Martirosyan, D. Analysis of Phenolic Acids in Propolis Using the High-Performance Liquid Chromatography Technique. *Curr. Nutr. Food Sci.* **2008**, *4*, 209–212. [CrossRef]
44. Yim, S.H.; Nam, S.H. Physiochemical, nutritional and functional characterization of 10 different pear cultivars (*Pyrus* spp.). *J. Appl. Bot. Food Qual.* **2016**, *89*, 73–81.
45. Babickaite, L.; Ramanauskiene, K.; Grigonis, A.; Ivaskiene, M.; Daunoras, G.; Klimiene, I.; Matusevicius, A.P. Determination of antimicrobial activity of chlorhexidine gel. *Acta Pol. Pharm.* **2016**, *73*, 1623–1630.
46. Uchil, R.R.; Kohli, G.S.; Katekhaye, V.M.; Swami, O.C. Strategies to combat antimicrobial resistance. *J. Clin. Diagn. Res.* **2014**, *8*, ME01–ME04.
47. Hobson, C.; Chan, A.N.; Wright, G.D. The Antibiotic Resistome: A Guide for the Discovery of Natural Products as Antimicrobial Agents. *Chem. Rev.* **2021**, *121*, 3464–3494. [CrossRef]
48. Kubkomawa, H.I.; Nafarnda, D.W.; Tizhe, M.A.; Daniel, T.K.; Shua, N.J.; Ugwu, C.C.; Opara, M.N.; Neils, J.S.; Okoli, I.C. Ethno-veterinary health management practices amongst livestock producers in Africa: A review. *Adv. Agric. Sci.* **2020**, *6*, 252–257.
49. Wynn, S.; Fougère, B. Clinical practice: Getting started. In *Veterinary Herbal Medicine*; Wynn, S., Fougère, B., Eds.; Elsevier: St. Louis, MO, USA, 2006; pp. 453–457.
50. Barth, S.A.; Preussger, D.; Pietschmann, J.; Feßler, A.T.; Heller, M.; Herbst, W.; Schnee, C.; Schwarz, S.; Kloss, F.; Berens, C.; et al. In Vitro Antibacterial Activity of Microbial Natural Products against Bacterial Pathogens of Veterinary and Zoonotic Relevance. *Antibiotics* **2024**, *13*, 135. [CrossRef]
51. Balcão, V.M.; Belline, B.G.; Silva, E.C.; Almeida, P.F.F.B.; Baldo, D.; Amorim, L.R.P.; Oliveira Júnior, J.M.; Vila, M.M.D.C.; Del Fiol, F.S. Isolation and Molecular Characterization of Two Novel Lytic Bacteriophages for the Biocontrol of *Escherichia coli* in Uterine Infections: In Vitro and Ex Vivo Preliminary Studies in Veterinary Medicine. *Pharmaceutics* **2022**, *14*, 2344. [CrossRef] [PubMed]
52. Meroni, G.; Cardin, E.; Rendina, C.; Millar, V.R.H.; Filipe, J.F.S.; Martino, P.A. In Vitro Efficacy of Essential Oils from Melaleuca Alternifolia and Rosmarinus Officinalis, Manuka Honey-Based Gel, and Propolis as Antibacterial Agents against Canine *Staphylococcus pseudintermedius* Strains. *Antibiotics* **2020**, *9*, 344. [CrossRef] [PubMed]
53. Sheppard, J.G.; McAleer, J.P.; Saralkar, P.; Geldenhuys, W.J.; Long, T.E. Allicin-Inspired Pyridyl Disulfides as Antimicrobial Agents for Multidrug-Resistant Staphylococcus Aureus. *Eur. J. Med. Chem.* **2018**, *143*, 1185–1195. [CrossRef]
54. Cui, H.; Zhang, C.; Su, K.; Fan, T.; Chen, L.; Yang, Z.; Zhang, M.; Li, J.; Zhang, Y.; Liu, J. Oregano Essential Oil in Livestock and Veterinary Medicine. *Animals* **2024**, *14*, 1532. [CrossRef]
55. Siddique, A.; Azim, S.; Ali, A.; Adnan, F.; Arif, M.; Imran, M.; Ganda, E.; Rahman, A. *Lactobacillus reuteri* and *Enterococcus faecium* from Poultry Gut Reduce Mucin Adhesion and Biofilm Formation of Cephalosporin and Fluoroquinolone-Resistant *Salmonella enterica*. *Animals* **2021**, *11*, 3435. [CrossRef] [PubMed]

56. Romero, B.; Susperregui, J.; Sahagún, A.M.; Diez, M.J.; Fernández, N.; García, J.J.; López, C.; Sierra, M.; Díez, R. Use of medicinal plants by veterinary practitioners in Spain: A cross-sectional survey. *Front. Vet. Sci.* **2022**, *9*, 1060738. [CrossRef]
57. Giral, T.; Hugues, B.; Soto, C.J. Suspensión oftálmica de propóleos-R: Una alternativa en el tratamiento de las oftalmopatias en animales afectivos. *Rev. Electrónica Clínica Vet.* **2007**, *2*, 1–6.
58. Alfaris, A.A.; Abdulsamadm, R.K.; Swadm, A.A. Comparative Studies between Propolis, Dexametason and Gentamycin Treatments of Induced Corneal Ulcer in Rabbits. *Iraqi J. Vet. Sci.* **2009**, *23*, S75–S80.
59. Zhang, Q.; Yang, A.; Tan, W.; Yang, W. Development, Physicochemical Properties, and Antibacterial Activity of Propolis Microcapsules. *Foods* **2023**, *12*, 3191. [CrossRef]
60. Sforcin, J.M.; Bankova, V. Propolis: Is there a potential for the development of new drugs? *J. Ethnopharmacol.* **2011**, *133*, 253–260. [CrossRef]
61. Woźniak, M.; Mrówczyńska, L.; Kwaśniewska-Sip, P.; Waskiewicz, A.; Nowak, P.; Ratajczak, I. Effect of the Solvent on Propolis Phenolic Profile and its Antifungal, Antioxidant, and In Vitro Cytoprotective Activity in Human Erythrocytes Under Oxidative Stress. *Molecules* **2020**, *25*, 4266. [CrossRef]
62. Radošević, K.; Ćurko, N.; Gaurina Srček, V.; Cvjetko Bubalo, M.; Tomašević, M.; Kovačević Ganić, K.; Radojčić Redovniković, I. Natural Deep Eutectic Solvents as Beneficial Extractants for Enhancement of Plant Extracts Bioactivity. *LWT Food Sci. Technol.* **2016**, *73*, 45–51. [CrossRef]
63. Mustafa, N.R.; Spelbos, V.S.; Witkamp, G.-J.; Verpoorte, R.; Choi, Y.H. Solubility and Stability of Some Pharmaceuticals in Natural Deep Eutectic Solvents-Based Formulations. *Molecules* **2021**, *26*, 2645. [CrossRef]
64. Bouchelaghem, S. Propolis characterization and antimicrobial activities against *Staphylococcus aureus* and *Candida albicans*: A review. *Saudi J. Biol. Sci.* **2022**, *29*, 1936–1946. [CrossRef]
65. Wojtacka, J. Propolis Against Pharmacological Interventions in Bees. *Molecules* **2022**, *27*, 4914. [CrossRef]
66. Stančiauskaitė, M.; Marksa, M.; Liaudanskas, M.; Ivanauskas, L.; Ivaskienė, M.; Ramanauskienė, K. Poplar buds (*Populus balsamifera* L., *Populus nigra* L.) and Lithuanian propolis extracts: Comparison of their composition and biological activity. *Plants* **2021**, *10*, 828. [CrossRef] [PubMed]
67. Woźniak, M.; Sip, A.; Mrówczyńska, L.; Broniarczyk, J.; Waśkiewicz, A.; Ratajczak, I. Biological Activity and Chemical Composition of Propolis from Various Regions of Poland. *Molecules* **2022**, *28*, 141. [CrossRef]
68. Kiliç, I.; Yeşiloğlu, Y. Spectroscopic Studies he the Antioxidant Activity of P-Coumaric Acid. *Spectrochim. Vineg. Part A Mol. Biomol. Spectrosc.* **2013**, *115*, 719–724. [CrossRef]
69. Naman, D.R.; Dilpreet, S. A critical appraisal on ferulic acid: Biological profile, biopharmaceutical challenges and nano formulations. *Health Sci. Rev.* **2022**, *5*, 100063.
70. Olatunde, A.; Aminu, M.; Ibrahim, M.A.; Tajuddeen, N.; Shuaibu, M.N. Vanillin: A food additive with multiple biological activities. *Eur. J. Med. Chem. Rep.* **2022**, *5*, 100055. [CrossRef]
71. Tanner, G.; Vardar, D.O.; Aydın, S.; Aytaç, Z.; Basaran, A.A.; Basaran, N. Use of in vitro assays to assess the potential cytotoxic, genotoxic and antigenotoxic effects of vanillic and cinnamic acid. *Drug Chem. Toxicol.* **2016**, *40*, 183–190. [CrossRef] [PubMed]
72. Kedar, S.B.; Singh, R.P. Genesis and development of DPPH method of antioxidant assay. *J. Food Sci. Technol.* **2011**, *48*, 412–422. [CrossRef] [PubMed]
73. Rasul, A.; Millimouno, F.M.; Eltayb, W.A.; Ali, M.; Lee, J.; Li, X. Pinocembrin: A Novel Natural Compound with Versatile Pharmacological and Biological Activities. *BioMed Res. Int.* **2013**, *2013*, 379850. [CrossRef]
74. Dezmirean, D.S.; Mărghita, S.L.A.; Chirila, F.; Copaciu, F.; Simonca, V.; Bobis, O.; Erler, S. Influence of geographic origin, plant source and polyphenolic substances on antimicrobial properties of propolis against human and honey bee pathogens. *J. Apic. Res.* **2017**, *56*, 588–597. [CrossRef]
75. Melo, J.G.A.D.; Soares, D.M.; Santos, S.C.D. In-vitro antibacterial activity of propolis extract against periodontopathogenic bacteria: A systematic review. *Rev. Port. Estomatol. Med. Dent. Cir. Maxilofac.* **2024**, *65*, 3–14. [CrossRef]
76. Vadillo-Rodríguez, V.; Fernández-Babiano, I.; Pérez-Giraldo, C.; Fernández-Calderón, M.C. Anti-Biofilm Perspectives of Propolis against *Staphylococcus epidermidis* Infections. *Biomolecules* **2024**, *14*, 779. [CrossRef] [PubMed]
77. Dezmirean, D.S.; Paşca, C.; Moise, A.R.; Bobiş, O. Plant Sources Responsible for the Chemical Composition and Main Bioactive Properties of Poplar-Type Propolis. *Plants* **2020**, *10*, 22. [CrossRef]
78. Kubiliene, L.; Laugaliene, V.; Pavilonis, A.; Maruska, A.; Majiene, D.; Barcauskaite, K.; Kubilius, R.; Kasparaviciene, G.; Savickas, A. Alternative preparation of propolis extracts: Comparison of their composition and biological activities. *BMC Complement. Altern. Med.* **2015**, *15*, 156. [CrossRef]
79. Brown, E.; Wright, G. Antibacterial friend discovery in the resistance era. *Nature* **2016**, *529*, 336–343. [CrossRef]
80. Jones, K.E.; Patel, N.G.; Levy, M.A.; Storeygard, A.; Balk, D.; Gittleman, J.L.; Daszak, P. Global trends in emerging infectious diseases. *Nature* **2008**, *451*, 990–993. [CrossRef]
81. WHO (World Health Organization). *Global Action Plan on Antimicrobial Resistance*; WHO Library Cataloguing-in-Publication Data; WHO: Geneva, Switzerland, 2015.
82. Betancourt, N.T.; García-Contreras, L.; Cruz Sánchez, T.A. Propolis in dogs: Clinical experiences and perspectives (A brief review). *Open J. Vet. Med.* **2015**, *5*, 53251. [CrossRef]

83. Cardoso, R.L.; Maboni, F.; Machado, G.; Hartz, A.S.; Castagna de Vargas, A. Antimicrobial activity of propolis extract against Staphylococcus coagulase positive and Malassezia pachydermatis of canine otitis. *Vet. Microbiol.* **2010**, *142*, 432–434. [CrossRef] [PubMed]
84. Freitas, A.S.; Cunha, A.; Oliveira, R.; Almeida-Aguiar, C. Propolis antibacterial and antioxidant synergisms with gentamicin and honey. *J. Appl. Microbiol.* **2022**, *132*, 2733–2745. [CrossRef] [PubMed]
85. Probst, I.S.; Sforcin, J.M.; VLM, R.; Fernandes, A.A.H.; Fernandes Júnior, A. Antimicrobial activity of propolis and essential oils and synergism between these natural products. *J. Venom. Animat. Toxins Incl. Trop. Dis.* **2011**, *17*, 159–167. [CrossRef]
86. Pobiega, K.; Kraśniewska, K.; Derewiaka, D.; Gniewosz, M. Comparison of the antimicrobial activity of propolis extracts obtained by means of various extraction methods. *J. Food Sci. Technol.* **2019**, *56*, 5386–5395. [CrossRef] [PubMed]
87. Bankova, V.; Boudourova-Krasteva, G.; Popov, S.; Sforcin, J.M.; Funari, S.R.C. Seasonal variations of the chemical composition of Brazilian propolis. *Apidologie* **1998**, *29*, 361–367. [CrossRef]
88. De Kraker, M.E.A.; Davey, P.G.; Grundmann, H. Mortality and Hospital Stay Associated with Resistant Staphylococcus aureus and Escherichia coli Bacteremia: Estimating the Burden of Antibiotic Resistance in Europe. *PLoS Med.* **2011**, *8*, e1001104. [CrossRef] [PubMed]
89. Tzani, A.; Pitterou, I.; Divani, F.; Tsiaka, T.; Sotiroudis, G.; Zoumpoulakis, P.; Detsi, A. Green Extraction of Greek Propolis Using Natural Deep Eutectic Solvents (NADES) and Incorporation of the NADES-Extracts in Cosmetic Formulation. *Sustain. Chem.* **2023**, *4*, 8–25. [CrossRef]
90. Bedair, H.M.; Samir, T.M.; Mansour, F.R. Antibacterial and antifungal activities of natural deep eutectic solvents. *Appl. Microbiol. Biotechnol.* **2024**, *108*, 198. [CrossRef] [PubMed]
91. Otręba, M.; Marek, Ł.; Tyczyńska, N.; Stojko, J.; Kurek-Górecka, A.; Górecki, M.; Olczyk, P.; Rzepecka-Stojko, A. Propolis as Natural Product in the Oral Cavity Bacterial Infections Treatment: A Systematic Review. *Appl. Sci.* **2022**, *12*, 10123. [CrossRef]
92. Grecka, K.; Kuś, P.M.; Okińczyc, P.; Worobo, R.W.; Walkusz, J.; Szweda, P. The Anti-Staphylococcal Potential of Ethanolic Polish Propolis Extracts. *Molecules* **2019**, *24*, 1732. [CrossRef]
93. Dziedzic, A.; Kubina, R.; Wojtyczka, R.D.; Kabała-Dzik, A.; Tanasiewicz, M.; Morawiec, T. The antibacterial effect of ethanol extract of polish propolis on mutans streptococci and lactobacilli isolated from saliva. *Evid. Based Complement Altern. Med.* **2013**, *2013*, 681891. [CrossRef] [PubMed]
94. Nichitoi, M.M.; Josceanu, A.M.; Isopescu, R.D.; Isopencu, G.O.; Geana, E.I.; Ciucure, C.T.; Lavric, V. Polyphenolics profile effects upon the antioxidant and antimicrobial activity of propolis extracts. *Sci. Rep.* **2021**, *11*, 20113. [CrossRef]
95. Boutoub, O.; El-Guendouz, S.; Matos, I.; El Ghadraoui, L.; Costa, M.C.; Carlier, J.D.; Faleiro, M.L.; Figueiredo, A.C.; Estevinho, L.M.; Miguel, M.G. Chemical Characterization and Biological Properties Assessment of *Euphorbia resinifera* and *Euphorbia officinarum* Moroccan Propolis. *Antibiotics* **2024**, *13*, 230. [CrossRef]
96. Kerek, Á.; Csanády, P.; Tuska-Szalay, B.; Kovács, L.; Jerzsele, Á. In Vitro Efficacy of Hungarian Propolis against Bacteria, Yeast, and *Trichomonas gallinae* Isolated from Pigeons—A Possible Antibiotic Alternative? *Resources* **2023**, *12*, 101. [CrossRef]
97. Dégi, J.; Herman, V.; Igna, V.; Dégi, D.M.; Hulea, A.; Muselin, F.; Cristina, R.T. Antibacterial Activity of Romanian Propolis against *Staphylococcus aureus* Isolated from Dogs with Superficial Pyoderma: In Vitro Test. *Vet. Sci.* **2022**, *9*, 299. [CrossRef]
98. Stepanović, S.; Antić, N.; Dakić, I.; Svabić-Vlahović, M. In vitro antimicrobial activity of propolis and synergism between propolis and antimicrobial drugs. *Microbiol. Res.* **2003**, *158*, 353–357. [CrossRef] [PubMed]
99. Zulhendri, F.; Chandrasekaran, K.; Kowacz, M.; Ravalia, M.; Kripal, K.; Fearnley, J.; Perera, C.O. Antiviral, Antibacterial, Antifungal, and Antiparasitic Properties of Propolis: A Review. *Foods* **2021**, *10*, 1360. [CrossRef]
100. Trusheva, B.; Petkov, H.; Popova, M.; Dimitrova, L.; Zaharieva, M.; Tsvetkova, I.; Bankova, V. "Green" approach to propolis extraction: Natural deep eutectic solvents. *Comptes Rendus L'academie Bulg. Sci.* **2019**, *72*, 897–905.
101. Deegan, K.R.; Fonseca, M.S.; Oliveira, D.C.P.; Santos, L.M.; Fernandez, C.C.; Hanna, S.A.; Machado, B.A.S.; Umsza-Guez, M.A.; Meyer, R.; Portela, R.W. Susceptibility of *Malassezia pachydermatis* Clinical Isolates to Allopathic Antifungals and Brazilian Red, Green, and Brown Propolis Extracts. *Front. Vet. Sci.* **2019**, *6*, 460. [CrossRef]
102. AL-Ani, I.; Zimmermann, S.; Reichling, J.; Wink, M. Antimicrobial Activities of European Propolis Collected from Various Geographic Origins Alone and in Combination with Antibiotics. *Medicines* **2018**, *5*, 2. [CrossRef] [PubMed]
103. Jorge, R.; Furtado, N.A.J.C.; Sousa, J.P.B.; da Silva Filho, A.A.; Gregório Junior, L.E.; Martins, C.H.G.; Soares, A.E.E.; Bastos, J.K.; Cunha, W.R.; Silva, M.L.A. Brazilian Propolis: Seasonal Variation of the Prenylated p-Coumaric Acids and Antimicrobial Activity. *Pharm. Biol.* **2008**, *46*, 889–893. [CrossRef]
104. Herald, P.J.; Davidson, P.M. Antibacterial activity of selected hydroxycinnamic acids. *J. Food Sci.* **1983**, *48*, 1378–1379. [CrossRef]
105. Tong, S.Y.; Davis, J.S.; Eichenberger, E.; Holland, T.L.; Fowler, V.G. *Staphylococcus aureus* infections: Epidemiology, pathophysiology, clinical manifestations, and management. *Clin. Microbiol. Rev.* **2015**, *28*, 603–661. [CrossRef] [PubMed]
106. Fuentes Esquivel, D.B.; Pérez, B.R.; Betancourt, N.T.; García Tovar, C.G.; Penieres Carrillo, J.G.; Galindo, F.H.; Flores, J.P.; Sánchez, T.A.C. In Vitro Antimycotic Activity and Structural Damage against Canine *Malassezia pachydermatis* Strains Caused by Mexican Stingless Bee Propolis. *Vet. Sci.* **2024**, *11*, 106. [CrossRef] [PubMed]
107. Cruz Sánchez, T.A.; Estrada García, P.A.; López Zamora, C.I.; Autran Martínez, M.; Pérez Valencia, V.; Londoño Orozco, A. Use of Propolis for Tipical Treatment of Dermatophytosis in Dog. *Open J. Vet. Med.* **2014**, *4*, 239–245. [CrossRef]

108. Rocha, D.; Freitas, D.S.; Magalhães, J.; Fernandes, M.; Silva, S.; Noro, J.; Ribeiro, A.; Cavaco-Paulo, A.; Martins, M.; Silva, C. NADES-Based Cork Extractives as Green Ingredients for Cosmetics and Textiles. *Processes* **2023**, *11*, 309. [CrossRef]
109. Kostoglou, D.; Vass, A.; Giaouris, E. Comparative Assessment of the Antibacterial and Antibiofilm Actions of Benzalkonium Chloride, Erythromycin, and L (+)-Lactic Acid against Raw Chicken Meat *Campylobacter* spp. Isolates. *Antibiotics* **2024**, *13*, 201. [CrossRef]

veterinary sciences

MDPI

Article

Current Situation of Bacterial Infections and Antimicrobial Resistance Profiles in Pet Rabbits in Spain

Mercedes Fernández [1], Biel Garcias [1], Inma Duran [2], Rafael A. Molina-López [3] and Laila Darwich [1,*]

[1] Departament de Sanitat i Anatomia Animal, Universitat Autònoma de Barcelona (UAB), 08193 Cerdanyola del Vallès, Spain; mariamercedes.fernandez2@autonoma.cat (M.F.); biel.garcias@uab.cat (B.G.)
[2] Departamento Veterinaria de Laboratorio Echevarne, S.A., 08037 Barcelona, Spain; idduran@laboratorioechevarne.com
[3] Catalan Wildlife Service, Centre de Fauna Salvatge de Torreferrussa, 08130 Santa Perpètua de Mogoda, Spain; rafael.molina@gencat.cat
* Correspondence: laila.darwich@uab.cat

Simple Summary: Rabbits are the second most common specialty pet among households in Europe and the USA. However, research on antimicrobial resistance (AMR) in pet rabbits is very scarce. Therefore, scientific data on AMR in pet rabbits is urgently needed as a guide for veterinarian clinicians to optimize antibiotic use in rabbits for reducing the selection of antibiotic resistance. In addition, antimicrobial stewardship programs should be conducted to educate rabbit owners not to misuse antibiotics on their pets as it may put their own health at risk. This paper aims to provide an overview of the current state of AMR in rabbits attended to in veterinary clinics distributed in Spain to highlight the importance of addressing AMR under the One Health approach.

Abstract: Research on antimicrobial resistance (AMR) in pet rabbits is very scarce. The aim of this study was to provide an overview of the current state of AMR in rabbits attended to in veterinary clinics distributed in Spain. Records of 3596 microbiological results of clinical cases submitted from 2010 to 2021 were analyzed. *Staphylococcus* spp. (15.8%), *Pseudomonas* spp. (12.7%), *Pasteurella* spp. (10%), *Bordetella* spp. (9.6%) and *Streptococcus* spp. (6.8%) were the most frequently diagnosed agents. Enterobacteriaceae, principally *Escherichia coli*, *Klebsiella pneumoniae* and *Enterobacter cloacae*, accounted for about 18% of the cases and showed the highest proportion of multi-drug resistance (MDR) isolates, with 48%, 57.5% and 36% of MDR, respectively. Regarding the antimicrobial susceptibility testing for a number of antimicrobial categories/families, the largest proportion of isolates showing resistance to a median of five antimicrobial categories was observed in *P. aeruginosa*, *Stenotrophomonas maltophilia* and *Burkolderia* spp. In contrast, infections caused by *Staphylococcus*, *Streptococcus* spp. and *Pasteurella multocida* were highly sensitive to conventional antimicrobials authorized for veterinary use (categories D and C). The emergence of AMR major nosocomial opportunistic pathogens such as *P. aeruginosa*, *S. maltophilia* and *K. pneumoniae* in pet rabbits can represent a serious public health challenge. Consequently, collaboration between veterinarians and human health professionals is crucial in the fight against antimicrobial resistance, to optimize, rationalize and prudently use antimicrobial therapies in domestic animals and humans.

Keywords: pet rabbits; antimicrobial resistance; One Health approach; zoonotic risk; Spain

Citation: Fernández, M.; Garcias, B.; Duran, I.; Molina-López, R.A.; Darwich, L. Current Situation of Bacterial Infections and Antimicrobial Resistance Profiles in Pet Rabbits in Spain. *Vet. Sci.* **2023**, *10*, 352. https://doi.org/10.3390/vetsci10050352

Academic Editor: George Valiakos

Received: 6 April 2023
Revised: 9 May 2023
Accepted: 12 May 2023
Published: 14 May 2023

1. Introduction

Antimicrobial resistance (AMR) is a growing global concern, with the emergence of multidrug-resistant bacteria representing a significant threat to human and animal health. The close interaction between pets and their owners can facilitate the transmission of pathogenic bacteria between humans and animals, especially multidrug-resistant (MDR) microorganisms, representing a serious threat for human and animal health. Moreover,

MDR infections complicate medical management, lengthen hospital stays and have a big economic impact [1].

Rabbits are the second most common specialty/exotic pet mammals among households, according to the American Veterinarian Association, and they are considered ideal pets for children in the USA and Europe [2]. Currently, rabbits are expanding in other regions, being extremely popular pets in Australia and in Asian countries such as Japan and Singapore [3]. Pet rabbits may also host parasites (*Encephalitozoon cuniculi*, *Cryptosporidium* spp., *Giardia* spp. and *Tricostrongylus* spp.), viruses (hepatitis E), bacteria (*Bartonella* spp., *Pasteurella* spp.) and fungi (dermatophytosis), which can be potential zoonotic pathogens for humans [4]. Elder people and children younger than 5 years, as well as immunocompromised persons and pregnant women, are particularly most susceptible to such pet-induced zoonoses [5]. However, related to AMR bacteria, most of the data published in pets are focused on dogs and cats [6–11] and very few are related to other pet species such as rabbits [4,12].

Thus, understanding the prevalence of AMR among pet rabbits is highly necessary from both veterinary and human medicine perspectives. Since the number of antibiotics available in veterinary medicine is limited, and there are many antibiotics contraindicated for oral administration in rabbits because of their toxicity (clindamycin, lincomycin, erythromycin, ampicillin, amoxicillin/clavulanic acid and cephalosporins), it is very important to select the best therapeutic option [13–15]. Thus, the use of antibiotics should be based on the results of susceptibility testing and the specific needs of each rabbit case. Empiric treatment should be administered only for urgent cases where the survival of the animal is compromised and should be based on scientific evidence. Therefore, scientific data on AMR in pet rabbits is urgently needed as a guide for veterinarian clinicians to optimize antibiotic use in rabbits for reducing the selection of antibiotic resistance. In addition, antimicrobial stewardship programs will also be conducted to educate rabbit owners not to misuse antibiotics on their pets as it may put their own health at risk.

This paper aims to provide an overview of the current state of AMR in rabbits attended to in veterinary clinics distributed in Spain and discuss the potential causes and consequences of this problem under the One Health approach.

2. Materials and Methods

2.1. Database Collection and Management

Retrospective data on microbiological results of clinical cases of pet rabbits submitted between 2010 and 2021 from Spain and Portugal were analyzed. The database was comprised of 3596 records. These records were provided by a private diagnostic laboratory in Barcelona (Spain), which has had the ISO-9001 quality management system certificate since 1998, and the ENAC (National Accreditation Entity) accreditation according to criteria included in the ISO standard/IEC 17025 defined in technical annexes 511/LE1947 for pharmaceutical toxicology and microbiology tests.

The first step was to filter and categorize the study variables to homogenize all the data for performing subsequent descriptive and quantitative statistical analyzes. The following variables were included in the study: geographical location of the sample; origin of the sample classified in categories as regards the pathological relevance in rabbits (abscesses, dental disease, dermatitis/skin disease, otitis, conjunctivitis, reproductive tract, respiratory tract, urinary tract infections); microbiological result (positive identification or negative/absence of bacterial growth); bacterial species (grouped by genus and species) and the antimicrobial sensitivity results (from the 84 antibiotics included in the study, the antibiotics most conventionally used in veterinary medicine and as a last resort for human medicine were selected).

2.2. Microbiological Diagnosis Techniques and Antimicrobial Susceptibility Testing

Bacterial identification was performed by means of the MALDI-TOF mass spectrometer, as previously described [9,10,12]. Gram-positive bacterial isolates were found by the antimicrobial susceptibility test (AST) using the standard disk diffusion method according to Performance Standards for Antimicrobial Susceptibility Testing for bacteria isolated from animals [16] and humans [17], for monitoring resistant microorganisms as a potential risk to public health. The panel included 21 antimicrobials corresponding to 9 classes or categories, and their respective disc concentrations: β-lactams (penicillin (PEN/10U), ampicillin (AMP/10 μg), cephalexin (LEX/30 μg), cefuroxime (CXM/30 μg), cefotaxime (CTX/30 μg), cefepime (FEP/30 μg), imipenem (IMI/10 μg), amoxicillin + clavulanic acid (AMC/30 μg) and aztreonam (AZT/30 μg)), fluoroquinolones (ciprofloxacin (CIP/5 μg), enrofloxacin (ENR/5 μg), marbofloxacin (MBF/5 μg)), aminoglycosides (amikacin (AMK/30 μg) and gentamicin (GEN/10 μg)), tetracyclines (doxycycline (DOX/30 μg)), polymyxins (polymyxin B (PMB/300 μg) and colistin (COL/10 μg)), trimethoprim/sulfonamides (trimethoprim + sulfametoxazol (TxS/25 μg)), glycopeptides (vancomycin (VAN/30 μg)), phosphonates (Fosfomycin (FOS/50 μg)) and phenicol's (chloramphenicol (CHL/10 μg)). In parallel, NM44 MicroScan (Beckman Coulter, Villepinte, France) system testing was performed to detect minimal inhibitory concentrations (MIC) [9,10]. Additionally, quality control for the AST was performed using internal controls in each automatic panel of the NM44 MicroScan (Beckman Coulter, Villepinte, France). In the case of manual antibiograms, McFarland standards were used as a reference, previously confirmed by a Densicheck (bioMérieux, Madrid, Spain).

Based on the lab testing readings, isolates were classified as susceptible, intermediate or resistant. For showing the AST histograms of antimicrobial categories, all isolates that exhibited intermediate resistance were grouped with the susceptible ones. Multidrug resistance (MDR) was defined as resistance to at least 1 agent in $\geq$3 antimicrobial categories and determined using R version 4.2.0 (R Core Team, 2022) [18], with the AMR package [19], as defined by Magiorakos et al. (2012), where intrinsic resistances were not considered in the analysis [20]. In the definitions proposed for MDR in this study, a bacterial isolate is considered resistant to an antimicrobial category when it is 'non-susceptible to at least one agent in a category' [20].

3. Results

The analysis of this study was conducted with 3596 records of clinical cases from different provinces of Spain. A microbiological identification was obtained in 2998 (83.4%) of the samples, and 598 samples were negative (no microbiological culture). According to the bacteriological identification, the most prevalent genera were *Staphylococcus* spp. (15.8%), *Pseudomonas* spp. (12.7%), *Pasteurella* spp. (10%), *Bordetella* spp. (9.6%) and *Streptococcus* spp. (6.8%). Enterobacteriaceae represented around 18% of the isolates, with *Enterobacter* spp., *Escherichia* spp. and *Klebsiella* spp. being the most frequent ones (Table 1).

Table 1. Frequencies of bacterial species identified in pet rabbits.

Bacteria Isolates	Number (% in spp.)	Overall % (N = 2998)
***Staphylococcus* spp.**	**n = 475**	**15.8**
S. aureus	171 (36)	5.70
S. xylosus	33 (7)	1.10
S. epidermidis	26 (5.5)	0.86
S. lugdunensis	14 (3)	0.46
S. pseudointermedius	13 (2.8)	0.43
S. sciuri	9 (1.9)	0.30
S. capitis	9 (1.9)	0.30
S. chromogenes	7 (1.5)	0.23
S. intermedius	6 (1.3)	0.20
S. simulans	6 (1.3)	0.20

Table 1. *Cont.*

Bacteria Isolates	Number (% in spp.)	Overall % (N = 2998)
S. chleiferi	6 (1.3)	0.20
S. cohnii	5 (1.1)	0.16
S. saprofhyticus	5 (1.1)	0.16
S. succinus	5 (1.1)	0.16
Others	160 (33.7)	5.33
***Pseudomonas* spp.**	**n = 382**	**12.7**
P. aeruginosa	264 (69)	8.80
P. putida	19 (5)	0.63
P. fluorescens	14 (3.7)	0.46
P. korensis	7 (1.8)	0.23
P. fulva	5 (1.3)	0.16
P. libaniensis	5 (1.3)	0.16
P. monteilii	5 (1.3)	0.16
Others	63 (16.5)	2.10
***Pasteurella* spp.**	**n = 302**	**10.1**
P. multocida	230 (76.2)	7.7
P. canis	10 (3.3)	0.3
Others	62 (20.5)	2.1
***Bordetella* spp.**	**n = 289**	**9.6**
B. bronchiseptica	278 (96.2)	9.3
Others	11 (3.8)	0.4
***Streptococcus* spp.**	**n = 204**	**6.8**
S. intermedius	32 (15.7)	1.1
S. anginosus	6 (3)	0.2
S. oralis	6 (3)	0.2
Others	160 (78.4)	5.3
***Enterobacter* spp.**	**n = 161**	**5.4**
E. cloacae	123 (76.39)	4.10
E. kobei	13 (8.07)	0.43
E. bugandensis	6 (3.72)	0.20
Others	19 (11.80)	0.63
***Escherichia* spp.**	**n = 153**	**5.1**
E. coli	141(92.15)	4.70
E. vulneris	7 (4.15)	0.23
Others	5 (3.26)	0.16
***Klebsiella* spp.**	**n = 131**	**4.4**
K. pneumoniae	75 (57.3)	2.5
K. oxytoca	43 (32.8)	1.4
Others	13 (9.9)	0.4
***Acinetobacter* spp.**	**n = 115**	**3.8**
A. iwoffii	28 (24.4)	0.9
A. baumannii	16 (13.9)	0.5
A. johnsonii	10 (8.7)	0.3
A. pitti	10 (8.7)	0.3
Others	51(44.3)	1.7
***Pantoea* spp.**	**n = 90**	**3.0**
P. agglomerans	69 (76.7)	2.3
Others	21 (23.3)	0.7

Table 1. *Cont.*

Bacteria Isolates	Number (% in spp.)	Overall % (N = 2998)
***Enterococcus* spp.**	**n = 77**	**2.6**
E. faecalis	40 (51.9)	1.3
Others	37 (48.05)	1.23
***Moraxella* spp.**	**n = 77**	**2.6**
M. branhamella	32 (41.6)	1.1
M. catarrhalis	25 (32.5)	0.8
Others	20 (26)	0.7
***Serratia* spp.**	**n = 54**	**1.8**
S. marcescens	39 (72.2)	1.3
S. liquefaciens	5 (9.3)	0.2
S. odorifera	5 (9.3)	0.2
Others	5 (9.3)	0.2
***Neisseria* spp.**	**n = 37**	**1.6**
N. animaloris/zoodegmatis	1 (2.7)	0.03
N. gonorrheae	1 (2.7)	0.03
N. species	1 (2.7)	0.03
Others	34 (92)	1.1
***Proteus* spp.**	**n = 34**	**1.1**
P. mirabilis	28 (82.4)	0.9
P. vulgaris	4 (11.7)	0.1
P. penneri	2 (5.9)	0.1
***Trueperella* spp.**	**n = 32**	**1.1**
T. pyogenes	32 (100)	1.1
***Strenotrophomonas* spp.**	**n = 27**	**0.9**
S. maltophilia	26 (96.3)	0.9
Others	1 (3.7)	0.0
***Burkholderia* spp.**	**n = 25**	**0.8**
B. cepacia	19 (76)	0.6
Others	6 (24)	0.2

The distribution of bacteria according to the origin of the samples showed that the most frequent origins were those coming from the respiratory tract (53%), followed by otitis (18%), abscesses, principally located in the head (16%), conjunctivitis (5%), reproductive tract (3%), skin disease/dermatitis (2%), urinary tract (2%) infections and dental disease (1%).

The most frequent pathogens involved in cases of abscesses (located mainly on the head), dental disease, dermatitis/skin disease, conjunctivitis and otitis were Gram-positive cocci (principally *Staphylococcus* spp., followed by *Streptococcus* spp.) and *Pseudomonas aeruginosa* (Figure 1). *Streptococcus* spp. was the primary agent responsible for reproductive tract infections, while *Enterococcus* spp. was the most frequently responsible for urinary infections. Gram-negative infections caused by *P. multocida* and *B. bronchiseptica* (33%), followed by *P. aeruginosa* (15%), were the most frequent causes of respiratory infections. Additionally, *Pasteurella* spp. was found in cases of abscesses, dermatitis/skin disease, conjunctivitis and otitis.

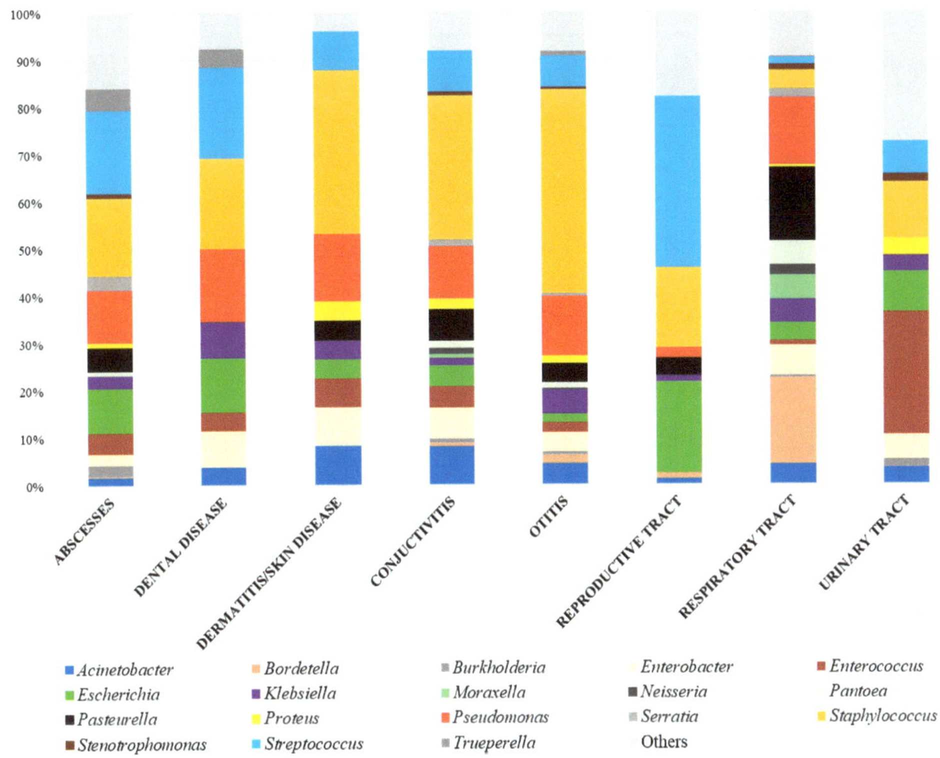

Figure 1. Distribution of bacteria genera regarding the sample origin in pet rabbits.

The Enterobacteriaceae family, principally represented by *E. coli*, *K. pneumoniae* and *E. cloacae*, was homogeneously distributed in all the pathological categories. *Acinetobacter* spp. was also isolated from diverse origins. Other less prevalent pathogens were *Stenotrophomonas maltophilia*, isolated from respiratory and urinary infections, conjunctivitis and otitis, *Burkholderia* spp., isolated from abscesses and the urinary tract, and *Trueperella pyogenes*, found in abscesses and dental disease (Figure 1).

As regards the AST results, *P. aeruginosa* was the most prevalent pathogen with the highest levels of AMR, presenting 80% of strains resistant to penicillins, inhibitors of β-lactamases (AMC), 1st and 2nd generation (1G/2G) cephalosporins, trimethoprim/sulfonamides and phenicols, and 60% of strains were resistant to 3rd and 4th generation (3G/4G) cephalosporins (Figure 2). As regards to the MDR profile, 8% (31/381) of *P. aeruginosa* strains were MDR, but the average number of antimicrobial categories or families that presented resistance was 5 (Table 2, Figure 3).

Other less representative bacteria but with the largest proportion of isolates showing resistance to a median of five antimicrobial categories were *Stenotrophomonas* spp., specifically *S. maltophilia* and *Burkolderia* spp. (Table 2). Both bacterial species were not considered MDR strains because of the intrinsic resistance to several families (Table 2). However, from the clinical point of view, it is interesting to remark that they presented high frequencies of resistance to β-lactams, with special attention to carbapenems (>80% *Stenotrophomonas* and 50% *Burkholderia*), also to polymyxins (>75% *Burkholderia* and 60% *Stenotrophomonas*) and fluoroquinolones (55% *Burkholderia* and 48% *Stenotrophomonas*) (Figure 2).

Table 2. Average number of AMR categories and frequencies of MDR bacterial species.

Genus	Isolates	Number of AMR Categories/Families		MDR * Profile
Gram-Negative	n	Mean	Median	%
Pseudomonas	381	5.0	5	8.1
Stenotrophomonas	26	5.2	5	0
Burkholderia	25	4.6	5	0
Acinetobacter	115	3.0	3	11.3
Bordetella	289	2.8	3	0
Pasteurella	299	0.7	0	0
Moraxella	77	1.2	1	0
Escherichia	129	2.7	3	47.9
Klebsiella	134	4.4	4	57.5
Enterobacter	157	3.6	3	35.7
Proteus	34	3.1	3	47.1
Serratia	54	3.3	3	33.3
Gram-Positive	n	Mean	Median	%
Staphylococcus	466	2.0	1	5
Streptococcus	204	1.6	1	0
Enterococcus	77	3.7	4	6.5

* According to [20].

Figure 2. *Cont.*

B

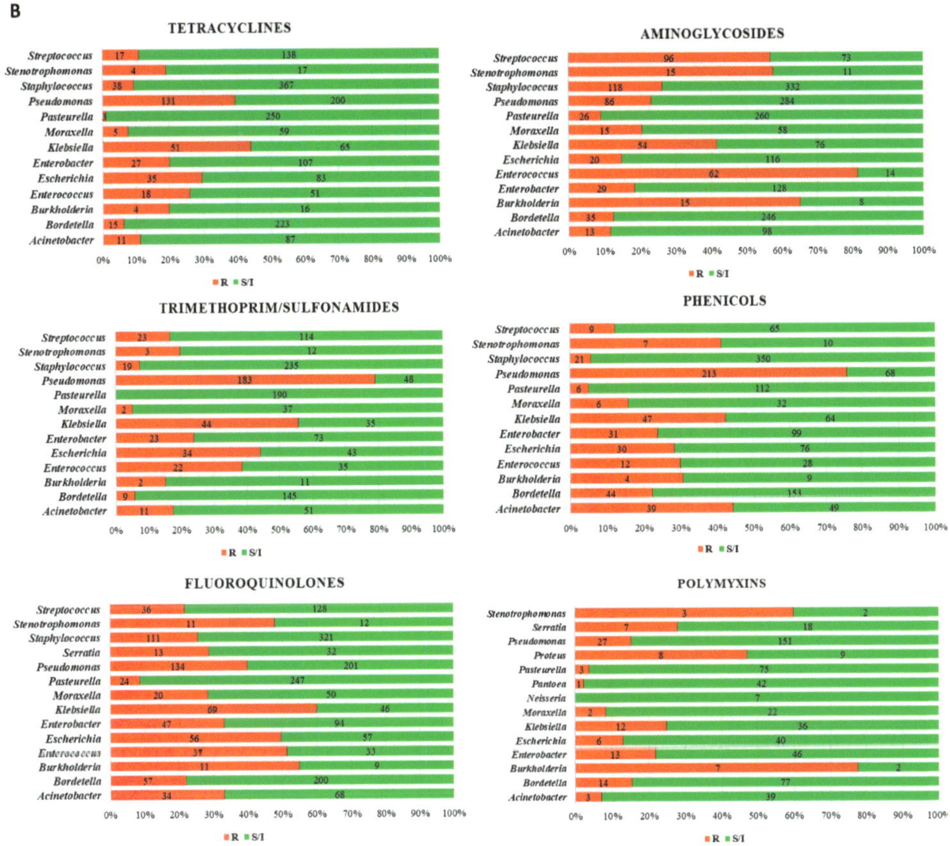

Figure 2. Percentage and number of isolates presenting resistant (red) or susceptible/intermediate (green) results in the AST for (**A**) the beta-lactams class and (**B**) other antimicrobial families.

The Enterobacteriaceae family, represented principally by *E. coli*, *K. pneumoniae* and *E. cloacae*, showed a high prevalence of AMR to β-lactams: penicillins (>80% *K. pneumoniae* and *E. cloacae*), AMC (>80% *E. cloacae*), 1G/2G cephalosporins (>50% *K. pneumoniae* and >70% *E. cloacae*) and 3G/4G cephalosporins (50% *K. pneumoniae*). Moreover, *K. pneumoniae* isolates showed resistance to trimethoprim/sulfonamides (50%) and to fluoroquinolones (60%) (Figure 2). Moreover, the percentage of MDR isolates was notable in enterobacteria isolates such as *K. pneumoniae* (58%), *E. coli* (48%), *Proteus* spp. (47%) and *E. cloacae* (36%) (Table 2). In addition, the average number of antimicrobial categories presenting resistance was three in almost all enterobacteria, except for *K. pneumoniae,* in which it was four (Figure 3).

Another bacterial spp. with a considerable resistance profile was *Acinetobacter* spp., with 11% of MDR (Table 2) and nearly 60% of the isolates presenting resistance to penicillins, AMC and 1G/2G cephalosporins (Figure 2). *Bordetella*, mainly *B. bronchiseptica*, was another pathogen with AMR resistance to 3 antimicrobial categories, finding 80% of resistance to penicillins and 1G/2G cephalosporins and 50% to 3G/4G cephalosporins (Figure 2). Additionally, *Enterococcus* spp., frequently isolated from UTI in rabbits, showed a high prevalence of AMR to aminoglycosides (>80%), 1G/2G cephalosporines (>70%), fluoroquinolones (>50%) and 3G/4G cephalosporines (>40%), with 6.5% of MDR strains (Figure 2,

Table 2). As regards the susceptibility to vancomycin, *Streptococcus* spp. (22%) presented the highest frequency of resistance, followed by *Enterococcus* (12%) and *Staphylococcus* (6%).

Finally, Gram-positive cocci (*Staphylococcus* and *Streptococcus*) and other Gram-negative bacteria, such as *Pasteurella multocida* and *Trueperella pyogenes*, were sensitive to a wide panel of conventional antimicrobial agents, including those classified in categories D and C (Figure 2).

Values of minimal inhibitory concentrations (MIC) can be found in Table S1. In general, *P. aeruginosa* presented the highest levels of MIC_{90} for a major portion of the antimicrobials tested.

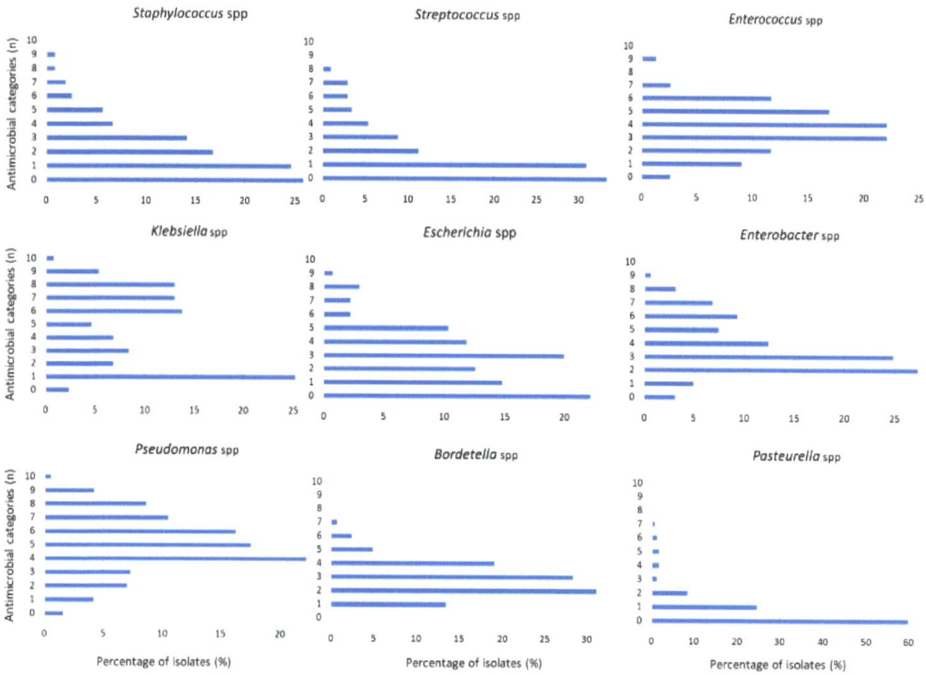

Figure 3. Distribution of the different bacterial isolates (%) according to the number of antimicrobial categories showing resistance.

4. Discussion

This study aimed to highlight the importance of addressing AMR in pet rabbits as a crucial step in the fight against antimicrobial resistance more broadly, enhancing the correct use of antibiotics to preserve their efficacy in the future to effectively control bacterial infections in people and pets.

The positive finding of these results is that the most common infections caused by Gram-positive cocci, basically *Staphylococcus* and *Streptococcus* spp. involved in abscesses, dental disease, dermatitis/skin disease, conjunctivitis and otitis, presented a low frequency of AMR, being sensitive to antimicrobials of categories D and C according to the EMA [21]. Additionally, *Pasteurella* (*P. multocida*), one of the most common bacteria of rabbits which colonizes the upper respiratory tract and the oro-pharynx, was found to be highly sensitive to conventional D and C class drugs. *Pasteurella multocida* can reside in the nasal flora of asymptomatic rabbits and spread to other sites during grooming, and it is also frequently isolated from abscesses because this bacterium has capsular polysaccharides that resist phagocytosis [22]. In pet rabbits, most abscesses occur around the head and face and are associated with dental disease. Another bacterial agent isolated from abscesses and dental disease was *Trueperella pyogenes*. This bacterium has been associated with sporadic cases of suppurative disorders in the lungs, liver, spleen and brain of rabbits [23]. Fortunately, and similar to *P. multocida*, *T. pyogenes* presented a highly sensitive pattern of AMR in our pet rabbits.

The zoonotic risk of *P. multocida* transmission to humans must be considered through bites, scratches or licks of companion animals, with the development of local inflammatory reactions and occasionally the occurrence of abscesses in people [5,24–26]. Moreover, in some patients, principally in immunocompromised people or persons with pulmonary disorders, pasteurellosis may result in more severe pathologies, such as pneumonia, endocarditis, meningitis and sepsis [27,28]. In a recent paper, *P. multocida* belonging to capsular type A was the type most often detected in humans, and although it was susceptible to the tested antibiotics, in agreement with our AST results, it was equipped with several virulence genes [4]. These findings are of particular interest because rabbits recovered from pasteurellosis very often become asymptomatic carriers of this infection and can represent a risk for the household members, especially for children and elder people [29].

On the other hand, Gram-negative infections caused by *P. multocida* and *B. bronchiseptica*, followed by *P. aeruginosa*, were principally involved in respiratory infections, in agreement with a pervious study conducted in pet rabbits in France [30]. In that study, the authors concluded that marbofloxacin was shown to be a potentially good treatment option for upper respiratory tract disease in pet rabbits. Although the use of fluoroquinolones is the most common therapeutic option in exotic animal medicine, the EMA recommendations appeal for the use of D and C categories in order to preserve the efficacy of critical antimicrobial classes such as fluoroquinolones (category B). For this reason, and considering the AST results of our study, for respiratory infections caused by *P. multocida* or *B. bronchiseptica*, trimethoprim/sulfonamides could be a good candidate for treatment in pet rabbits.

Non-fermenting Gram-negative bacilli, such as *P. aeruginosa* and *Acinetobacter baumannii*, are among the major opportunistic pathogens involved in the global antibiotic resistance epidemic in human medicine [31]. Data on pet rabbits showed that the antimicrobial treatment of *P. aeruginosa* can be more complicated, since a high percentage of the isolates presented a resistance profile, including antimicrobials of category B (3G/4G cephalosporins and fluoroquinolones). This pathogen is also found in a wide spectrum of rabbit pathologies (abscesses, dental disease, dermatitis/skin disease, conjunctivitis, otitis and respiratory infections), and the treatment options are very few, limited to carbapenems and polymyxins, which are antimicrobials of category A (reserved for critical use in human medicine), but also to aminoglycosides. Since this former family is classified in category C, aminoglycosides could be the best option for treating pseudomonal infections in rabbits. It is important to note that polymyxins can be highly toxic to rabbits and should be avoided for treatments. However, this antimicrobial class was added in this study for its relevance as a last-resort drug for human medicine.

As regards the Enterobacteriaceae family, *E. coli*, *K. pneumoniae* and *E. cloacae* represented the most frequent species isolated from a large diversity of pathologies. *Escherichia coli* infections can cause enteritis, sepsis and urinary tract infections in rabbits. Although *E. coli* was the most prevalent enterobacteria, the frequency of MDR was lower compared to *K. pneumoniae*, as observed in other pet studies in Spain [9,12]. According to our results, good candidates for treating infections caused by *E. coli* could be aminoglycosides. On the other hand, *K. pneumoniae* showed high resistance to most of the antimicrobial classes of conventional use in veterinary medicine, leaving carbapenems as the best therapeutic option even though it is a category A drug. Considering other antimicrobials authorized for veterinary medicine, the best options were aminoglycosides, chloramphenicol or doxycycline, although more than 40% of the isolates presented resistance to these drugs. As a result, the increasing occurrence of *K. pneumoniae* as a MDR infection and a zoonotic agent represents a real threat to both animal and human health [32,33]. In addition, *E. cloacae* is another emerging pathogen recognized as a nosocomial bacterium contributing to septic arthritis, skin/soft tissue infections, bacteremia, lower respiratory tract and urinary tract infection, endocarditis, osteomyelitis and intra-abdominal infections in humans [34].

Other less representative bacteria, but with a proportion of resistance to several antimicrobial categories (five as a median), were *S. maltophilia* and *Burkolderia* spp. Both bacterial species presented high frequencies of resistant isolates to β-lactams (including carbapenems), as well as to category A (polymyxins) and B (fluoroquinolones) drugs. *S. maltophilia* is an emerging nosocomial pathogen, with intrinsic resistance to beta-lactams, capable of causing healthcare-associated infections in intensive care units, life-threatening diseases in immunocompromised patients and severe pulmonary infections in cystic fibrosis and COVID-19-infected individuals [35,36].

Lastly, it was interesting to note that 12% of the Enterococcus isolates were resistant to vancomycin, more than 80% to aminoglycosides, around 70% to 1G/2G cephalosporines, half of them to fluoroquinolones and 40% to 3G/4G cephalosporines. With these AMR profiles, the treatment of UTI caused by this bacterium in rabbits can be difficult to plan without a previous susceptibility testing.

Overall, the emergence of AMR strains such as *P. aeruginosa*, *A. baumannii*, *S. maltophilia* and *K. pneumoniae* in pet rabbits can represent a serious health threat for the owners, since they are among the major opportunistic pathogens with significant contributions to mortality in hospitals worldwide [31,37]. Moreover, these pathogens are designated as urgent/serious threats by the Centers for Disease Control and Prevention and are part of the World Health Organization's list of critical priority pathogens [38].

It is important to remember that the list of antimicrobial therapeutic options for treating bacterial infections in rabbits is not exhaustive and the use of antibiotics should be based on the results of susceptibility testing, the specific needs of each animal case and the risk of toxicity of these drugs in rabbits. However, for urgent cases, when the severity of the clinical process requires immediate antimicrobial therapy with no time for AST analysis, the data reported in the present study can be useful for veterinary practitioners to apply empirical therapy. It is crucial to keep in mind that the best way to proceed for reducing AMR selection is to perform a proper antimicrobial diagnosis with the corresponding AST. Then, antimicrobials with a sensitive result must be prioritized according to the EMA categories, mainly D and C.

Finally, the results of this study provided objective data on the microbiological results in pet rabbits in Spain. The high levels of AMR to critically important antibiotics in human medicine found in pet rabbits are of great concern since potential transmission of resistance genes from rabbits to humans or other pets can occur. Considering that the predominant bacteria in this study are among the top pathogens directly attributed to human deaths due to AMR, it is critical that veterinarians and physicians work together to optimize, rationalize and prudently use antimicrobial therapies in domestic animals and humans under the One Health approach.

Supplementary Materials: The following supporting information can be downloaded at: https://www.mdpi.com/article/10.3390/vetsci10050352/s1, Table S1: Minimal Inhibitory Concentrations (MIC) in different bacterial isolates and antimicrobials.

Author Contributions: Conceptualization, L.D.; methodology, M.F., B.G. and I.D.; software, M.F. and B.G.; validation, B.G., R.A.M.-L. and L.D.; formal analysis, M.F.; investigation, L.D.; resources, I.D.; data curation, M.F.; writing—original draft preparation, M.F. and L.D.; writing—review and editing, B.G., I.D., R.A.M.-L. and L.D.; visualization, M.F.; supervision, L.D.; project administration, L.D.; funding acquisition, L.D. All authors have read and agreed to the published version of the manuscript.

Funding: M.F. was granted by the Paraguay Program Don Carlos Antonio López and B.G. was supported by Departament de Recerca i Universitats de la Generalitat de Catalunya (FI-SDUR 2020).

Institutional Review Board Statement: Not applicable.

Informed Consent Statement: Written informed consent was obtained from all participants involved in the study.

Data Availability Statement: Data will be made available upon reasonable request to the corresponding author.

Conflicts of Interest: The authors declare no conflict of interest.

References

1. Giraldi, G.; Montesano, M.; Napoli, C.; Frati, P.; La Russa, R.; Santurro, A.; Scopetti, M.; Orsi, G.B. Healthcare-Associated Infections Due to Multidrug-Resistant Organisms: A Surveillance Study on Extra Hospital Stay and Direct Costs. *Curr. Pharm. Biotechnol.* **2019**, *20*, 643–652. [CrossRef]
2. Available online: https://www.avma.org/resources-tools/reports-statistics/us-pet-ownership-statistics (accessed on 26 April 2023).
3. DeMello, M. Rabbits Multiplying Like Rabbits: The Rise in the Worldwide Popularity of Rabbits as Pets. In *Companion Animals in Everyday Life*; Pręgowski, M., Ed.; Palgrave Macmillan: New York, NY, USA, 2016.
4. D'Amico, F.; Casalino, G.; Bozzo, G.; Camarda, A.; Lombardi, R.; Dimuccio, M.M.; Circella, E. Spreading of Pasteurella multocida Infection in a Pet Rabbit Breeding and Possible Implications on Healed Bunnies. *Vet. Sci.* **2022**, *9*, 301. [CrossRef] [PubMed]
5. Centers of Disease Control and Prevention. Available online: https://www.cdc.gov/onehealth/basics/zoonotic-diseases.html (accessed on 31 March 2023).
6. Guardabassi, L.; Schwarz, S.; Lloyd, D.H. Pet animals as reservoirs of antimicrobial-resistant bacteria. *J. Antimicrob. Chemother.* **2004**, *66*, 700–709. [CrossRef] [PubMed]
7. Lloyd, D.H. Reservoirs of antimicrobial resistance in pet animals. *Clin. Infect. Dis.* **2007**, *45*, S148–S152. [CrossRef] [PubMed]
8. Wedley, A.L.; Dawson, S.; Maddox, T.W.; Coyne, K.P.; Pinchbeck, G.L.; Clegg, P.; Nuttall, T.; Kirchner, M.; Williams, N.J. Carriage of antimicrobial resistant Escherichia coli in dogs: Prevalence, associated risk factors and molecular characteristics. *Vet. Microbiol.* **2017**, *199*, 23–30. [CrossRef] [PubMed]
9. Li, Y.; Fernández, R.; Durán, I.; Molina-López, R.A.; Darwich, L. Antimicrobial Resistance in Bacteria Isolated from Cats and Dogs from the Iberian Peninsula. *Front. Microbiol.* **2021**, *11*, 621597. [CrossRef]
10. Darwich, L.; Seminati, C.; Burballa, A.; Nieto, A.; Durán, I.; Tarradas, N.; Molina-López, R.A. Antimicrobial susceptibility of bacterial isolates from urinary tract infections in companion animals in Spain. *Vet. Rec.* **2021**, *188*, e60. [CrossRef]
11. Schirò, G.; Gambino, D.; Mira, F.; Vitale, M.; Guercio, A.; Purpari, G.; Antoci, F.; Licitra, F.; Chiaramonte, G.; La Giglia, M.; et al. Antimicrobial Resistance (AMR) of Bacteria Isolated from Dogs with Canine Parvovirus (CPV) Infection: The Need for a Rational Use of Antibiotics in Companion Animal Health. *Antibiotics* **2022**, *11*, 142. [CrossRef]
12. Muñoz-Ibarra, E.; Molina-López, R.A.; Durán, I.; Garcias, B.; Martín, M.; Darwich, L. Antimicrobial Resistance in Bacteria Isolated from Exotic Pets: The Situation in the Iberian Peninsula. *Animals* **2022**, *12*, 1912. [CrossRef]
13. Carpenter, J.W. (Ed.) *Exotic Animal Formulary*, 3rd ed.; Elsevier Publishing: St. Louis, MO, USA, 2005; 564p.
14. Ivey, E.S.; Morrisey, J.K. Therapeutics for rabbits. *Vet. Clin. N. Am.* **2000**, *3*, 183 220. [CrossRef]
15. Quesenberry, K.E.; Carpenter, J.W. (Eds.) *Ferrets Rabbits, and Rodents: Clinical Medicine and Surgery*; WB Saunders Co.: Philadelphia, PA, USA, 2004; 461p.
16. *M31-A3*; Performance Standards for Antimicrobial Disk and Dilution Susceptibility Tests for Bacteria Isolated from Animals. 3rd ed. CLSI: Wayne, PA, USA, 2008.
17. *M100-S26*; Performance Standards for Antimicrobial Susceptibility Testing: Twenty-Sixth Informational Supplement. CLSI: Wayne, PA, USA, 2016.
18. R Core Team. *R: A Language and Environment for Statistical Computing*; R Foundation for Statistical Computing: Vienna, Austria, 2022. Available online: https://www.R-project.org/ (accessed on 26 April 2023).
19. Berends, M.S.; Luz, C.F.; Friedrich, A.W.; Sinha, B.N.M.; Albers, C.J.; Glasner, C. AMR: An R Package for Working with Antimicrobial Resistance Data. *J. Stat. Softw.* **2022**, *104*, 1–31. [CrossRef]
20. Magiorakos, A.-P.; Srinivasan, A.; Carey, R.B.; Carmeli, Y.; Falagas, M.E.; Giske, C.G.; Harbarth, S.; Hindler, J.F.; Kahlmeter, G.; Olsson-Liljequist, B.; et al. Multidrug-Resistant, Extensively Drug-Resistant and Pandrug-Resistant Bacteria: An International Expert Proposal for Interim Standard Definitions for Acquired Resistance. *Clin. Microbiol. Infect.* **2012**, *18*, 268–281. [CrossRef] [PubMed]
21. EMA/CVMP/CHMP. *Categorisation of Antibiotics in the European Union*; European Medicines Agency: Amsterdam, The Netherlands, 2019.
22. Deeb, B. Update for veterinary practitioners on pasteurellosis in rabbits. *J. Small Exotic Anim. Med.* **1993**, *2*, 112–113.
23. Rzewuska, M.; Kwiecień, E.; Chrobak-Chmiel, D.; Kizerwetter-Świda, M.; Stefańska, I.; Gieryńska, M. Pathogenicity and Virulence of Trueperella pyogenes: A Review. *Int. J. Mol. Sci.* **2019**, *20*, 2737. [CrossRef] [PubMed]
24. Kimura, R.; Hayashi, Y.; Takeuchi, T.; Shimizu, M.; Iwata, M.; Tanahashi, J.; Makoto, I. Pasteurella multocida septicemia caused by close contact with a domestic cat: Case report and literature review. *J. Infect. Chemother.* **2004**, *10*, 250–252. [CrossRef]
25. Weber, D.J.; Wolfson, J.S.; Swartz, M.N.; Hooper, D.C. Pasteurella multocida infections. Report of 34 cases and review of the literature. *Medicine* **1984**, *63*, 133–154. [CrossRef] [PubMed]
26. Wilson, B.A.; Ho, M. Pasteurella multocida: From zoonosis to cellular microbiology. *Clin. Microbiol. Rev.* **2013**, *26*, 631–655. [CrossRef]

27. Abrahamian, F.M.; Goldstein, E.J. Microbiology of animal bite wound infections. *Clin. Microbiol. Rev.* **2011**, *24*, 231–246. [CrossRef]
28. Wilkie, I.W.; Harper, M.; Boyce, J.D.; Adler, B. *Pasteurella Multocida: Diseases and Pathogenesis*; Springer: Heidelberg/Berlin, Germany, 2012; pp. 1–22.
29. Körmöndi, S.; Terhes, G.; Pál, Z.; Varga, E.; Harmati, M.; Buzás, K.; Urbán, E. Human Pasteurellosis Health Risk for Elderly Persons Living with Companion Animals. *Emerg. Infect. Dis.* **2019**, *25*, 229–235. [CrossRef]
30. Rougier, S.; Galland, D.; Boucher, S.; Boussarie, D.; Vallé, M. Epidemiology and susceptibility of pathogenic bacteria responsible for upper respiratory tract infections in pet rabbits. *Vet. Microbiol.* **2006**, *115*, 192–198. [CrossRef]
31. Wisplinghoff, H. Pseudomonas *spp.*, Acinetobacter *spp. and Miscellaneous Gram-Negative Bacilli*; Elsevier: Amsterdam, The Netherlands, 2017; pp. 1579–1599.e2.
32. Wareth, G.; Neubauer, H. The Animal-Foods-Environment Interface of Klebsiella Pneumoniae in Germany: An Observational Study on Pathogenicity, Resistance Development and the Current Situation. *Vet. Res.* **2021**, *52*, 16. [CrossRef]
33. Schmiedel, J.; Falgenhauer, L.; Domann, E.; Bauerfeind, R.; Prenger-Berninghoff, E.; Imirzalioglu, C.; Chakraborty, T. Multiresistant Extended-Spectrum β-Lactamase-Producing Enterobacteriaceae from Humans, Companion Animals and Horses in Central Hesse, Germany. *BMC Microbiol.* **2014**, *14*, 187. [CrossRef]
34. Soteras, A.R.; Mackay, A.S.; Borràs-Maixenchs, N.; Martí, J.A. Catheter related necrotizing fascitiis in haematological patients: Case report and implications for nursing. *Ann. Oncol.* **2019**, *30*, v843. [CrossRef]
35. Al-Hamad, A.; Burnie, J.; Upton, M. Enhancement of antibiotic susceptibility of Stenotrophomonas maltophilia using a polyclonal antibody developed against an ABC multidrug efflux pump. *Can. J. Microbiol.* **2011**, *57*, 820–828. [CrossRef] [PubMed]
36. Peykov, S.; Strateva, T. Whole-Genome Sequencing-Based Resistome Analysis of Nosocomial Multidrug-Resistant Non-Fermenting Gram-Negative Pathogens from the Balkans. *Microorganisms* **2023**, *11*, 651. [CrossRef] [PubMed]
37. De Oliveira, D.M.P.; Forde, B.M.; Kidd, T.J.; Harris, P.N.A.; Schembri, M.A.; Beatson, S.A.; Paterson, D.L.; Walker, M.J. Antimicrobial resistance in ESKAPE pathogens. *Clin. Microbiol. Rev.* **2020**, *33*, e00181-19. [CrossRef] [PubMed]
38. WHO Publishes List of Bacteria for Which New Antibiotics Are Urgently Needed. Available online: https://www.who.int/en/news-room/detail/27-02-2017-who-publishes-list-of-bacteria-for-which-new-antibiotics-are-urgently-needed (accessed on 31 March 2023).

veterinary sciences

MDPI

Review

Immune Activated Cellular Therapy for Drug Resistant Infections: Rationale, Mechanisms, and Implications for Veterinary Medicine

Lynn M. Pezzanite [1], Lyndah Chow [1], Alyssa Strumpf [2], Valerie Johnson [3] and Steven W. Dow [1,4,*]

1 Department of Clinical Sciences, College of Veterinary Medicine and Biomedical Sciences, Colorado State University, Fort Collins, CO 80523, USA
2 Department of Clinical Sciences, College of Veterinary Medicine and Biomedical Sciences, The Ohio State University, Columbus, OH 43210, USA
3 Department of Clinical Sciences, College of Veterinary Medicine and Biomedical Sciences, Michigan State University, Lansing, MI 48824, USA
4 Department of Microbiology, Immunology, and Pathology, College of Veterinary Medicine and Biomedical Sciences, Colorado State University, Fort Collins, CO 80523, USA
* Correspondence: steven.dow@colostate.edu

Simple Summary: Mesenchymal stromal/stem cells have intrinsic antimicrobial properties, thus making them attractive as an alternative treatment strategy in chronic, drug-resistant bacterial infections. Recent evidence has suggested that these antimicrobial effects can be significantly enhanced by immune activation just prior to injection. This review examines the potential role for cellular therapies in treatment of drug resistant infections in veterinary medicine, drawing on insights across species and discussing the therapeutic potential of this approach overall in today's veterinary patients.

check for updates

Citation: Pezzanite, L.M.; Chow, L.; Strumpf, A.; Johnson, V.; Dow, S.W. Immune Activated Cellular Therapy for Drug Resistant Infections: Rationale, Mechanisms, and Implications for Veterinary Medicine. *Vet. Sci.* **2022**, *9*, 610. https://doi.org/10.3390/vetsci9110610

Academic Editor: George Valiakos

Received: 29 September 2022
Accepted: 31 October 2022
Published: 4 November 2022

Abstract: Antimicrobial resistance and biofilm formation both present challenges to treatment of bacterial infections with conventional antibiotic therapy and serve as the impetus for development of improved therapeutic approaches. Mesenchymal stromal cell (MSC) therapy exerts an antimicrobial effect as demonstrated in multiple acute bacterial infection models. This effect can be enhanced by pre-conditioning the MSC with Toll or Nod-like receptor stimulation, termed activated cellular therapy (ACT). The purpose of this review is to summarize the current literature on mechanisms of antimicrobial activity of MSC with emphasis on enhanced effects through receptor agonism, and data supporting use of ACT in treatment of bacterial infections in veterinary species including dogs, cats, and horses with implications for further treatment applications. This review will advance the field's understanding of the use of activated antimicrobial cellular therapy to treat infection, including mechanisms of action and potential therapeutic applications.

Keywords: mesenchymal; stromal; stem; cell; antimicrobial; antibiotic resistance

1. Introduction

Selection of antibiotic resistant bacteria in both human and veterinary medicine necessitates novel therapeutic approaches for successful management. Chronic infections, particularly those involving biofilms and multi-drug resistant organisms, evade most attempts at effective treatment. Recent reports by the National Institutes of Health (NIH), National Institute of Allergy and Infectious Diseases (NIAID), Centers for Disease Control and Prevention (CDC), World Health Organization (WHO), and Natural Resources Defense Council (NRDC) reflect the magnitude of the problem in healthcare [1–8]. In 2013, the CDC reported that an estimated two million people developed antibiotic-resistant infections annually, with greater than 23,000 cases resulting in death [1]. Similarly, antimicrobial resistance has been extensively recently documented in veterinary medicine, and considered one of the most important issues threatening animal health worldwide [9]. Conventional approaches to treatment of bacterial infections (i.e., the development of new antibiotics), are not able to keep pace with the increasing incidence of multi-drug resistant infections [3].

Antimicrobial cellular therapy (ACT) represents a new approach to address the growing issue of chronic, drug-resistant infection. This approach employs living cells, mesenchymal stromal or 'stem' cells (MSC), to augment the activity of conventional antibiotic therapy. Recent work has focused on optimizing cellular therapeutic strategies to focus on use of ACT as an adjunctive therapy for multi-drug resistant (MDR) bacterial infections, including both acute and chronic cases, as will be discussed in this review. This work builds off the use of MSC for treatment of bacterial infections, previously reported in the lung or peritoneal cavity [10–13] and particularly in biofilms [14–27] and previous work by other groups demonstrating that pre-activation of MSC with inflammatory licensing agents enhances the antibacterial and immunomodulatory abilities of MSC which may enhance their effect in treatment of infection [16,17,24–26,28–46]. Summary of the studies detailing the antimicrobial effects of mesenchymal stromal cell therapy in treatment of bacterial biofilms and that activation of MSC enhances their innate antibacterial and immunomodulatory effects are detailed in Tables 1 and 2, respectively.

Several key features distinguish the current version of ACT from other forms of cellular therapy for treating infections. First, the use of allogeneic MSC that have been activated with toll or nod-like receptors prior to administration. Pre-activation takes advantage of receptors that are commonly present in inflammation and infection to enhance the migratory properties of MSC and activate host innate immune defenses against infection [16,17,19,24–26,28–30,32–34,36–39,41,44,45]. A second defining characteristic of this approach in ACT is the use of repeated cell infusions for optimal effect. In addition, both intravenous and local routes of delivery were explored [43]. Systemic administration ensures that activated MSC will reach sites of deep-seated infection via chemokine-mediated migration and interact fully with the host immune response to stimulate effective antibacterial immune responses. However, intra-articular administration in an equine model of septic arthritis demonstrated a beneficial effect in localized disease processes such as those isolated to synovial structures suggesting that route of administration may be tailored to the specific disease process [25]. Finally, the concurrent administration of conventional antibiotics with ACT enhances the effect in an additive or synergistic manner, which we will discuss further.

Evidence for the effectiveness of the ACT approach has been generated in both mouse models [17,24,36,45], pet dogs with spontaneous chronic, drug-resistant bacterial infections involving soft tissues and bones [26], and an induced case–control study modelling septic arthritis in horses [25]. Thus, there is compelling preclinical evidence that ACT may be an effective means of stimulating clearance of recalcitrant, drug-resistant infections. In this article, we will review the evidence supporting use of TLR agonism to improve cellular therapy in treatment of bacterial infections in murine, canine, and equine disease models and further discuss mechanisms of action by which ACT exerts an effect. Finally, we will discuss the implications of these studies in the clinical application of cellular therapy to manage patients with intractable MDR infections.

Table 1. Summary of studies demonstrating efficacy of (MSC) and conditioned medium (MSC-CM) in treatment of bacterial biofilms.

Investigator	Reference	Species	Culture Conditions or Lesion	Cell Source	Cell Dose	Protocol	Route of Administration	Outcome Parameters	Main Findings
Yuan et al. (2014)	[14]	Rat	Subcutaneous infection MRSA	Bone marrow	2×10^7, 2×10^6, or 2×10^5 cells/rat	Dosed daily for 4 doses	Intravenous	Quantitative cultures Immunoassays cytokines	MSC reduced bacterial colonies. MSC reduced cytokine expression (IL1β, IL6, IL10, CCL5).
Criman et al. (2016)	[15]	Rat	Subcutaneous *E.coli* inoculated meshes	Bone marrow	7.5×10^5 MSC/mesh	MSC seeded meshes vs non-seeded meshes	Seeded in meshes	Microbiologic mesh evaluation Histologic mesh evaluation	Augmentation of bioprosthetic materials with MSC enhanced resistance to bacterial infection.
Johnson et al. (2017)	[16]	Murine	*Staphylococcus aureus* implant infection model	Adipose	1×10^6 cells/injection	TLR-3 poly I:C activated or not with or without antibiotics Dosed every 3 days, 3 doses	Intravenous	IVIS luminescence imaging to determine bacterial burden Wound tissue histology	Activated MSC co-administered with antibiotics was most effective to reduce bacterial bioburden.
		Canine	Naturally occurring wounds	Adipose	2×10^6 cells/kg	TLR-3 poly I:C activated + antibiotics Dosed every 2 weeks, 3 doses	Intravenous	Quantitative cultures Clinical signs Phone follow-up	Repeated MSC injection resulted in clearance of bacteria and wound healing.
Asami et al. (2018)	[17]	Murine	*Streptococcus pneumoniae* pulmonary infection	Bone marrow	1×10^6 cells/injection	Once1 hour after bacterial inoculation	Intravenous	Bacteria bronchoalveolar lavage Myeloperoxidase activity assay Bichinchoninic acid protein assay Histopathologic examination	MSC-CM modulates TNFα, IL-6, IL-10 after stimulation with TLR2, TLR4, TLR9 ligands. MSC-CM suppresses CXCL1, CXCL2 production after stimulation with TLR2 and TLR9 ligands. MSC IV decreased total cells, neutrophils, and myeloperoxidase activity during pulmonary infection. MSC IV decreased BALF cytokine levels TNFα, IL-6, IFN-γ, CCL2, GM-CSF during pulmonary infection.

Table 1. *Cont.*

Investigator	Reference	Species	Culture Conditions or Lesion	Cell Source	Cell Dose	Protocol	Route of Administration	Outcome Parameters	Main Findings
Wood et al. (2018)	[18]	Human	In vitro *Staphylococcus aureus*, Pseudomonas co-culture	Adipose	N/A		In vitro	Scanning electron microscopy Colony forming units Biofilm assay	MSC inhibited *P. aeruginosa* biofilm formation due to bacterial adhesion, engulfment/phagocytosis and secretion of antibacterial factors.
Chow et al. (2019)	[19]	Human	*Staphylococcus aureus* In vitro biofilm assay Mouse mesh implant model	Bone marrow	N/A 1×10^6 cells/injection	TLR and Nod-like receptor agonists TLR-3 poly I:C activated with antibiotics dosed every 3 days for 4 doses	In vitro Intravenous	Live/dead biofilms confocal microscopy bacterial density via IVIS live imaging	MSC secreted factors disrupted *MRSA* biofilm formation. Activated MSC treatment decreases bacterial bioburden in mouse chronic biofilm infection model.
Bujnakova et al. (2020)	[20]	Canine	In vitro biofilm *Staphylococcus aureus* *Escherichia coli*	Bone marrow	N/A	In vitro coculture *S. aureus*, *E.coli* biofilms	In vitro	Disc diffusion test Spectrophotometric crystal violet assay Bioluminescence assay	MSC-CM inhibited biofilm formation and quorum sensing.
Bahroudi et al. (2020)	[21]	Human	In vitro *Vibrio cholerae* co-culture with MSC secretome	Bone marrow	N/A	MSC secretome coculture V. cholerae 1:8 to 1:128	In vitro	Plate crystal violet assay	MSC secretome prevented biofilm formation of *Vibrio cholerae* in a dose-dependent manner.
Marx et al. (2020)	[22]	Equine	In vitro *Pseudomonas, Staphylococcus* biofilms	Peripheral blood	N/A	In vitro co-culture with *Pseudomonas* and *Staphylococcus* biofilms	In vitro	Protease array Confocal microscopy biofilm composition Western blot analysis	MSC secretome inhibits biofilm formation and mature biofilms of Pseudomonas and Staphylococcus spp. MSC secrete cysteine proteases that destabilize MRSA biofilms increasing efficacy of antibiotics.
Marx et al. (2021)	[23]	Equine	*Ex vivo* equine skin biofilm explant model	Peripheral blood	N/A	In vitro co-culture MSC-CM with MRSA and MSSA	In vitro explant	Immunofluorescence activity Biofilm live/dead staining	MSC decreased MRSA viability in mature biofilms. Equine MSCs secrete CCL2 that increased antimicrobial peptide secretion by equine keratinocytes.

Table 1. *Cont.*

Investigator	Reference	Species	Culture Conditions or Lesion	Cell Source	Cell Dose	Protocol	Route of Administration	Outcome Parameters	Main Findings
Pezzanite et al. (2021)	[24]	Equine	In vitro MRSA biofilm assays	Bone marrow	N/A	TLR-3, TLR-4 NOD activated MSC	In vitro biofilms	Bactericidal activity Neutrophil bacterial phagocytosis Cytokine analysis Antimicrobial peptide secretion	MSC stimulation TLR3 poly I:C suppressed biofilm formation enhanced neutrophil phagocytosis increased MCP-1 secretion, enhanced antimicrobial peptide production.
Pezzanite et al. (2022)	[25]	Equine	In vivo MRSA septic arthritis	Bone marrow	20×10^6 cells/joint	TLR-3 poly I:C activated MSC Dosed every 3 days for 3 doses	Intra-articular	Clinical pain scoring Quantitative bacterial cultures Complete blood counts Cytokines synovial fluid, plasma Imaging (radiographs, ultrasound, MRI) Macroscopic joint scoring Histologic changes	Activated MSC therapy resulted in improved pain scores, ultrasound and MRI scoring, quantittative bacterial counts, systemic neutrophil and serum amyloid A, synovial fluid lactate and serum amyloid A synovial fluid IL-6 and IL-18.
Johnson et al. (2022)	[26]	Canine	Naturally occurring chronic multidrug resistant infections	Adipose	2×10^6 cells/kg	TLR-3 poly I:C activated with antibiotics Dosed every 2 weeks for 3 doses	Intravenous	Quantitative cultures Clinical signs Phone follow-up	Repeated delivery of activated allogeneic MSC resulted in infection clearance and wound healing.
Yang et al. (2022)	[27]	Human	*Pseudomonas aeruginosa* inoculated tracheal tubes	Umbilical cord	N/A	In vitro co-culture, 8 MSC concentrations	In vitro biofilms	Titration MSC concentration Anti-biofilm experiment Bacterial motility assay DNA microarray experiment	Antibacterial peptides from MSC affected biofim formation by downregulating polysaccharide biosynthesis protein which correlated to MSC concentration.

Table 2. Summary of studies demonstrating evidence that activation of MSC enhances their innate antibacterial and immunomodulatory properties.

Investigator	Reference	Species	Culture Conditions or Lesion	Cell Source	Cell Dose	Protocol	Route	Outcome Parameters	Main Findings
Liotta et al. (2008)	[28]	Human	In vitro TLR activation	Bone marrow	N/A	TLR-3 poly I:C or TLR-4 LPS activation	In vitro	Flow cytometric evaluation	BM-MSCs expressed high levels TLR3 and 4 which induce nuclear factor k-$\mathcal{B}$ activity, IL6, IL8, CXCL10
			T-cell co-culture					MSC differentiation assays	Ligation TLR3 and TLR4 on MSCs inhibited ability of MSC to suppress T-cell proliferation without
								T-cell proliferation assays	influencing immunophenotype or differentiation potential
								ELISA cytokines/chemokines analysis	TLR-triggering was related to impaired Notch receptor signaling in T cells
								IDO activity measures	TLR3 and TLR4 expression on MSCs provide effective mechanisms to block immunosuppressive activities and restore efficient T-cell response to infection such as viruses or Gram-negative bacteria
								Confocal microscopy	
								Quantitative analysis NFK-$\mathcal{B}$ translocation	
								RNA extraction and rtPCR	
Opitz et al. (2009)	[29]	Human	In vitro co-culture	Bone marrow	N/A	MSC T-cells in mixed leukocyte reactions	In vitro	Karyotype analysis of MSC	TLR ligation activates innate and adaptive immune response pathways to protect against pathogens
			MSC with T-cells			TLR-3 poly I:C or TLR-4 LPS activation		Flow cytometric analysis MSC	TLR expressed on human bm-MSC enhanced immunosuppressive phenotype of MSC
								Mixed leukocyte reactions	Immnunosuppression mediated by TLR was dependent on production of IDO1

Table 2. *Cont.*

Investigator	Reference	Species	Culture Conditions or Lesion	Cell Source	Cell Dose	Protocol	Route	Outcome Parameters	Main Findings
								Quantitative rt-PCR	Induction of IDO1 by TLR involved autocrine interferon signaling loop which depended on protein kinase R
								Liquid chromatography Western blot analysis, siRNA ELISA cell culture supernatants	
Romieu-mourez et al. (2009)	[30]	Human	In vitro activation	Bone marrow	N/A	TLR-3 poly I:C or TLR-4 LPS activation	In vitro	Flow cytometric analysis	Human MSC and macrophages expressed TLR3 and TLR4 at comparable levels
			cytokines, TLR agonists					real-time RT-PCR	TLR-mediated activation of MSC resulted in production inflammatory mediators IL-1β, IL-6, IL-8/CXCL8, CCL5
								Immunoblot analysis	IFN priming combined with TLR activation increases immune responses induced by Ag-presenting MSC
								Growth response to TNF-α, IFN-α, IFN-γ	TLR activation resulted in inflammatory site attracting innate immune cells
								Immune effector infiltration analysis Neutrophil chemotaxis assay	
Cassatella et al. (2011)	[32]	Human	In vitro activated	Bone marrow	N/A	TLR-3 poly I:C or TLR-4 LPS activation	In vitro	Cytofluorometric analysis	TLR-3 MSC activation enhanced anti-apoptosis of neutrophils more than TLR-4
			MSC neutrophil coculture					ELISA immunoassays	TLR-3 and TLR-4 activation enhanced respiratory burst ability and CD11b expression by PMN

Table 2. *Cont.*

Investigator	Reference	Species	Culture Conditions or Lesion	Cell Source	Cell Dose	Protocol	Route	Outcome Parameters	Main Findings
								Respiratory burst cytochrome C reduction	TLR-3 activation effects mediated by IL-6, IFN-*B* and GM-CSF
									TLR-4 activation effects mediated by GM-CSF
Lei et al. (2011)	[33]	Murine	In vitro TLR activation	Bone marrow	N/A	TLR-2 or TLR-4 activation	In vitro	MSC migration	TLR2 ligation (but not TLR4) inhibited MSC migration, MSC mediated immunosuppression on allo-MLR,
								Allogeneic mixed lymphocyte reaction	and reduced MSC mediated expansion of Treg cells
								Induction Treg cell	TLR2 activation induced lower CXCL10 mRNA and protein expressions
									TLR2 and TLR4 had different effects on immunomodulatory capacity of MSC
Giuliani et al. (2014)	[34]	Human	In vitro MSC NK cell coculture	Bone marrow	N/A	TLR-3 or TLR-4 activation	In vitro	Flow cytometry CD107 degranulation	TLR primed MSC are more resistant than unprimed MSC to IL-2 activated NK-induced killing
				Embryonic		NK cell MSC coculture		ELISA culture supernatants	TLR-primed MSC modulated naturall killer group 2D ligands MHC class I chain A, ULBP3, DNAM-1 ligands
								Chromium release assay	MSC adapt their immunobehavior in inflammatory context, decreasing susceptibility to NK killing
									TLR3 but not TL4 primed MSC enhance suppressive functionns against NK cells
Johnson et al. (2017)	[16]	Murine	*Staphylococcus aureus*	Adipose	1×10^6 cells/	TLR-3 poly I:C activation +/- antibiotics	Intravenous	Bacterial burden IVIS imaging	Activated MSC co-administered with antibiotics was most effective to reduce bacterial bioburden

Table 2. *Cont.*

Investigator	Reference	Species	Culture Conditions or Lesion	Cell Source	Cell Dose	Protocol	Route	Outcome Parameters	Main Findings
		Canine	implant infection model Naturally occurring wounds	Adipose	/injection 2×10^6 cells/kg	dosed every 3 days for 3 doses TLR-3 poly I:C activated with antibiotics dosed every 2 weeks for 3 doses	Intravenous	Wound tissue histology Quantitative cultures Clinical signs, Phone follow-up	Clearance of bacteria and wound healing following repeated IV injection
Gorskaya et al. (2017)	[36]	Murine	Intraperitoneal injection NLR, TLR, *S. typhimurium*	Bone marrow	NLR/TLR ligands 10 µg/mouse	NLR2 and TLR (LPS, flagellin, CpG, poly I:C) and *S. typhimurium* antigenic complex	Intraperitoneal	Efficiency bone marrow MSC colony formation	NLR, TLR and *S. typhimurium* antigenic complex increase efficiency of MSC cloning and content by 1 hr
Rashedi et al. (2017)	[37]	Human	In vitro activation TLR ligands	Bone marrow	N/A	TLR-3, TLR-4 effect on MSC Treg induction	In vitro	MSC, CD4+ lymphocyte co-culture assays Gene and protein expression analysis Flow cytometric analysis Quantification cytokines culture medium	TLR3/4 activation MSC enhanced Treg generation in CD4+ lymphocyte/MSC cultures TLR3/4 activation augmented Treg induction via Notch pathway
Petri et al. (2017)	[38]	Human	In vitro coculture TLR-3 TLR-3 activated MSCs and NK cells	Nasal mucosa	N/A	TLR-3 activated MSC effect on NK cells	In vitro	ELISA immunoassays Flow cytometric analysis Surface/intracellular staining	Early time points TLR3-activated MSC secrete type I interferon to enhance NK cell effector function Later time points NK cell function limited by TGF-β and IL-6 Feedback regulatory NK cells to MSCs promote survival, proliferation, pro-angiogenic properties

Table 2. *Cont.*

Investigator	Reference	Species	Culture Conditions or Lesion	Cell Source	Cell Dose	Protocol	Route	Outcome Parameters	Main Findings
								Cytotoxicity assays Degranulation assays NK cell proliferation assays MSC invasion and proliferation assays	
Cassano et al. (2018)	[39]	Equine	In vitro co-culture TLR ligands	Bone marrow	N/A	TLR-3 or TLR-4 activation	In vitro	T-cell proliferation via flow cytometry	TLR3/4 priming increased MSC expression IL6, CCL2, CXCL10
						MSC co-culture inflammatory macrophages Suppression T-cell proliiferation assay		Macrophage RNA gene expression	TLR3/4 priming or exposure to inflammatory macrophages enhanced immunomodulatory function demonstrated by decreased T-cell proliferation
Cortes-Araya et al. (2018)	[41]	Equine	In vitro comparison MSC tissue sources	Endometrium	N/A	TLR-4 primed MSC versus unprimed	In vitro	Antimicrobial peptide immunocytochemistry	Lipocalin-2 was expressed at higher levels in EM-MSC than AD or BMD
			In vitro activation with TLR4 ligand	Adipose				Cytokine secretion via ELISA	TLR-4 stimulated lipocalin-2 production by all three cell types
				Bone marrow				Gene expression analyses	TLR-4 induced expression IL-6, IL-8, MCP-1, chemokine ligand-5, TLR4 by all three cell types
Asami et al. (2018)	[17]	Murine	In vitro activation with TLR ligands	Bone marrow	1×10^6 cells	1 injection 1 hour after bacterial inoculation	Intravenous	Bacteria bronchoalveolar lavage	MSC-CM modulates TNFα, IL-6, IL-10 after
			Streptococcus pneumoniae pulmonary infection		/injection			Myeloperoxidase activity assay Bichinchoninic acid protein assay	stimulation with TLR2, TLR4, TLR9 ligands. MSC-CM suppresses CXCL1, CXCL2 production

Table 2. *Cont.*

Investigator	Reference	Species	Culture Conditions or Lesion	Cell Source	Cell Dose	Protocol	Route	Outcome Parameters	Main Findings
								Histopathologic examination	after stimulation with TLR2 and TLR9 ligands. MSC IV decreased total cells, neutrophils, and myeloperoxidase activity during pulmonary infection. MSC IV decreased BALF cytokine levels TNFα, IL-6, IFN-γ, CCL2, GM-CSF during pulmonary infection.
Chow et al. (2019)	[19]	Human	In vitro *Staphylococcus aureus* biofilm assay	Bone marrow	N/A	Comparison TLR, NLR receptor agonists	In vitro	Live/dead biofilms via confocal microscopy	MSC secreted factors disrupted *MRSA* biofilm formation
			Mice with mesh implant biofilm animal model		1×10^6 cells /injection	TLR-3 poly I:C activated with antibiotics Dosed every 3 days for 4 doses	Intravenous	bacterial density by IVIS live imaging	Activated MSC treatment decreases bacterial bioburden in mouse chronic biofilm infection model
Kurte et al. (2020)	[44]	Murine	In vitro splenocyte and MSC and Tcell	Bone marrow	N/A		In vitro	Quantitative real-time PCR	Time dependent LPS activation regulate IL6 and iNOS expression in MSCs.
			and MSC co-cultures				Subcutaneous	Flow cytometry	Immunosuppressive activity of MSCs on T cell proliferation depends on time dependent LPS activation.
			Murine autoimmune encephalomyelitis (EAE)					Immunosuppression assay	Long exposure to LPS enhances MSC therapeutic potential in EAE.
								Treg, Th17, Th1 differentiation assay	TLR4 expression involved in immunosuppressive capacity of MSCs in vitro.
								Thelper analysis in treated mouse lymph nodes	TLR4 inhibition disrupts capacity of MSCs to inhibit Th1 and Th17 cells in vitro.

Table 2. *Cont.*

Investigator	Reference	Species	Culture Conditions or Lesion	Cell Source	Cell Dose	Protocol	Route	Outcome Parameters	Main Findings
									TLR4 deficiency reduces therapeutic effect of MSCs in EAE.
Aqdas et al. (2021)	[45]	Murine	In vitro co-culture MSC with Mycobacterium tuberculosis (Mtb)	Bone marrow	N/A	TLR-4 or NOD-2 activated MSC	In vitro	Cytokine secretion ELISA (IL-6, IL-10, IL-12, TNF-α) RT-qPCR (IL-6, IL-12, IL-10, iNOS, TNF-α, TGF-$\mathcal{B}$) Phenotypic charactization of MSC markers Evaluation MSC differentiation Bacterial load determination post-infection Bacterial tracking into autolysosomes	TLR4/NOD-2 augmented pro-inflammatory cytokine secretion. TLR4/NOD-2 co-localized Mtb in lysosomes. TLR4-NOD-2 induced autophagy. TLR4-NOD-2 enhanced NF-κ$\mathcal{B}$ activity via p38 MAPK. TLR4-NOD-2 reduced intracellular Mtb survival. Triggering TLR4-NOD-2 pathway may be future immunotherapy.
Pezzanite et al. (2021)	[24]	Equine	In vitro MRSA biofilm assays	Bone marrow	N/A	TLR-3, TLR-4 and NOD activated MSC	In vitro	Bactericidal activity Neutrophil bacterial phagocytosis Cytokine analysis Antimicrobial peptide secretion	MSC stimulation with TLR3 poly I:C suppressed biofilm formation, enhanced neutrophil phagocytosis, increased MCP-1 secretion and enhanced antimicrobial peptide cathetlicidin production
Johnson et al. (2022)	[26]	Canine	Naturally occurring chronic multidrug resistant infections	Adipose	2×10^6 cells/kg	TLR-3 poly I:C activated with antibiotics dosed every 2 weeks for 3 doses	Intravenous	Quantitative cultures Clinical signs, Phone follow-up	Repeated delivery of activated allogeneic MSC resulted in infection clearance and wound healing

Table 2. *Cont.*

Investigator	Reference	Species	Culture Conditions or Lesion	Cell Source	Cell Dose	Protocol	Route	Outcome Parameters	Main Findings
Pezzanite et al. (2022)	[25]	Equine	MRSA inoculated septic arthritis	Bone marrow	20×10^6 cells/joint	TLR-3 poly I:C activated MSC dosed every 3 days for 3 doses	Intra-articular	Clinical pain scoring Quantitative bacterial cultures Complete blood counts Cytokine analyses (blood, synovial fluid) Imaging (radiographs, ultrasound, MRI)	Activated MSC therapy resulted in improved pain scores, ultrasound and MRI scoring, quantitative bacterial counts, systemic neutrophil and serum amyloid A, and synovial fluid lactate, serum

2. Principles of Cellular Therapy to Treat Bacterial Infection

2.1. Mechanisms of MSC Antimicrobial and Immunomodulatory Action

Direct antimicrobial activity of MSC from multiple species and tissue sources has been reported, primarily through secretion of antimicrobial peptides that potentiate the activity of conventional antibiotics by increasing drug permeability of bacterial cell walls [13,16,43,47–57]. In addition, while MSC themselves express low immunogenicity, MSC are immunologically active, suppressing inflammation associated with infection by both direct cell-to-cell contact and secreted factors [57–63] including immune suppressive cytokines (e.g., IL-10, TGF-ß), metabolites (e.g., IDO, PGE2, adenosine), and matrix factors (e.g., galectins) [19,57,62,64–69]. MSC secreted factors not only suppressed biofilm formation but further disrupted formed biofilms in vitro [23,70]. MSC embedded implants have previously been demonstrated to have enhanced bacterial clearance and be more resistant to biofilm formation [15]. As biofilms are a defining feature of chronic bacterial infections, including those involving bone, synovial structures, and implants [15,71–73], the biofilm dispersing properties displayed by MSC are key to their role in treatment of chronic infection. The rationale for and approach to ACT takes advantage of and optimizes these innate properties of MSC for enhanced treatment of MSC [31,42,57,74] (Figure 1).

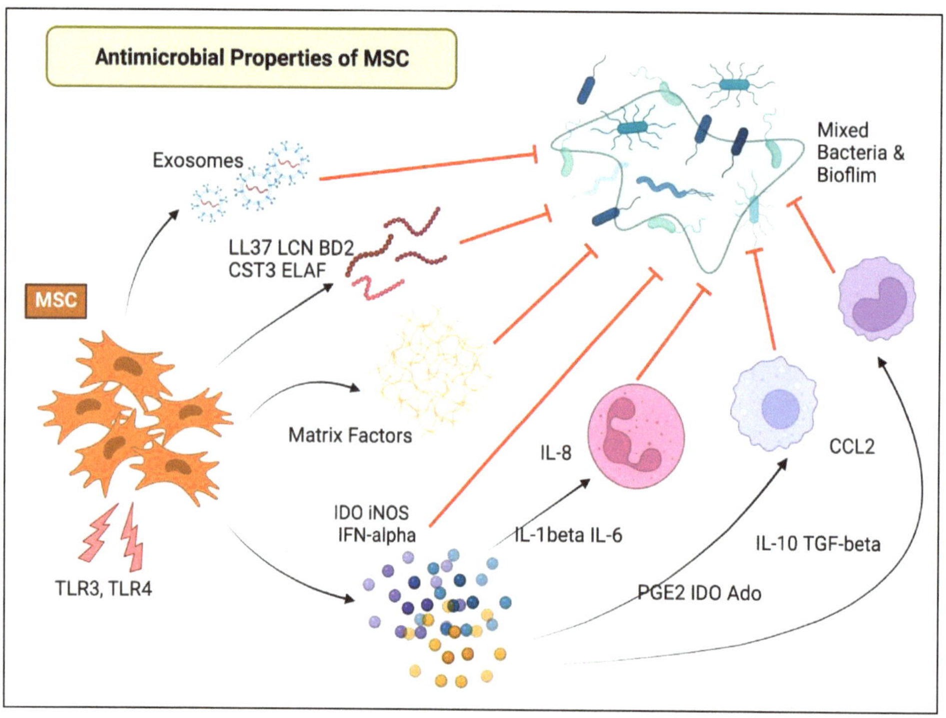

Figure 1. Immune mechanisms for antimicrobial properties of MSC against biofilms. Direct antimicrobial activity of MSC via secreted factors including antimicrobial peptides and indirect immunomodulatory activity of MSC are illustrated. Directly, cationic antimicrobial peptides (e.g., cathelicidin, lipocalin-2, ß-defensin 2), induce damage to bacterial membranes or alter bacterial function either directly or indirectly. Indirectly, MSC activate host immune cells, modulate local inflammation and induce angiogenesis and fibrogenesis, targeting several different cell types including T cells, macrophages, neutrophils, and dendritic cells. This activity is primarily mediated by up-regulation or inhibition of immunomodulatory cytokines and chemokines that in turn augment the immune system either to a pro-inflammatory or an anti-inflammatory state.

2.2. Cellular Activation Techniques

The functional properties of MSCs can be modified through activation of Toll-like receptors (TLR), nucleotide-binding oligomerization domain (NOD-like receptors or NLRs), or RIG-I-like receptors (RLR) [75]. Toll-like receptors (TLRs) specifically have been recognized as regulators of stromal cell functions, including survival, differentiation, and growth [35], with thirteen different TLRs identified to date in mammalian species [35]. TLRs are expressed either on intracellular membranes of the endoplasmic reticulum, lysosomes, and endosomes (TLRs 3, 7, 8 and 9) or on the cell surface (TLRs 1, 2, 4, 5, and 6) [42]. MSC derived from multiple tissue sources and species express TLRs (e.g., TLR2, TLR3, TLR4, and TLR9), which play an important role in their regulatory effects in immune modulation and response to inflammation in infection [33,76], and signaling through TLR pathways is regulated at multiple levels from transcriptional to post-translational [42]. Furthermore, interactions between TLR pathways and micro-RNAs (miRNAs) dictate either suppression or activation of the TLR signaling and downstream responses in MSCs [42]. Differences in TLR stimuli used, culture conditions or MSC source have been shown to play a role in resultant action following MSC priming, leading to inconsistent findings reported with TLR activation of MSC [31]. MSCs activated with TLRs have been demonstrated to exhibit immunosuppressive properties through induction of indoleamine-2,3-dioxygenase-1 via protein kinase R and interferon-ß [29] and to recruit immune inflammatory cells, through upregulation of secretion of immunomodulatory cytokines (CCL5, IL1ß, IL-6, IL-8) [30]. In vivo injection of various ligands (NLR2, TLR3,4 and 5) further enhanced proliferation of MSCs, increased cloning efficiency, and affected cell differentiation [36].

Importantly, activation with different TLR ligands have resulted in differential effects [46]. For example, TLR4 activation was found to induce a pro-inflammatory phenotype in MSC, termed *MSC1*, whereas TLR3 activation resulted in an *MSC2* phenotype with upregulation of more immunosuppressive pathways [77–79]. TLR3 but not TLR4 primed MSC enhanced their immune-suppressive activity again natural killer cells, through modulation of natural killer group 2D ligand major histocompatibility complex class I chain A and ULBP3 and DNAM-1 ligands, which was also found to be context dependent to the site of inflammation [34]. Ligation of TLR3 and TLR4 further inhibited MSCs' ability to suppress T-cell proliferation by affecting Notch signaling pathways, which are transmembrane receptor proteins important in cell–cell communication, solidifying MSCs' role in immunosuppression [28,37]. In addition, TLR4 activation can stimulate the release of cytokines, especially immunomodulatory chemokines such as MCP-1 and IL-8 that recruit monocytes and neutrophils, respectively [41]. Priming of equine MSC with both TLR3 and TLR4 increased expression of CXCL10, CCL2, and IL-6 and resulted in decreased T cell proliferation (TLR3 to a greater extent than TLR4) [39]. TLR3 agonist polyinosinic:polycytidylic acid (poly I:C) stimulation of MSC further regulated key innate immune cells known to be important to anti-viral immunity in a time-dependent fashion where early activated MSC secrete type I interferon to enhance NK cell effector function and at later time points produce greater amounts of IL-6 and TGF-ß to induce senescence in NK cells and terminate inflammatory responses [38].

Furthermore, ligation of specific TLR agonists (eg., TLR2 versus TLR4 activation) can actually inhibit MSC migration, MSC-mediated immunosuppression, and reduce expansion of regulatory T cells, diminishing MSC potential effect in treating inflammatory disease [33]. In another study, inhibition of TLR4 resulted in reduced proliferation and osteogenic differentiation of adipose derived MSC. These findings indicate that TLR receptors also regulate cell differentiation pathways, which may be relevant in the setting of bacterial infections where multiple different TLR and NLR ligands are expressed.

In a study evaluating the effect of TLR activation of murine MSC in the treatment of pulmonary infection, activation with TLR 2, 4 and 9 resulted in significantly decreased production of pro-inflammatory cytokines IL-6 and TNF- α [17]. Finally, multiple aspects of culture techniques, including time of TLR agonist exposure, concentration of TLR agonist, and MSC concentration during cell activation have all been demonstrated to affect both the immunosuppressive and the antibacterial activity of MSC [24,44]. These studies provide some explanation for the previously conflicting reports regarding overall net effects of TLR stimulation, suggesting MSC polarization and ligand selection are important aspects to consider in application of TLR agonists to activation of MSC in clinical scenarios. Specifically, MSC polarization refers to the process by which MSCs may be polarized by downstream TLR signaling into two relatively homogeneous phenotypes previously classified as MSC1 and MSC2, providing both a mechanism by which to reduce heterogeneity in cellular populations and potentially improve efficacy of current cell-based therapies [77]. Taken together, these findings support the concept that MSCs' immunomodulatory and antimicrobial function can be significantly upregulated just prior to injection by priming or 'licensing' with innate immune ligands such as TLR agonists, and that selection of these agonists can significantly impact the quality and the magnitude of the downstream pathways that are activated.

Activation of MSC with TLR ligands stimulates production of antimicrobial peptides, including lipocalin-2, hepcidin, and beta-defensin-2, and cathelicidin [11,32,48,51,80]. Stimulation of MSC with IFN- γ, as would typically be found in an inflammatory microenvironment as in bacterial infection, resulted in enhanced mRNA expression of TLR3 as well as IDO1, and increased secretion of immunomodulatory cytokines including IL-10 [81]. When Toll-like receptor (TLR) activation was compared to that of nucleotide-binding oligomerization domain (NOD)-like receptor (NLR) ligand stimulation of MSC specifically to enhance antimicrobial properties and immunomodulation, activation with TLR3 ligand poly I:C increased bactericidal activity, suppressed biofilm formation, enhanced neutrophil bacterial phagocytosis and increased immunomodulatory cytokine secretion (MCP-1) by equine MSC compared to nonstimulated MSC and activation with other TLR and NLR agonists [24]. Of all ligands evaluated, MSCs treated with TLR3 ligand poly I:C, of all ligands evaluated, resulted in greater production of indoleamine 2,3-dioxygenase (IDO), a clinically relevant therapeutic factor, and attenuated pathology in a mouse model of dextran sodium sulfate (DSS) induced colitis [82]. In an additional in vivo mouse model of chronic wound infection, mice treated with TLR3 activated MSC demonstrated migration to the site of infection, which was mechanistically shown to be mediated in part by upregulation of CXCR4 expression [16]. For example, activated MSC migrated more efficiently to an SDF-1 stimulus in vitro, and to sites of wound infection in vivo. Thus, pre-activation with a TLR ligand such as pIC was demonstrated to augment MSC antimicrobial activity through a variety of indirect mechanisms and was moved forward in clinical studies in dogs with naturally occurring wounds and horses with septic arthritis involving multidrug resistant organisms.

2.3. *Route of Administration, Dosing, and Number of Injections*

Both systemic and local intraperitoneal or intrasynovial injection of MSC have resulted in successful treatment of infection in animal models [25,82,83] and supports previous studies demonstrating that priming of MSC induces population-normalizing effects that can standardize what would otherwise be heterogenous cell populations [83]. Doses of 2×10^6 cells/kg and up to 1×10^9 cell/kg, which have previously been reported as optimal for immunomodulation in humans and large animals [84], were injected intravenously in mice with chronic *Staphylococcus aureus* impregnated implant infections and dogs with chronic naturally occurring wounds [16]. Mechanistically, when administered systemically via intravenous administration, MSC have been shown to interact with host innate immune cells, principally neutrophils and monocytes, at multiple sites, including lungs, spleen, liver, and sites of infection [64,65,85]. For example, these effects resulted in enhanced bacterial phagocytosis, mediated by MSC-secreted cytokines such as interleukin-18 (IL-8) and stimulation of neutrophil extracellular trap (NET) formation, leading to enhanced bacterial killing and neutrophil survival [16,26,78]. Recruitment of monocytes to sites of inflammation, such as bacterial infection, is mediated by chemokine CCL2 (MCP-1) produced by MSC, which mobilizes release of inflammatory monocytes from bone marrow and recruitment to sites of high CCL2 production (i.e., infection) [34]. Once recruited to wound tissues, monocytes rapidly differentiate to macrophages; important to the mechanism of ACT, TLR-3 activated MSC induce differentiation of wound macrophages from an M1 (pro-inflammatory) to M2 (reparative) phenotype [16]. This response is consistent with the anti-inflammatory phenotype of TLR-3 activated MSC previously reported [77–79].

When ACT was further explored in a large animal model of septic arthritis, local administration was investigated to minimize the need for larger numbers of MSC when dose was extrapolated to increased body mass [25], with positive results in reduction of local and systemic inflammation, decreased bacterial burden within joints and improved pain scores [25]. Furthermore, in a mouse model of induced colitis, intraperitoneal but not intravenous injection of TLR3 activated MSC was found to attenuate disease severity [82]. In previous studies, local injection of MSC at sites of wound infection have not been appreciated to be as effective as systemic administration [16], indicating that further investigation and comparison of routes of administration is warranted and the optimal route for a particular clinical scenario may depend on a number of factors. These studies illustrate the pros and cons of different routes of administration depending on the size of the patient, cost considerations, and condition for and accessibility of the lesion for which MSC are being administered.

Multiple versus single administrations may further improve eradication of chronic infections, theoretically due to a cumulative impact on activation of host defenses [25]. In studies performed in pet dogs with chronic MDR infections, some animals received up to 10 MSC infusions via intravenous administration [16]. A potential concern with the use of repeated injections of allogeneic MSC is the potential for induction of harmful host adaptive immune reactions to infused MSC; however, no adverse events were seen in dogs or horses receiving multiple MSC administrations for chronic infections, which may reflect the high level of systemic and local inflammation already present in multidrug resistant infections [16,25]. Future studies may employ recently investigated techniques to reduce immunogenicity when injecting allogeneic MSC such as major histocompatibility (MHC) haplotyping and matching or TGFß2 stimulation to reduce immunogenicity to MSC-mismatched stromal cell donors [86,87]. (Tables 1 and 2).

2.4. Combination of MSC with Antibiotics for Enhanced Bacterial Killing

Co administration of antibiotics with activated MSC has been a key feature of ACT for optimal bactericidal effect. Based on our studies, all major classes of antibiotics including beta-lactam drugs (penicillins, cephalosporins, carbapenems), aminoglycosides, fluoroquinolones, glycopeptide (vancomycin), and cyclic lipopeptide (daptomycin) antibiotics exhibit synergistic or additive activity with MSC secreted factors in vitro [70]. In support of this concept, the most effective treatment protocol for mice with chronic biofilm infections was activated MSC in combination with antibiotics compared to antibiotics alone, or activated or non-activated MSC alone [16]. Furthermore, canine clinical studies with spontaneous MDR infections demonstrated that administration of antibiotics to which the infecting bacteria are resistant can still be combined effectively with activated MSC treatment.

3. Evidence for Antimicrobial Activity in Animal Models

3.1. Rodent Models of Infection

Multiple rodent studies have supported both the antimicrobial effects of MSC in treatment of infection at various sites (e.g., thoracic and peritoneal cavities, subcutaneous chronic implant) [17] as well as the benefits of priming of MSC in culture prior to administration [17]. Mice with *Streptococcus pneumoniae* pulmonary infection treated with MSC exhibited reduced myeloperoxidase activity in the lungs, decreased neutrophil number in bronchoalveolar lavage fluid and lower levels of pro-inflammatory cytokines as well as bacterial load in the lungs following treatment [17]. In this model, activation of the murine MSC with TLR agonists 2,4,9 or live *S. pneumoniae* bacteria resulted in reduced production of IL-6 and TNF- α [17]. Intraperitoneal administration of TLR3 polyI:C activated MSC further reduced disease severity in mice with DSS-induced colitis through enhanced immunosuppressive activity by stimulating MSCs to increase production of indoleamine 2,3-dioxygenase (IDO) [82]. MSC can also be combined with various substrates or polymers to increase immune modulation ability [88]. In an acute model of bacterial wound infection, Kudinov et al. demonstrated that the combination of proteins secreted from MSC along with chitosan gel was able to ameliorate the presence of microorganisms in the burn wound area [89].

3.2. Naturally Occurring Canine Model of Chronic Infection

Dogs represent a translational model for orthopedic implant infection in humans as they develop naturally occurring implant infections in similar body sites which involve similar bacterial pathogens and antibiotic resistance patterns as chronic infections in humans. As infections were naturally occurring, induction in laboratory species could be avoided. Therefore, using the dog as a realistic, translational chronic infection model, activated allogeneic MSC were administered repeatedly intravenously without negative side effects, and in many cases, resolved infections that had resisted prolonged treatment (i.e., weeks to months) with conventional antibiotics. The canine model also addresses key issues regarding the scalability of ACT for treatment of chronic infection, as dogs in these studies have been treated with comparable doses of activated MSC (typically 2×10^6 cells per kg body weight) that have also been used for systemic MSC infusion in humans [35,90]. Moreover, dogs as an outbred species also address the safety issue of repeated intravenous delivery of fully allogeneic MSC, as the donor source for MSC in all the dog studies reported by our group were adipose tissues of unrelated dogs [16]. Adverse events associated with multiple repeated infusions of activated canine allogeneic MSC over periods of up to six months were not observed, and clinical study animals have now been followed for at least two years with no subsequent adverse events noted.

3.3. Induced Equine Model of Septic Arthritis

The encouraging findings demonstrated with TLR activation of MSC in vitro and in murine and canine models of infection prompted further evaluation of ACT in a large animal (equine) model of septic arthritis. The equine preclinical model is a clinically and translationally relevant model for human infection for several reasons. Development of infectious arthritis as a naturally occurring disease process in horses is well-documented, their large joint volume allows for repeated collection of synovial fluid to analyze a larger number of outcome parameters and their cartilage thickness, joint volume and loading forces more closely replicates that of people than many other veterinary species [91–96]. In this work, multi-drug resistant *Staphylococcal* septic arthritis was treated with three intra-articular injections of TLR3-activated MSC and antibiotics or antibiotics alone. Horse pain scores, diagnostic imaging findings (ultrasound, magnetic resonance imaging), quantitative bacterial counts, systemic parameters of inflammation (neutrophil counts and acute phase marker serum amyloid A), and intra-synovial cytokine levels of pro-inflammatory cytokines interleukin-6 and interleukin-18 were improved in MSC + antibiotic treated horses and no adverse events were noted (Figure 2). These studies serve as strong evidence that the use of ACT has considerable promise as a new approach to management of chronic and/or multidrug resistant infections.

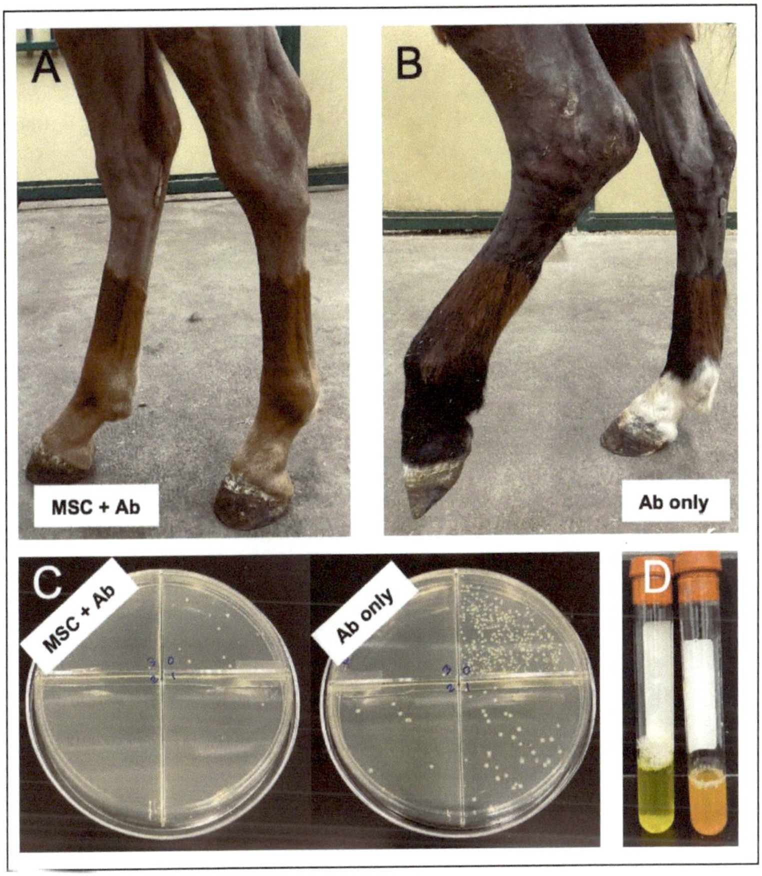

Figure 2. Evaluation of TLR poly I:C activated bone marrow derived MSC therapy in an equine model of multi-drug resistant USA300 methicillin resistant *Staphylococcus aureus* [25]; original unpublished images presented with permission from the authors]. Representative images of horses at day 7 following intra-articular inoculation of the left tarsocrural joint treated with three intra-articular injections of (**A**) MSC and antibiotics, or (**B**) antibiotics alone. Quantitative bacterial cultures were significantly reduced in horses treated with (**C**) MSC and antibiotics versus (**D**) antibiotics alone. Synovial fluid parameters serum amyloid A, lactate, and inflammatory biomarkers IL-6 and IL-18 were significantly improved in horses treated with MSC and antibiotics (left) versus antibiotics alone (right) (**D**).

4. Discussion

Cellular therapy is emerging as a promising adjunctive therapy to combat the growing problem of drug-resistant bacterial infections and those involving biofilms, and investigation of strategies to improve potency of MSCs in an ongoing area of research [42]. While there remains an incomplete understanding of the underlying mechanisms of action of TLR agonism in ACT, as well as the demonstrated additive and synergistic effects with specific antibiotics, it is apparent from these studies that TLR-activated cellular therapy for treatment of infection is well-tolerated, effective, and can be readily implemented using allogeneic sources (i.e., bone marrow or adipose tissue derived MSC obtained from young, healthy, unrelated donors) and in a variety of chronic inflammatory disease states [74]. The site of infection also does not appear to be a limiting factor, as intravenous delivery of cells was sufficient to home to sites of infection in mice and dog models and intrasynovial injection was used to effectively treat localized infections in horses. Moreover, specific resistance patterns or bacterial strains do not seem to reduce the antimicrobial effect of MSC, as activity of ACT has been observed against a variety of different Gram-positive and -negative bacterial isolates, many displaying multiple antibiotic resistances and for which development of resistance is very different. Further characterization of the effect of TLRs in biological regulation of stromal cell function could improve MSC-based cellular immunotherapies in treatment of infection [74].

Despite promising pre-clinical studies, potential obstacles to clinical implementation of ACT still must be addressed. Regulatory pathways for approval of veterinary cellular therapies in the United States by the Food and Drug Administration (FDA) is a lengthy and expensive process, with none approved to date despite greater than ten years of development efforts. Furthermore, the primary target for the majority of cellular therapies is osteoarthritis, as the market for infections in veterinary medicine may not justify development costs. In addition, there is generally a lack of spontaneous animal models of chronic infection in which to evaluate activated cellular therapies and therefore to use for FDA approval. Finally, the use of cellular therapy specifically to treat chronic drug resistant infections was not reported until 2017 by Johnson et al., so therapy for this specific indication is relataively early in the development process. As a result, a more complete understanding of the mechanisms of action of cellular activation and optimal combinations with various antibiotics is indicated. Recent evidence suggests that long noncoding RNAs (lncRNAs) regulate a wide range of biological processes and are differentially expressed in TLR3 activated MSC, providing some framework for better understanding the molecular mechanisms by which TLR activation modulates MSCs' functions [35]. Another potential issue is donor-to-donor MSC variability as MSCs from different genetic backgrounds have been shown to exhibit distinct antibacterial phenotypes [83], which at present has been addressed by using MSC derived from young, healthy donor animals and avoiding extensive MSC passaging. Hirakawa et al. recently demonstrated that CRISPR-based gene modulation could be used to engineer MSCs with enhanced antibacterial properties through upregulation of CD14, and further investigation of these methods is indicated [83]. The relative impact of the host immune status on response to ACT is also a potential treatment variable, which may limit improvement following ACT therapy in elderly or immunocompromised patients. The optimal number of ACT treatments has also not been established, nor is it clear which clinical parameters (i.e., biomarkers) are best suited to monitor treatment responses, or time frame at which to assess treatment impact as response may take weeks to months to manifest in the case of persistent, chronic bacterial infections. Finally, recent studies have begun to investigate the application of MSC derived exosomes as an acellular therapy capable of reparation [97], immunomodulation and drug-delivery, specifically in the context of treating sepsis, which may represent a promising future direction for anti-infective cellular therapies.

5. Conclusions

In summary, the use of activated cellular therapy to manage refractory or drug resistant bacterial infections is promising as an innovative option to augment antibiotic therapy. Further evaluation of mechanisms of action and investigation of ACT in randomized controlled clinical trials is indicated.

6. Patents

Provisional patents have been filed covering immune activated MSC technology described herein (S.D., L.P., L.C.).

Author Contributions: L.M.P., L.C., A.S., V.J. and S.W.D. contributed to conception and design and drafted the manuscript. All authors have read and agreed to the published version of the manuscript.

Funding: This research was funded in part by the Charlies Shipley Family Foundation, Grayson Jockey Club Research Foundation, ACVS Zoetis Dual Training Grant, NIH/NCATS CCTSI CTSA 5TL1TR002533-02, NIH 5T32OD010437-19, Carolyn Quan and Porter Bennett, Gates Center for Regenerative Medicine (Anschutz Medical Campus, University of Colorado Denver).

Institutional Review Board Statement: Not applicable.

Informed Consent Statement: Not applicable.

Data Availability Statement: Not applicable.

Acknowledgments: The authors would like to thank and acknowledge members of the Center for Immune and Regenerative Medicine at Colorado State University for their contributions to the antimicrobial cellular therapy work described herein.

Conflicts of Interest: The co-authors declare commercial affiliations with Validus Cellular Therapeutics (S.D. co-founder) and eQcell Inc. (L.P., S.D. advisory board and stock options).

References

1. CDC. Antibiotic Resistance Threats in the United States. 2013. Available online: https://www.cdc.gov/drugresistance/threat-report-2013/pdf/ar-threats-2013-508.pdf (accessed on 4 April 2022).
2. World Health Organization (WHO). Antimicrobial Resistance. 2018. Available online: https://www.who.int/news-room/fact-sheets/detail/antimicrobial-resistance (accessed on 4 April 2022).
3. Ventola, C.L. The Antibiotic Resistance Crisis: Part 1: Causes and threats. *Pharm. Ther.* **2015**, *40*, 277–283.
4. Aslam, B.; Wang, W.; Arshad, M.I.; Khurshid, M.; Muzammil, S.; Nisar, M.A.; Alvi, R.F.; Aslam, M.A.; Qamar, M.U.; Salamat, M.K.F.; et al. Antibiotic resistance: A rundown of a global crisis. *Infect. Drug Resist.* **2018**, *11*, 1645–1658. [CrossRef] [PubMed]
5. Bin Zaman, S.; Hussain, M.A.; Nye, R.; Mehta, V.; Mamun, K.T.; Hossain, N. A Review on Antibiotic Resistance: Alarm Bells are Ringing. *Cureus* **2017**, *9*, e1403. [CrossRef] [PubMed]
6. Cumper, C. The Antibiotic Resistance Crisis. 2018. Available online: http://eureka.criver.com/the-antibiotic-resistance-crisis/ (accessed on 8 April 2022).
7. NRDC. Next FDA Commissioner Must Address Antibiotic Resistance Crisis. 2019. Available online: https://www.nrdc.org/media/2019/190305 (accessed on 8 April 2022).
8. NIHMedline Plus. The End Antibiotics? 2018. Available online: https://medlineplus.gov/magazine/issues/winter18/articles/winter18pg8-11.html (accessed on 8 April 2022).
9. Palma, E.; Tilocca, B.; Roncada, P. Antimicrobial Resistance in Veterinary Medicine: An Overview. *Int. J. Mol. Sci.* **2020**, *21*, 1914. [CrossRef]
10. Sutton, M.T.; Fletcher, D.; Ghosh, S.K.; Weinberg, A.; van Heeckeren, R.; Kaur, S.; Sadeghi, Z.; Hijaz, A.; Reese, J.; Lazarus, H.M.; et al. Antimicrobial Properties of Mesenchymal Stem Cells: Therapeutic Potential for Cystic Fibrosis Infection, and Treatment. *Stem Cells Int.* **2016**, *2016*, 5303048. [CrossRef]
11. Gupta, N.; Krasnodembskaya, A.; Kapetanaki, M.; Mouded, M.; Tan, X.; Serikov, V.; Matthay, M.A. Mesenchymal stem cells enhance survival and bacterial clearance in murine Escherichia coli pneumonia. *Thorax* **2012**, *67*, 533–539. [CrossRef]
12. Krasnodembskaya, A.; Samarani, G.; Song, Y.; Zhuo, H.; Su, X.; Lee, J.-W.; Gupta, N.; Petrini, M.; Matthay, M.A. Human mesenchymal stem cells reduce mortality and bacteremia in gram-negative sepsis in mice in part by enhancing the phagocytic activity of blood monocytes. *Am. J. Physiol. Lung Cell. Mol. Physiol.* **2012**, *302*, L1003–L1013. [CrossRef]
13. Alcayaga-Miranda, F.; Cuenca, J.; Khoury, M. Antimicrobial Activity of Mesenchymal Stem Cells: Current Status and New Perspectives of Antimicrobial Peptide-Based Therapies. *Front. Immunol.* **2017**, *8*, 339. [CrossRef]
14. Yuan, Y.; Lin, S.; Guo, N.; Zhao, C.; Shen, S.; Bu, X.; Ye, H. Marrow mesenchymal stromal cells reduce methicillin-resistant Staphylococcus aureus infection in rat models. *Cytotherapy* **2014**, *16*, 56–63. [CrossRef]
15. Criman, E.T.; Kurata, W.E.; Matsumoto, K.W.; Aubin, H.T.; Campbell, C.E.; Pierce, L.M. Bone Marrow–Derived Mesenchymal Stem Cells Enhance Bacterial Clearance and Preserve Bioprosthetic Integrity in a Model of Mesh Infection. *Plast. Reconstr. Surg. Glob. Open* **2016**, *4*, e751. [CrossRef]
16. Johnson, V.; Webb, T.; Norman, A.; Coy, J.; Kurihara, J.; Regan, D.; Dow, S. Activated Mesenchymal Stem Cells Interact with Antibiotics and Host Innate Immune Responses to Control Chronic Bacterial Infections. *Sci. Rep.* **2017**, *7*, 9575. [CrossRef] [PubMed]
17. Asami, T.; Ishii, M.; Namkoong, H.; Yagi, K.; Tasaka, S.; Asakura, T.; Suzuki, S.; Kamo, T.; Okamori, S.; Kamata, H. Anti-inflammatory roles of mesenchymal stromal cells during acute *Streptococcus pneumoniae* pulmonary infection in mice. *Cytotherapy* **2018**, *20*, 302–313. [CrossRef] [PubMed]
18. Wood, C.R.; Al Dhahri, D.; Al-Delfi, I.; Pickles, N.A.; Sammons, R.L.; Worthington, T.; Wright, K.T.; Johnson, W.E.B. Human adipose tissue-derived mesenchymal stem/stromal cells adhere to and inhibit the growth of Staphylococcus aureus and Pseudomonas aeruginosa. *J. Med. Microbiol.* **2018**, *67*, 1789–1795. [CrossRef] [PubMed]

19. Chow, L.; Johnson, V.; Coy, J.; Regan, D.; Dow, S. Mechanisms of Immune Suppression Utilized by Canine Adipose and Bone Marrow-Derived Mesenchymal Stem Cells. *Stem Cells Dev.* **2017**, *26*, 374–389. [CrossRef]
20. Bujňáková, D.; Čuvalová, A.; Čížek, M.; Humenik, F.; Salzet, M.; Čížková, D. Canine Bone Marrow Mesenchymal Stem Cell Conditioned Media Affect Bacterial Growth, Biofilm-Associated Staphylococcus aureus and AHL-Dependent Quorum Sensing. *Microorganisms* **2020**, *8*, 1478. [CrossRef]
21. Bahroudi, M.; Bakhshi, B.; Soudi, S.; Najar-Peerayeh, S. Antibacterial and antibiofilm activity of bone marrow-derived human mesenchymal stem cells secretome against Vibrio cholerae. *Microb. Pathog.* **2020**, *139*, 103867. [CrossRef]
22. Marx, C.; Gardner, S.; Harman, R.M.; Van de Walle, G.R. The mesenchymal stromal cell secretome impairs methicillin-resistant Staphylococcus aureus biofilms via cysteine protease activity in the equine model. *Stem Cells Transl. Med.* **2020**, *9*, 746–757. [CrossRef]
23. Marx, C.; Gardner, S.; Harman, R.M.; Wagner, B.; Van de Walle, G.R. Mesenchymal Stromal Cell-Secreted CCL2 Promotes Antibacterial Defense Mechanisms Through Increased Antimicrobial Peptide Expression in Keratinocytes. *Stem Cells Transl. Med.* **2021**, *10*, 1666–1679. [CrossRef]
24. Pezzanite, L.M.; Chow, L.; Johnson, V.; Griffenhagen, G.M.; Goodrich, L.; Dow, S. Toll-like receptor activation of equine mesenchymal stromal cells to enhance antibacterial activity and immunomodulatory cytokine secretion. *Vet. Surg.* **2021**, *50*, 858–871. [CrossRef]
25. Pezzanite, L.M.; Chow, L.; Phillips, J.; Griffenhagen, G.M.; Moore, A.R.; Schaer, T.P.; Engiles, J.B.; Werpy, N.; Gilbertie, J.; Schnabel, L.V.; et al. TLR-activated mesenchymal stromal cell therapy and antibiotics to treat multi-drug resistant *Staphyloccocal* septic arthritis in an equine model. *Ann. Trans. Med.* **2022**, *in press*.
26. Johnson, V.; Chow, L.; Harrison, J.; Soontararak, S.; Dow, S. Activated Mesenchymal Stromal Cell Therapy for Treatment of Multi-Drug Resistant Bacterial Infections in Dogs. *Front. Vet. Sci.* **2022**, *9*, 925701. [CrossRef] [PubMed]
27. Yang, H.; Xu, F.; Zheng, X.; Yang, S.; Ren, Z.; Yang, J. Human Umbilical Cord Mesenchymal Stem Cells Prevent Bacterial Biofilm Formation. *BioMed Res. Int.* **2022**, *2022*, 1530525. [CrossRef] [PubMed]
28. Liotta, F.; Angeli, R.; Cosmi, L.; Filì, L.; Manuelli, C.; Frosali, F.; Mazzinghi, B.; Maggi, L.; Pasini, A.; Lisi, V.; et al. Toll-Like Receptors 3 and 4 Are Expressed by Human Bone Marrow-Derived Mesenchymal Stem Cells and Can Inhibit Their T-Cell Modulatory Activity by Impairing Notch Signaling. *Stem Cells* **2008**, *26*, 279–289. [CrossRef] [PubMed]
29. Opitz, C.A.; Litzenburger, U.M.; Lutz, C.; Lanz, T.V.; Tritschler, I.; Köppel, A.; Tolosa, E.; Hoberg, M.; Anderl, J.; Aicher, W.K.; et al. Toll-Like Receptor Engagement Enhances the Immunosuppressive Properties of Human Bone Marrow-Derived Mesenchymal Stem Cells by Inducing Indoleamine-2,3-dioxygenase-1 via Interferon-β and Protein Kinase R. *Stem Cells* **2009**, *27*, 909–919. [CrossRef]
30. Romieu-Mourez, R.; Francois, M.; Boivin, M.-N.; Bouchentouf, M.; Spaner, D.E.; Galipeau, J. Cytokine Modulation of TLR Expression and Activation in Mesenchymal Stromal Cells Leads to a Proinflammatory Phenotype. *J. Immunol.* **2009**, *182*, 7963–7973. [CrossRef]
31. DelaRosa, O.; Lombardo, E. Modulation of Adult Mesenchymal Stem Cells Activity by Toll-Like Receptors: Implications on Therapeutic Potential. *Mediat. Inflamm.* **2010**, *2010*, 865601. [CrossRef]
32. Cassatella, M.A.; Mosna, F.; Micheletti, A.; Lisi, V.; Tamassia, N.; Cont, C.; Calzetti, F.; Pelletier, M.; Pizzolo, G.; Krampera, M. Toll-Like Receptor-3-Activated Human Mesenchymal Stromal Cells Significantly Prolong the Survival and Function of Neutrophils. *Stem Cells* **2011**, *29*, 1001–1011. [CrossRef]
33. Lei, J.; Wang, Z.; Hui, D.; Yu, W.; Zhou, D.; Xia, W.; Chen, C.; Zhang, Q.; Wang, Z.; Zhang, Q.; et al. Ligation of TLR2 and TLR4 on murine bone marrow-derived mesenchymal stem cells triggers differential effects on their immunosuppressive activity. *Cell. Immunol.* **2011**, *271*, 147–156. [CrossRef]
34. Giuliani, M.; Bennaceur-Griscelli, A.; Nanbakhsh, A.; Oudrhiri, N.; Chouaib, S.; Azzarone, B.; Durrbach, A.; Lataillade, J.-J. TLR Ligands Stimulation Protects MSC from NK Killing. *Stem Cells* **2014**, *32*, 290–300. [CrossRef]
35. Wang, L.; Zhu, C.-Y.; Ma, D.-X.; Gu, Z.-Y.; Xu, C.-C.; Wang, F.-Y.; Chen, J.-G.; Liu, C.-J.; Guan, L.-X.; Gao, R.; et al. Efficacy and safety of mesenchymal stromal cells for the prophylaxis of chronic graft-versus-host disease after allogeneic hematopoietic stem cell transplantation: A meta-analysis of randomized controlled trials. *Ann. Hematol.* **2018**, *97*, 1941–1950. [CrossRef]
36. Gorskaya, Y.F.; Tukhvatulin, A.I.; Nesterenko, V.G. NNLR2 and TLR3, TLR4, TLR5 ligands, injected in vivo, improve after 1 h the efficiency of cloning and proliferative activity of bone marrow multipotent stromal cells and reduce the content of osteogenic multipotent stromal cells in CBA mice. *Microbiol. Immunol.* **2017**, *163*, 356–360. [CrossRef]
37. Rashedi, I.; Gómez-Aristizábal, A.; Wang, X.-H.; Viswanathan, S.; Keating, A. TLR3 or TLR4 Activation Enhances Mesenchymal Stromal Cell-Mediated Treg Induction via Notch Signaling. *Stem Cells* **2017**, *35*, 265–275. [CrossRef] [PubMed]
38. Petri, R.M.; Hackel, A.; Hahnel, K.; Dumitru, C.A.; Bruderek, K.; Flohe, S.B.; Paschen, A.; Lang, S.; Brandau, S. Activated Tissue-Resident Mesenchymal Stromal Cells Regulate Natural Killer Cell Immune and Tissue-Regenerative Function. *Stem Cell Rep.* **2017**, *9*, 985–998. [CrossRef] [PubMed]
39. Cassano, J.M.; Schnabel, L.V.; Goodale, M.B.; Fortier, L.A. The immunomodulatory function of equine MSCs is enhanced by priming through an inflammatory microenvironment or TLR3 ligand. *Vet. Immunol. Immunopathol.* **2018**, *195*, 33–39. [CrossRef] [PubMed]
40. Cassano, J.M.; Schnabel, L.V.; Goodale, M.B.; Fortier, L.A. Inflammatory licensed equine MSCs are chondroprotective and exhibit enhanced immunomodulation in an inflammatory environment. *Stem Cell Res. Ther.* **2018**, *9*, 82. [CrossRef]

41. Cortés-Araya, Y.; Amilon, K.; Rink, B.E.; Black, G.; Lisowski, Z.; Donadeu, F.X.; Esteves, C.L. Comparison of Antibacterial and Immunological Properties of Mesenchymal Stem/Stromal Cells from Equine Bone Marrow, Endometrium, and Adipose Tissue. *Stem Cells Dev.* **2018**, *27*, 1518–1525. [CrossRef]
42. Abdi, J.; Rashedi, I.; Keating, A. Concise review: TLR pathway-miRNA interplay in mesenchymal stromal cells: Regulatory roles and therapeutic directions. *Cells.* **2018**, *36*, 1655–1662. [CrossRef]
43. Olsen, A.; Johnson, V.; Webb, T.; Santangelo, K.S.; Dow, S.; Duerr, F.M. Evaluation of Intravenously Delivered Allogeneic Mesenchymal Stem Cells for Treatment of Elbow Osteoarthritis in Dogs: A Pilot Study. *Vet. Comp. Orthop. Traumatol.* **2019**, *32*, 173–181. [CrossRef]
44. Kurte, M.; Vega-Letter, A.M.; Luz-Crawford, P.; Djouad, F.; Noël, D.; Khoury, M.; Carrión, F. Time-dependent LPS exposure commands MSC immunoplasticity through TLR4 activation leading to opposite therapeutic outcome in EAE. *Stem Cell Res. Ther.* **2020**, *11*, 416. [CrossRef]
45. Aqdas, M.; Singh, S.; Amir, M.; Maurya, S.K.; Pahari, S.; Agrewala, J.N. Cumulative Signaling Through NOD-2 and TLR-4 Eliminates the Mycobacterium Tuberculosis Concealed Inside the Mesenchymal Stem Cells. *Front. Cell. Infect. Microbiol.* **2021**, *11*, 669168. [CrossRef]
46. Aziz, S.G.-G.; Alipour, S.; Ranjbarvan, P.; Azari, A.; Babaei, G.; Golchin, A. Critical roles of TLRs on the polarization of mesenchymal stem cells for cell therapy of viral infections: A notice for COVID-19 treatment. *Comp. Clin. Pathol.* **2021**, *30*, 119–128. [CrossRef] [PubMed]
47. Krasnodembskaya, A.; Song, Y.; Fang, X.; Gupta, N.; Serikov, V.; Lee, J.-W.; Matthay, M.A. Antibacterial Effect of Human Mesenchymal Stem Cells Is Mediated in Part from Secretion of the Antimicrobial Peptide LL-37. *Stem Cells* **2010**, *28*, 2229–2238. [CrossRef] [PubMed]
48. Sung, D.K.; Chang, Y.S.; Sung, S.I.; Yoo, H.S.; Ahn, S.Y.; Park, W.S. Antibacterial effect of mesenchymal stem cells against Escherichia coli is mediated by secretion of beta- defensin- 2 via toll- like receptor 4 signalling. *Cell. Microbiol.* **2016**, *18*, 424–436. [CrossRef] [PubMed]
49. Kang, H.-K.; Kim, C.; Seo, C.H.; Park, Y. The therapeutic applications of antimicrobial peptides (AMPs): A patent review. *J. Microbiol.* **2017**, *55*, 1–12. [CrossRef] [PubMed]
50. Fabisiak, A.; Murawska, N.; Fichna, J. LL-37: Cathelicidin-related antimicrobial peptide with pleiotropic activity. *Pharmacol. Rep.* **2016**, *68*, 802–808. [CrossRef]
51. Doss, M.; White, M.R.; Tecle, T.; Hartshorn, K.L. Human defensins and LL-37 in mucosal immunity. *J. Leukoc. Biol.* **2010**, *87*, 79–92. [CrossRef]
52. Chakraborty, K.; Ghosh, S.; Koley, H.; Mukhopadhyay, A.K.; Ramamurthy, T.; Saha, D.R.; Mukhopadhyay, D.; Roychowdhury, S.; Hamabata, T.; Takeda, Y.; et al. Bacterial exotoxins downregulate cathelicidin (hCAP-18/LL-37) and human beta-defensin 1 (HBD-1) expression in the intestinal epithelial cells. *Cell Microbiol.* **2008**, *10*, 2520–2537. [CrossRef]
53. Armitage, A.E.; Eddowes, L.A.; Gileadi, U.; Cole, S.; Spottiswoode, N.; Selvakumar, T.A.; Ho, L.-P.; Townsend, A.R.M.; Drakesmith, H. Hepcidin regulation by innate immune and infectious stimuli. *Blood* **2011**, *118*, 4129–4139. [CrossRef]
54. Michels, K.; Nemeth, E., Ganz, T.; Mehrad, B. Hepcidin and Host Defense against Infectious Diseases. *PLoS Pathog.* **2015**, *11*, e1004998. [CrossRef]
55. Kaundal, U.; Bagai, U.; Rakha, A. Immunomodulatory plasticity of mesenchymal stem cells: A potential key to successful solid organ transplantation. *J. Transl. Med.* **2018**, *16*, 31. [CrossRef]
56. Harman, R.M.; Yang, S.; He, M.K.; Van de Walle, G.R. Antimicrobial peptides secreted by equine mesenchymal stromal cells inhibit the growth of bacteria commonly found in skin wounds. *Stem Cell Res. Ther.* **2017**, *8*, 157. [CrossRef] [PubMed]
57. Russell, K.A.; Garbin, L.C.; Wong, J.M.; Koch, T.G. Mesenchymal Stromal Cells as Potential Antimicrobial for Veterinary Use—A Comprehensive Review. *Front. Microbiol.* **2020**, *11*, 606404. [CrossRef] [PubMed]
58. Kyurkchiev, D.; Bochev, I.; Ivanova-Todorova, E.; Mourdjeva, M.; Oreshkova, T.; Belemezova, K.; Kyurkchiev, S. Secretion of immunoregulatory cytokines by mesenchymal stem cells. *World J. Stem Cells* **2014**, *6*, 552–570. [CrossRef] [PubMed]
59. Gonzalez-Rey, E.; Gonzalez, M.A.; Varela, N.; O'Valle, F.; Hernandez-Cortes, P.; Rico, L.; Büscher, D.; Delgado, M. Human adipose-derived mesenchymal stem cells reduce inflammatory and T cell responses and induce regulatory T cells in vitro in rheumatoid arthritis. *Ann. Rheum. Dis.* **2010**, *69*, 241–248. [CrossRef] [PubMed]
60. Davies Lindsay, C.; Heldring, N.; Kadri, N.; Le Blanc, K. Mesenchymal stromal cell secretion of programmed death-1 ligands regulates t cell mediated immunosuppression. *Stem Cells* **2016**, *35*, 766–776. [CrossRef]
61. Kadle, R.L.; Abdou, S.A.; Villarreal-Ponce, A.; Soares, M.A.; Sultan, D.L.; David, J.A.; Massie, J.; Rifkin, W.J.; Rabbani, P.; Ceradini, D.J. Microenvironmental cues enhance mesenchymal stem cell-mediated immunomodulation and regulatory T-cell expansion. *PLoS ONE* **2018**, *13*, e0193178. [CrossRef]
62. Zhou, K.; Guo, S.; Tong, S.; Sun, Q.; Li, F.; Zhang, X.; Qiao, Y.; Liang, G. Immunosuppression of human adipose-derived stem cells on t-cell subsets via the reduction of nf-kappab activation mediated by pd-l1/pd-1 and gal-9/tim-3 pathways. *Stem Cells Dev.* **2018**, *27*, 1191–1202. [CrossRef]
63. Rozenberg, A.; Rezk, A.; Boivin, M.-N.; Darlington, P.J.; Nyirenda, M.; Li, R.; Jalili, F.; Winer, R.; Artsy, E.A.; Uccelli, A.; et al. Human Mesenchymal Stem Cells Impact Th17 and Th1 Responses Through a Prostaglandin E2 and Myeloid-Dependent Mechanism. *Stem Cells Transl. Med.* **2016**, *5*, 1506–1514. [CrossRef]

64. Takeda, K.; Webb, T.L.; Ning, F.; Shiraishi, Y.; Regan, D.P.; Chow, L.; Smith, M.J.; Ashino, S.; Guth, A.M.; Hopkins, S.; et al. Mesenchymal Stem Cells Recruit CCR2+ Monocytes to Suppress Allergic Airway Inflammation. *J. Immunol.* **2018**, *200*, 1261–1269. [CrossRef]
65. Brandau, S.; Jakob, M.; Bruderek, K.; Bootz, F.; Giebel, B.; Radtke, S.; Mauel, K.; Jäger, M.; Flohe, S.; Lang, S. Mesenchymal Stem Cells Augment the Anti-Bacterial Activity of Neutrophil Granulocytes. *PLoS ONE* **2014**, *9*, e106903. [CrossRef]
66. de Witte, S.F.H.; Luk, F.; Parraga, J.M.S.; Gargesha, M.; Merino, A.; Korevaar, S.S.; Shankar, A.S.; O'Flynn, L.; Elliman, S.J.; Roy, D.; et al. Immunomodulation by therapeutic mesenchymal stromal cells (MSC) is triggered through phagocytosis of MSC by monocytic cells. *Stem Cells* **2018**, *36*, 602–615. [CrossRef] [PubMed]
67. Salami, F.; Tavassoli, A.; Mehrzad, J.; Parham, A. Immunomodulatory effects of mesenchymal stem cells on leukocytes with emphasis on neutrophils. *Immunobiology* **2018**, *223*, 786–791. [CrossRef] [PubMed]
68. Kim, S.-N.; Lee, H.-J.; Jeon, M.-S.; Yi, T.; Song, S.U. Galectin-9 is involved in immunosuppression mediated by human bone marrow-derived clonal mesenchymal stem cells. *Immune Netw.* **2015**, *15*, 241–251. [CrossRef] [PubMed]
69. Yagi, H.; Soto-Gutierrez, A.; Parekkadan, B.; Kitagawa, Y.; Tompkins, R.G.; Kobayashi, N.; Yarmush, M.L. Mesenchymal Stem Cells: Mechanisms of Immunomodulation and Homing. *Cell Transplant.* **2010**, *19*, 667–679. [CrossRef]
70. Chow, L.; Johnson, V.; Impastato, R.; Coy, J.; Strumpf, A.; Dow, S. Antibacterial activity of human mesenchymal stem cells mediated directly by constitutively secreted factors and indirectly by activation of innate immune effector cells. *Stem Cells Transl. Med.* **2020**, *9*, 235–249. [CrossRef]
71. Bjarnsholt, T. The role of bacterial biofilms in chronic infections. *APMIS* **2013**, *121*, 1–58. [CrossRef]
72. Bjarnsholt, T.; Ciofu, O.; Molin, S.; Givskov, M.; Høiby, N. Applying insights from biofilm biology to drug development—Can a new approach be developed? *Nat. Rev. Drug Discov.* **2013**, *12*, 791–808. [CrossRef]
73. Walton, K.D.; Lord, A.; Kendall, L.V.; Dow, S.W. Comparison of 3 real-time, quantitative murine models of staphylococcal biofilm infection by using in vivo bioluminescent imaging. *Comp. Med.* **2014**, *64*, 25–33.
74. Shirjang, S.; Mansoori, B.; Solali, S.; Hagh, M.F.; Shamsasenjan, K. Toll-like receptors as a key regulator of mesenchymal stem cell function: An up-to-date review. *Cell. Immunol.* **2017**, *315*, 1–10. [CrossRef]
75. Chen, X.; Zhang, Z.; Zhou, H.; Zhou, G. Characterization of mesenchymal stem cells under the stimulation of Toll-like receptor agonists. *Dev. Growth Differ.* **2014**, *56*, 233–244. [CrossRef]
76. Saeedi, P.; Halabian, R.; Fooladi, A.A.I. Antimicrobial effects of mesenchymal stem cells primed by modified LPS on bacterial clearance in sepsis. *J. Cell. Physiol.* **2018**, *234*, 4970–4986. [CrossRef] [PubMed]
77. Waterman, R.S.; Tomchuck, S.L.; Henkle, S.L.; Betancourt, A.M. A New Mesenchymal Stem Cell (MSC) Paradigm: Polarization into a Pro-Inflammatory MSC1 or an Immunosuppressive MSC2 Phenotype. *PLoS ONE* **2010**, *5*, e10088. [CrossRef] [PubMed]
78. Ward, C.L.; Sanchez, C.J., Jr.; Pollot, B.E.; Romano, D.R.; Hardy, S.K.; Becerra, S.C.; Rathbone, C.R.; Wenke, J.C. Soluble factors from biofilms of wound pathogens modulate human bone marrow-derived stromal cell differentiation, migration, angiogenesis, and cytokine secretion. *BMC Microbiol.* **2015**, *15*, 75. [CrossRef] [PubMed]
79. Hsu, S.-C.; Wang, L.-T.; Yao, C.-L.; Lai, H.-Y.; Chan, K.-Y.; Liu, B.-S.; Chong, P.; Lee, O.K.-S.; Chen, H.-W. Mesenchymal stem cells promote neutrophil activation by inducing IL-17 production in CD4+ CD45RO+ T cells. *Immunobiology* **2013**, *218*, 90–95. [CrossRef]
80. Pezzanite, L.; Chow, L.; Griffenhagen, G.; Dow, S.; Goodrich, L. Impact of Three Different Serum Sources on Functional Properties of Equine Mesenchymal Stromal Cells. *Front. Vet. Sci.* **2021**, *8*, 634064. [CrossRef]
81. Deng, J.; Li, D.; Huang, X.; Li, W.; Zhao, F.; Gu, C.; Shen, L.; Cao, S.; Ren, Z.; Zuo, Z.; et al. Interferon-γ enhances the immunosuppressive ability of canine bone marrow-derived mesenchymal stem cells by activating the TLR3-dependent IDO/kynurenine pathway. *Mol. Biol. Rep.* **2022**, *49*, 8337–8347. [CrossRef] [PubMed]
82. Ryu, D.-B.; Lim, J.-Y.; Lee, S.-E.; Park, G.; Min, C.-K. Induction of Indoleamine 2,3-dioxygenase by Pre-treatment with Poly(I:C) May Enhance the Efficacy of MSC Treatment in DSS-induced Colitis. *Immune Netw.* **2016**, *16*, 358–365. [CrossRef] [PubMed]
83. Hirakawa, M.P.; Tjahjono, N.; Light, Y.K.; Celebi, A.N.; Celebi, N.N.; Chintalapudi, P.; Butler, K.S.; Branda, S.S.; Krishnakumar, R. Upregulation of CD14 in mesenchymal stromal cells accelerates lipopolysaccharide-induced response and enhances antibacterial properties. *iScience* **2022**, *25*, 103759. [CrossRef] [PubMed]
84. Schallmoser, K.; Rohde, E.; Reinisch, A.; Bartmann, C.; Thaler, D.; Drexler, C.; Obenauf, A.; Lanzer, G.; Linkesch, W.; Strunk, D. Rapid Large-Scale Expansion of Functional Mesenchymal Stem Cells from Unmanipulated Bone Marrow Without Animal Serum. *Tissue Eng. Part C Methods* **2008**, *14*, 185–196. [CrossRef] [PubMed]
85. Pedrazza, L.; Cunha, A.A.; Luft, C.; Nunes, N.K.; Schimitz, F.; Gassen, R.B.; Breda, R.V.; Donadio, M.V.F.; de Souza Wyse, A.T.; Pitrez, P.M.C.; et al. Mesenchymal stem cells improves survival in LPS-induced acute lung injury acting through inhibition of NETs formation. *J. Cell. Physiol.* **2017**, *232*, 3552–3564. [CrossRef] [PubMed]
86. Berglund, A.K.; Long, J.M.; Robertson, J.B.; Schnabel, L.V. TGF-ß2 reduces the cell-mediated immunogenicity of equine MHC-mismatched bone marrow-derived mesenchymal stem cells without altering immunomodulatory properties. *Front. Cell Dev. Biol.* **2021**, *9*, 628382. [CrossRef] [PubMed]
87. Rowland, A.L.; Miller, D.; Berglund, A.; Schnabel, L.V.; Levine, G.J.; Antczak, D.F.; Watts, A.E. Cross-matching of allogeneic mesenchymal stromal cells eliminates recipient immune targeting. *Stem Cells Transl. Med.* **2021**, *10*, 694–710. [CrossRef] [PubMed]
88. Arifka, M.; Wilar, G.; Elamin, K.M.; Wathoni, N. Polymeric Hydrogels as Mesenchymal Stem Cell Secretome Delivery System in Biomedical Applications. *Polymers* **2022**, *14*, 1218. [CrossRef] [PubMed]

89. Kudinov, V.A.; Artyushev, R.I.; Zurina, I.M.; Lapshin, R.D.; Snopova, L.B.; Mukhina, I.V.; Grinakovskaya, O.S.; Saburina, I.N. Anntimicrobial and regenerative effects of placental multipotent mesenchymal stromal cell secretome-based chitosan gel on infected burns in rats. *Pharmceuticals* **2021**, *14*, 1263. [CrossRef] [PubMed]
90. Fisher, S.A.; Cutler, A.; Doree, C.; Brunskill, S.J.; Stanworth, S.J.; Navarrete, C.; Girdlestone, J. Mesenchymal stromal cells as treatment or prophylaxis for acute or chronic graft-versus-host disease in haematopoietic stem cell transplant (HSCT) recipients with a haematological condition. *Cochrane Database Syst. Rev.* **2019**, *2019*, CD009768. [CrossRef]
91. Frisbie, D.D.; Cross, M.W.; McIlwraith, C.W. A comparative study of articular cartilage thickness in the stifle of animal species used in human pre-clinical studies compared to articular cartilage thickness in the human knee. *Vet. Comp. Orthop. Traumatol.* **2006**, *19*, 142–146. [CrossRef]
92. McIlwraith, C.W.; Fortier, L.A.; Frisbie, D.D.; Nixon, A.J. Equine Models of Articular Cartilage Repair. *Cartilage* **2011**, *2*, 317–326. [CrossRef]
93. McIlwraith, C.W.; Frisbie, D.D.; Kawcak, C.E. The horse as a model of naturally occurring osteoarthritis. *Bone Jt. Res.* **2012**, *1*, 297–309. [CrossRef]
94. Chu, C.R.; Szczodry, M.; Bruno, S. Animal Models for Cartilage Regeneration and Repair. *Tissue Eng. Part B Rev.* **2010**, *16*, 105–115. [CrossRef]
95. Reesink, H.; Watts, A.; Mohammed, H.; Jay, G.; Nixon, A. Lubricin/proteoglycan 4 increases in both experimental and naturally occurring equine osteoarthritis. *Osteoarthr. Cartil.* **2017**, *25*, 128–137. [CrossRef]
96. Frisbie, D.D.; Kawcak, C.E.; Trotter, G.W.; Powers, B.E.; Walton, R.M.; McIlwraith, C.W. Effects of triamcinolone acetonide on an in vivo equine osteochondral fragment exercise model. *Equine Vet. J.* **1997**, *29*, 349–359. [CrossRef] [PubMed]
97. Wu, X.; Jin, S.; Ding, C.; Wang, Y.; He, D.; Liu, Y. Mesenchymal Stem Cell-Derived Exosome Therapy of Microbial Diseases: From Bench to Bed. *Front. Microbiol.* **2022**, *12*, 804813. [CrossRef] [PubMed]

MDPI AG
Grosspeteranlage 5
4052 Basel
Switzerland
Tel.: +41 61 683 77 34

Veterinary Sciences Editorial Office
E-mail: vetsci@mdpi.com
www.mdpi.com/journal/vetsci

www.ingramcontent.com/pod-product-compliance
Lightning Source LLC
Chambersburg PA
CBHW041347210926
51850CB00013B/3011
* 9 7 8 3 7 2 5 8 2 7 6 0 2 *